The Many Worlds of Hugh Everett III

The Many Worlds of Hugh Everett III

Multiple Universes, Mutual Assured
Destruction, and the Meltdown of a
Nuclear Family

Peter Byrne

OXFORD
UNIVERSITY PRESS

OXFORD
UNIVERSITY PRESS

Great Clarendon Street, Oxford, OX2 6DP,
United Kingdom

Oxford University Press is a department of the University of Oxford.
It furthers the University's objective of excellence in research, scholarship,
and education by publishing worldwide. Oxford is a registered trade mark of
Oxford University Press in the UK and in certain other countries

The moral rights of the authors have been asserted

First published 2010
First published in paperback 2012
Reprinted 2017

Published in the United States of America by Oxford University Press
198 Madison Avenue, New York, NY 10016, United States of America

British Library Cataloguing in Publication Data
Data available

Library of Congress Cataloging in Publication Data
Data available

ISBN 978-0-19-965924-1

Dedicated to Stacey L. Evans
and our son, Miles Patrick Byrne.

These necromantic books are heavenly,
Lines, circles, scenes, letters and characters:
Ay, these are those that Faustus most desires.
Oh, what a world of profit and delight,
Of power, of honour, of omnipotence,
Is promised to the studious artizan!
All things that move between the quiet poles
Shall be at my command. Emperors and kings
Are but obeyed in their several provinces.
Nor can they raise the wind or rend the clouds.
But his dominion that exceeds in this
Stretcheth as far as doth the mind of man:
A sound magician is a demi-god.

Christopher Marlowe, *Doctor Faustus*, Act One, Scene One, 1592.

What is that little Devil's pitchfork?

Mark Everett, 2007. Upon seeing ψ, the Greek letter (psi)
symbolizing the quantum mechanical wave function.

Contents

Book 12: Everett's Legacy

Forewords

The Boxes

Growing up in my family once was odd enough. I had no desire to do it again. As a means of survival, I decided I had to always be moving forward. I ran to California and made a new life for myself. After the deaths of my father, mother and sister, I was left with the grizzly task of going back to the family house in Virginia and cleaning it out. I only had a few days to go through 35 years of boxes that had accumulated during the family's time in the house, as well as decade upon decade's worth of my grandparents', great grandparents', and so on's boxes.

Back in California, I unceremoniously stacked box after box of my family's past onto shelves in the dirt-floored crawl-space section of my basement. They sat there gathering dust for nearly another decade while I made music in the room directly on the other side of the wall.

I knew the day was coming when the boxes would have to be opened. I just didn't want to be the one to do it. Although I've been lucky enough to end up being happy with my life (part hard work, part miracle) and feeling at peace with my family history, I still don't relish going back to that world. If I play a concert in the Washington, D.C. area, the moment I step off the plane I can smell death in the air. I was sure those boxes held the same smell.

Luckily Peter Byrne came along to smell those boxes for me. The boxes have now become this book and I've learned a lot from what Peter found in them. Family secrets and secrets to the universe written on legal pads, diary pages, check stubs, and napkins. Peter managed to dig through the smell and bring the people buried in the boxes back to life. Alternately enlightening and troubling, like any good book should be.

It's an endlessly strange feeling being the lone survivor of the family. I'm pretty busy these days with my own job. It's been a great pleasure to have this new part-time job of helping my father get the attention he didn't get while he was alive. I've learned to forgive him for his shortcomings as a father by identifying with him in some ways. I recommend it, if you haven't tried it.

Mark Oliver Everett

Los Feliz, California

My Friend, Hugh

Hugh Everett III and I were fellow graduate students at Princeton, roommates one semester and close friends throughout. We each completed PhD dissertations in 1957, which were published that year in *Reviews of Modern Physics*, but our research relations with John Archibald Wheeler, who endorsed these degrees, were very different. My work was to add some fancy mathematics to an idea that Wheeler suggested. Hugh's work and the arguments to support it were all originated by Hugh. Wheeler served as his public relations manager, trying to negotiate peace between Hugh's upstart (but logical) ideas and the widely accepted (and hardly contested) interpretation of quantum mechanics attributed to Niels Bohr, Wheeler's mentor.

Hugh's thesis topic was much influenced by his personality. As Peter Byrne says, after extensive conversations with Hugh's acquaintances, "He loved to argue." I think it was his favorite sport, closely related to what was then called "oneupmanship." So when Niels Bohr visited Princeton, and his young assistant tried to explain Bohr's views on quantum mechanics, Hugh found it medieval: While mathematically formulated physics applied to everything when no one was looking, as soon as the results were to be unveiled God threw the dice (which Einstein doubted) and reset the equations to a probabilistically chosen result. So Hugh looked at what would be predicted if the mathematical formulation (the Schrödinger equation) were assumed to work all the time. He found, to his delight, that this implied a preposterous view of the world that was just as unintuitive as Copernicus' 16th century view that we, sitting comfortably in our chairs, are moving at tremendous speed through the solar system. So Hugh obtained a psychological prize in that no one could fault his logic, even if they couldn't stomach his conclusions. The most common reaction to this dilemma was just to ignore Hugh's work.

Quantum physicists had their hands full around 1957 with exciting research that found Bohr's viewpoint adequate. New elementary particles were being discovered and their relations systematized; mirror symmetry violation led to a Nobel prize; nuclear structure was beginning to make sense; the maser was about to be upgraded to the laser; the relativistic quantum theory of light and electrons was being widely understood, as was the source of energy in the Sun; superconductivity had just been explained, and condensed matter theory was flowering supported by the success of the transistor. None of this would benefit by using Hugh's view of the quantum instead of Bohr's. To speak of the wave function of the Universe, as Wheeler and I were willing to do, implied a viewpoint different from Bohr's, but serious consideration of quantum effects in the (then still questionable) Big Bang would not occur until a couple of decades later, and gravity and cosmology were, in 1957, not generally considered ripe for attention. So Hugh had a long wait before his dissertation began to be appreciated, all of which this book describes.

Hugh would, of course, have been happy if his quantum ideas were noticed and applauded, but when they were mostly ignored he was instead chagrined and perplexed. He could not understand why a perfectly logical idea had so little impact. But he had more important things to do than help the world properly understand quantum theory. He needed a job that would make lots of money and keep him out of the post-Korean War draft. As I learned from this book, he (with George Pugh) may have helped cool down the Cold War by analyzing the global effects of radioactive fallout.

The present situation for the interpretation of quantum mechanics is that several important fields of physics are uncomfortable with Bohr's viewpoint. Some participants work with Everett's view, others try to develop alternatives to both Bohr and Everett without the philosophically troubling "many worlds" picture that Everett proposed. The most important problems that demand an alternative to the "Copenhagen interpretation" preached by Bohr are: (a) modeling the early stages of the Big Bang Universe and (b) designing a quantum computer. Fundamental research useful toward solving these two problems are studies in the field of mesoscopic physics, which means physics involving few enough atoms that quantum effects can be expected, but in large enough numbers that they should tend toward classical (non-quantum) behaviors. These studies also go by the names "quantum-classical transition," or "decoherence." The researchers in these fields all know of Everett's ideas. Some buy his views enthusiastically, others look for other ways to let the Schrödinger equation operate steadily (with no mystical interruptions for the "collapse of the wave function"), but without the Universe branching into Hugh's continually multiplying "many worlds".

It may take many decades for mathematical progress to be matched by philosophical understanding. Hugh Everett proposed that we not search for remedies for the implausible "collapse of the wave function" by changing the mathematics of the Schrödinger equation (or its relativistic field theory upgrades), but instead just look hard at what would be predicted if we let the equations show us how they think Nature behaves. Now, over 50 years later, there is a strong effort to do just that, but the broad picture is not yet clear. Thus my guess for the outcome of the active search for a satisfactory understanding of quantum mechanics, is that some different "big picture" will arise, which is not "many worlds," but which still upholds Hugh Everett's conviction that paranormal influences will not overrule the Schrödinger equation.

Charles W. Misner

University of Maryland

BOOK 1
BEGINNINGS

Introduction:
The Story of Q

We can believe that we will first understand how simple the
universe is when we recognize how strange it is.

J. A. Wheeler, 1973[1]

On January 2, 1971 two U.S. Air Marshals were stewarding a White House courier from Dulles Airport in Virginia to Los Angeles. The courier was en route to the "western White House" in San Clemente, California, carrying classified information to Dr. Henry Kissinger, national security advisor to President Richard Nixon. Their flight was delayed due to bad weather, so the courier and his bodyguards lounged at the bar in the airline's Admiral Club. They chatted about the nature of their jobs, complaining about having to work over the holidays.

Nearby, a heavy-set, goateed man wearing a black suit was drinking gin and chain-smoking Kent cigarettes stuck in a long filter. He listened intently to the idle talk, from which he deduced the couriers' profession.

On the night flight to California, the goateed man enjoyed several more drinks. Returning to his seat after visiting the rest room, he suddenly pulled out a miniature camera and photographed one of the federal marshals. Startled, the marshal demanded, "Why did you do that?" The man with the camera replied, enigmatically, "For my files." Concerned, the marshal ducked into the cockpit. He radioed ahead asking for an FBI agent to meet the plane so that the interloper could be detained. But the agent was tardy in arriving, and the mystery man disappeared into the crowd.

Fearful that the nation's security had been compromised, the agent telexed a report to top FBI officials. Inquiries were made, and agents determined that the goateed man was Dr. Hugh Everett III, president of Lambda Corporation of Arlington, Virginia, which designed "computer modules." They tracked Everett to his room at a Holiday Inn near Santa Barbara, where he was attending a conference on advanced techniques in data processing. Sheepishly, he explained to FBI interrogators that he had been "affected" by several drinks, and that

---

[1] Wheeler, J. A. (1973). 245. Wheeler often wrote this slogan on his classroom chalk boards.

taking the photograph, "was merely a stupid act that he did on the spur of the moment." He had "no ulterior motive," and his "for my file" remark was said to get a "reaction" from the marshals. "He was extremely apologetic and embarrassed over the incident."[2]

The U.S. Attorney decided that Everett had not violated any statutes, and the matter was consigned to FBI files. But what the agents did not record, if they knew it at all, was that the goateed man had an incredibly high national security clearance, a "Q" clearance, which allowed him access to some of the Pentagon's most precious secrets, including the software he had designed for targeting cities with nuclear weapons in a hot war. A programming genius, Everett had also designed algorithms for the National Security Agency, the code-breaking "puzzle palace," the very existence of which was a state secret. If FBI officials knew that the man behind the miniature spy camera was a Q, they may well have wondered just how safe the nation's military and intelligence jewels were in the hands of such an impetuous inebriate. But even though excessive drinking was technically a cause for stripping a Q of his security clearance, the culture of the national security bureaucracy easily tolerated alcoholics.

Everett was not only a prankster and a software guru, he was also a master game theorist whose work epitomized the ethos of "rationality" worshipped by the Cold War technocracy. His specialty was designing and running computer simulations of nuclear wars, testing America's ability to launch the real thing. One high-ranking general considered him to be "worth his weight in Plutonium 239."[3] And, outside the top secret world of intelligence and operations research, at which he excelled, his revolutionary work in theoretical physics, known as the "many worlds interpretation" of quantum mechanics, was just beginning to be recognized as "one of the most daring and most ambitious theories ever constructed in the history of science."[4]

To his neighbors in suburban McLean, Virginia, the dad wearing the black suits studded with cigarette burns was just another government worker. Little did they know that his day job was planning the end of history. Nor did they know that many physicists and philosophers were subscribing to his theory that all physically probable events—from summer snow to nuclear winter—*actually happen inside a multiplicity of branching universes that contain countless copies of everything and everybody in every possible configuration.*

Many worlds

In his 1957 PhD thesis at Princeton University, Everett formulated a solution to a vexing problem in quantum theory, the measurement problem. Simply put, the problem arises because, logically, an atomic particle can move through

---

[2] FBI. (1971).
[3] Wheeler to Everett, 10/30/57.
[4] Jammer, M. (1974). 509.

space and time in a multiplicity of directions at once—as if it is an expanding, spherical *wave* simultaneously passing through all possible trajectories. But when we interact with the particle—when we measure it—we always find it at one place, not many. This fact does not accord with the fundamental rule of quantum mechanics, the Schrödinger equation.

Everett showed that it is mathematically consistent to say that when a scientist measures the position of an atomic particle, he *splits* into numerous copies of himself. Each copy resides in a different universe. And each copy sees the particle in a different position. The set of all copies covers the set of all possible particle positions inside a *multiverse*. According to Everett, each universe inside the multiverse is constantly branching, like a tree, into separate but parallel worlds that cannot communicate with each other. Each parallel universe records a self-consistent history drawn from a range of physically possible histories. No one of these universes is any more or less real than another. Importantly, this does not mean that *anything* is possible: physical reality exercises certain constraints on what is probable.

According to physicist, Bryce DeWitt, Everett showed that

> This universe is constantly splitting into a stupendous number of branches, all resulting from the measurement like interactions between its myriad of components. Moreover, every quantum transition taking place on every star, in every galaxy, in every remote corner of the universe is splitting our local world on earth into myriads of copies of itself....Here is schizophrenia with a vengeance.[5]

A consequence of the "many worlds" logic is that there are universes in which dinosaurs survived and humans remained shrew-like; universes in which *you* win the state lottery every week; universes in which Wall Street does not exist and global resources are equally shared. Sure, it seems like an improbable idea, but the many worlds theory is widely recognized as a major contender for interpreting how quantum theory links to physical reality. Today, quantum cosmologists use Everett's "universal wave function" to view the cosmos from inside the cosmos, as opposed to trying to view it from outside, which is obviously impossible. And some scientists claim that recent discoveries made by satellites mapping the microwave residue of the Big Bang might be evidence validating Everett's theory![6]

In fact, so seriously is this extremely counter-intuitive idea taken that there is a school of philosophy based at University of Oxford known as "Everettian." And if you look in the index of just about any book on the foundations of quantum mechanics, you will see the entry, "Everett, Hugh III." In July 2007, the prestigious scientific journal, *Nature*, featured his theory on its cover, celebrating the half-centenary of its publication.

---

[5] DeWitt, B. S. (1970). 161.
[6] See Epilogue for claims that there may be physical evidence for multiple universes in the cosmic microwave background radiation.

Whether you believe in it or not, understanding the argument of Everett's many worlds model is of central importance to any attempt to rationalize the mysteries of the quantum world.

Sadly, homage to the revolutionary theory came largely after the untimely death of the theorist in 1982. Originally fascinated by quantum mechanics, Everett had left physics behind after his published thesis was dismissed by most of the physics establishment. Two decades later, his appreciation of its burgeoning popularity was over-shadowed by the collapse of his personal world. Despite enjoying a challenging career in military operations research, and having enough money to indulge himself with rich foods, fine wines, sexual escapades, and Caribbean cruises, the joke-loving mathematician was inwardly fey.

At the time of the drunken incident on the airplane, Everett's consulting business was starting to disintegrate due to cut-backs in military research budgets and his mis-management of the business end. His marriage had long been troubled, and his self-effacing wife, Nancy, was resigned to accepting his affairs with employees and prostitutes (while engaging in a few affairs of her own). Within a decade, the scientist was ruined financially and personally—his body wracked by compulsive smoking, drinking, and eating. Only Nancy knew of the black depression that fuelled his self-destruction, and she felt powerless to intervene. Gradually, the demons of addiction and despair overcame him, sucking the pleasure out of both work and play.

Ironically, Everett had devoted his life to making models of reality, but he was largely oblivious to the harm he caused those closest to him. He barely acknowledged the existence of the pair of troubled children, Mark and Liz, who yearned for parental attention and acted out their frustration when it was not forthcoming. When Everett died suddenly of a heart attack, his teenage son, Mark, trying to revive his corpse, reflected that he could not recall ever having touched his father in life.

Archeology

After his father died, Mark moved to Los Angeles and became a successful songwriter known to his fans as "E." Years later, he learned that the man he had considered to be a "robot" capable only of sitting at the dining room table drinking martinis and smoking Kent cigarettes and scribbling line after line of computer code (it looked like "gibberish" to him) was a world-renowned quantum theorist.

On a scorching hot day in June 2007, Mark and I descended into the sepulcher-cool basement of his house in Los Angeles. We were followed by a camera crew from the British Broadcasting Corporation, which was shooting a documentary about Mark's journey to discover his father, *Parallel Worlds, Parallel Lives*. Much of the sloping, dirt-floored basement was filled with guitars and amplifiers and bicycles and bags of cast-off clothing and broken chairs.

One wall was lined with wooden shelves holding the family saga—two dozen cardboard boxes bursting with memorabilia and paper trails left by Hugh III, Nancy, daughter Elizabeth ("Liz"), and Everett's parents.

Several cartons were stuffed with photographs. Other boxes held birth and marriage and death certificates, expired passports, financial spreadsheets, tax returns, military medals, personal diaries, and hundreds of letters home documenting the lives of three generations of Everetts: their secret hopes, private joys and heartbreaks, their public celebrations and mournings—the intimate history of an emotionally cloistered family gathering dust in a dark room.

Everett's boxes sequestered old textbooks, physics and operations research papers, stacks of letters, used airplane tickets, cancelled checks to liquor stores, stained hotel receipts, a deck of playing cards festooned with naked, peroxide blondes, a super-8 porno film with no pretension toward plot, and a scrap of paper on which Everett had scrawled a *fallacious* "proof" of the existence of God. Several were stuffed with thousands of sheets of yellow legal paper covered with algorithms variously designed to track ballistic missiles with nuclear warheads or to outwit the housing and stock markets or to generate financial spreadsheets on first generation personal computers.

Another box held artwork made by the kids for Father's Day. Laying beneath the childish art were letters from some of the most renowned quantum physicists of the day, John Wheeler, Niels Bohr, Norbert Wiener. And there was an ancient Panasonic Dictaphone containing the only known tape recording of Everett: a cocktail party conversation between him and his friend, Charles Misner, made in 1977. (There is a touching scene in *Parallel Worlds, Parallel Lives*, where the adult son listens to the tape for the first time, hearing his father's voice as if back from the dead. In the background, an adolescent Mark is heard pounding on a trap set, gradually drowning out the conversation.)

After E and the film crew left on a journey across America to confab with his father's classmates and colleagues, I began sorting through the boxed treasure. Festooned with clip lights illuminating a folding table set with bottles of water, magic markers, and a laptop computer, the basement became a kind of archeological dig as box by box, layer by layer, page by page, scientific and biographical gems emerged from the higgledy-piggledy. Chief among the finds was the original, handwritten draft of Everett's dissertation. Letters and other papers showed that he had followed the rehabilitation of his theory with great interest, though galled by what he perceived as the failure of great physicists to understand the core of his theory. There were several writings that addressed the burning question of whether or not he had viewed his multiple universes as totally abstract, or as physically real.

In other cartons, Everett's mother, Katharine Kennedy Everett, left behind a poignant record of her life in letters and poems. Waves of emotional pain seemed to radiate from boxes containing certain items, particularly Nancy's intensely intimate diaries, and the alcoholic, depressed Liz's desperate letters home. And then there were the photographs of the broken family pasted in three funeral books kept safely buried in the basement by the sole-surviving Everett.

This is a book about anti-heroes. It's about a tragically dysfunctional American family as reconstructed from intimate records and memories of the living. It's about the technocratic mindset that waged the Cold War, bringing humanity to the brink of destruction under the banner of rationality. It's about the seemingly intractable problem of modeling and understanding a complex system from within that same system—be that a quantum system of multiple universes, or a political system composed of mirrored superpowers facing off with hydrogen bombs, or a sad and confused family stumbling toward destruction while surrounded by socio-economic privilege. These three strands—quantum mechanics, computerized war gaming, and the fate of a small, nuclear family shaped during the Cold War—gradually weave together as we tell the story of a powerfully intelligent, but morally conflicted man who significantly affected our world.

And we take our time doing it, as it is necessary to explain the scientific, historical, and cultural context of Everett's remarkable ideas, so that we may comprehend why his accomplishments were so vital. At the heart of the tale is the puzzle of probability: that foundation of decision-making that seems so commonsensical and, yet, cannot be explained. Probing probability, Everett attempted to account for how the world of our experience emerges from an indeterministic quantum netherworld. This journey led him to invent one of the strangest, most important ideas in science. It is not just that his nearly impeccable logic leads to multiple universes—it leads to a vision of a completely deterministic reality in which probability itself becomes an illusion. Yet, in his influential job as a war-planner, he relied upon probabilistic equations to build weapons systems assuring destruction. A genius at statistical analysis in both physics and operations research, he invented powerful computer algorithms for use by the military that, if his interpretation of quantum mechanics is correct, precipitated the explosion of city-vaporizing hydrogen bombs in worlds beyond number.

Tracking Everett's rise and fall as a military consultant provides a glimpse of the top secret paths trod by many Cold War scientists who, despite horrible misgivings about the nature of their work, did it anyway. But in Everett's case, professional success inside the Beltway did not translate into psychological ease, and the toys of the affluent middle-class did not substitute for the lost nurturing that his children so dearly desired. So why do we care about this man, Hugh Everett III? Is it possible to feel compassion for someone who was so ready to blow up the Earth? And what about the multiple universes? Should we care about them even if we cannot prove that they—and multiple we's—exist?

My answer is—yes—we should care about Everett as a person, because his life reflects America's collective personality during the Cold War and beyond. His story transcends a personal travail—it is about the creeping militarization of science and civil society and the failure of consumerism to provide happiness. And we should care about his theory of many worlds because it sheds some light on how the ancient riddle of probability—coupled to the physicality of information and consciousness—unites physics and philosophy in modeling a purely quantum mechanical reality. But, perhaps, most of all, we should care about Everett not just because of the contribution he made to quantum mechanics and cosmology—whether or not his worlds exist, it is one of the most influential theories in the history of physics—but because the unvarnished story of what happened to his family, as recorded in the boxes in the basement, is an American tragedy—and, yet, it is also a lesson in the quality of forgiveness.

1

Family Origins: a Sketch

> *We gather most of our information by intercepting a small fraction of [the] environment. Different observers agree about reality based on consensus reached in this fashion.*
> Wojciech H. Zurek, 1998[1]

Cosmic wonder

In this universe, Hugh Everett III was born in Washington, D.C. on November 11, 1930: Armistice Day. That national holiday marked the signing of an unsteady peace by the global powers that fought the First World War. It was celebrated all across America with parades, flag-waving, and patriotic speeches.

Also on the day that Everett was born, Albert Einstein published "Religion and Science" in the *Berliner Tageblatt*. The founder of relativity and quantum theory was to pop up in Everett's life at several pivotal moments. But on this day, Einstein wrote that the best scientists possess a "cosmic religious feeling" and "a deep conviction of the rationality of the universe." He prayed that a "priestly caste" of scientists would learn to conjure the machinery of peace, not war, and that political leaders would mirror their rationality.

Unfortunately, in 1939, world war erupted again. This time, the industrial powers mobilized legions of scientists to produce weapons of mass destruction far more lethal than the previous war's mustard gas, mortars, biplanes, and machine guns. The Second World War concluded with the United States exploding atomic bombs over Hiroshima and Nagasaki. After the incineration of the Japanese cities, Einstein reflected, "By painful experience we have learned that rational thinking does not suffice to solve the problems of our social life." Pure science was not, in Einstein's view, an end in itself. He warned,

> Penetrating research and keen scientific work have often had tragic implications for mankind, producing, on the one hand, inventions which liberated man from exhausting physical labor, making his life richer and easier, but on the other hand, making him a slave to his technological environment, and—most catastrophic of all—creating the means for his own mass destruction. This, indeed, is a tragedy of overwhelming poignancy![2]

---

[1] Zurek, W. H. (1998).
[2] "A Message to Intellectuals," in Einstein, A. (1954). 149–150.

When Einstein wrote those words in 1948, Everett was novitiate to a "priestly caste" of scientists serving military-industrialism. His career was typical of the careers of many physicists swept up in the post-war marriage of the academy to the national security state: propelled by a talent for mathematics, it segued seamlessly from high school to college to graduate school to the top secret world inhabited by legions of Cold Warriors idolizing reason.

Like many of his compatriots, Everett was irreligious and cynical about human nature. But, deep down, he held on to a sense of cosmic wonder: He burned to reduce the complexity of the universe to rational formulae. This drive to extract order, logic, and beauty from seeming chaos was inherited from both of his parents.

Family tree (with nicknames)

Everett's great-grandfather, Charles Everett, a medical doctor, emigrated from Scotland. He served in the Confederate Army during the Civil War before marrying Virginia Haynes of Virginia. Their son, Hugh Everett, was born in Denton, Texas, a frontier town. Just before the turn of the century, Hugh married Laura Katherine Clardy, also of Virginia, later known to her grandchildren as "Ma Maw." She was "hard rock" Baptist and a staunch member of the Women's Christian Temperance Union. Sharp of tongue, she was unforgiving toward what she perceived as the failings of her adult children.[3]

Hugh, nicknamed "Day Day" by his children, worked most of his career as a compositor for the Government Printing Office and *The Washington Post* newspaper in Washington D.C. He briefly owned a small printing company, Terminal Press, during the Great Depression. And he smoked cigarettes.

Hugh became Hugh Everett, Sr. when Hugh Everett, Jr. was born in 1903 in the District of Columbia, the eldest of four children: Charles Edward ("Jiggs"),

Everett's paternal grandmother, Laura Clardy.

[3] Information in this section is from interviews with members of the extended Everett family and documents in the basement archives. For instance, in a September 23, 1959 letter to Hugh Jr. and Sara, who were on a world cruise, Nancy wrote that Ma Maw "has always felt that she has missed out on life somewhere—and not being able to blame herself—takes it out on those closest."

of Washington D.C. (who also became a printer), and two sisters, Virginia ("Ginny"), a housewife, and Kathryn ("Kaka"), a school teaching "spinster," as unmarried women were called in those days. Everett, Jr.'s siblings and their families congregated during the holidays, but they were not a close-knit group. "Day Day" died of a heart attack or stroke in 1958; "Ma Maw" passed away quietly in 1973 at the age of 94.

Hugh Everett, Jr., circa 1922.

Mr and Mrs Hugh Everett, Sr., circa 1950.

Katharine Lucille Kennedy, of Baltimore, Maryland, married Hugh Everett, Jr. of Washington, D.C. in 1924. The newlyweds were 20 years of age. She was a school teacher. He was employed by Southern Railway Company as a payroll clerk. After earning a degree in civil engineering at George Washington University, he designed railway bridges. He was also an officer in the National Guard, and he held the "world" rifle record at 1000 yards from 1928 to 1936.[4]

He was laid-off by the railroad after the stock market crash of 1929, but weathered the Great Depression by freelancing as an engineer for government and military agencies. He earned a Masters in Patent Law and a Doctorate in Juridical Science from local colleges.

Meanwhile, the marriage was falling apart.

Socially flamboyant and politically liberal, Katharine was definitely not suited to living with her engineer-soldier husband. Repulsed by housewifery, Katharine set her sights on a writing career. Before marriage, she had attended American University and the Corcoran School of Art (Sculpture) in the District of Columbia. After Hugh III was born, she earned a Masters degree in education from George Washington University. But by 1935, the

---

[4] *Washington Post*, 1/8/55; according to Everett, Jr.'s high school yearbook in 1920, "Hugh is a marksman. He is also a handsome lady-killer."

Katharine with baby Hugh (III), circa 1930.

school teacher was transitioning into the life of a professional magazine writer—selling tales of love gone wrong, children's stories, science fiction, and reams of purple poetry. Soon thereafter, she left her husband, toddler in tow.

Everett, Jr. fell in love with Sarah Thrift, a secretary at the National Park Service. In late 1935, he drove to Nuevo Laredo, Mexico, to purchase divorce papers for himself and Sarah, whose early marriage had also disintegrated. Crossing the border, he wrote her: "Now I am on my way home to you with two very legal looking documents in my suitcase and a song of love in my heart."[5]

The next year found the newly divorced newlyweds in the Panama Canal Zone where Everett, Jr. worked as an engineer. He was rehired by Southern Railroad after the economy perked up a bit in 1937. They moved into a colonial-style house in Bethesda, Maryland, a woody suburb of Washington, and were well-suited to each other. With another world war looming, he signed up for active duty as an Army officer.

Meanwhile, Katharine's career as a pulp-fiction writer took off, but she did not have the time nor the financial wherewithal to properly care for her son: so, Everett, age seven, moved in with his father and step-mother, who were loving, but ran a household in which the rule of thumb was "sink or swim." Indeed, Everett learned to swim by being peremptorily tossed into a lake.[6]

As the Second World War ignited in Europe, Everett, Jr. was commissioned Lieutenant Colonel in the Army. He served on the general staff in the battles for Sicily, Rome-Arno, and the Rhineland. His son remained in Bethesda with Sarah for the duration.[7]

In 1940, Katharine sued for a legal divorce under the laws of Maryland. The court granted her a "divorce a vinculo matrimonii," a legal term signifying that if one of the spouses had been adulterous during the marriage, that spouse could not legally marry his or her paramour. (The thicket of Mexican and Maryland divorce laws caused Everett, Jr. difficulties many years later when the Veteran's Administration did not want to recognize Sarah as his lawful wife for pension purposes.) The divorce court awarded formal custody of her son to Katharine, although he continued to live in his father's house, and under his supervision, as she watched from afar.

Katharine never remarried. She was the distant, dark star that periodically perturbed the motion of Everett's orbit around the planet of his paternal namesake. Long as he might for his elusive mother, it was his father to whom he clung.

---

[5] Everett, Jr. to Thrift, 10/11/35, basement.
[6] Everett, M. (2008). 13.
[7] After the war, a beautiful Sicilian woman that Colonel Everett had befriended during the occupation of Sicily wrote to him thanking him and Sarah for some dresses they had sent to her, and requesting an overcoat for her father. The letter was intimate, without revealing the exact nature of their relationship. She sent a photo of herself and her teenage daughter and requested a picture of the Colonel's son, who "must be a big boy by now." Larregsoffore to Everett, Jr., 2/14/48.

"Pudge" for life

Clearly, Everett inherited his left-brained, logical capabilities from his father, who was an accomplished statistician. From his right-brainy mother, he was blessed with intuition, creativity, and a streak of rebelliousness. He partook of their talents—indeed, he far transcended his father's analytical abilities—but he was also saddled with their less desirable qualities: an addictive personality from his father; depression from his mother.

Father and son shared much more than a name. Both were compulsive photographers, happy to peer at the world through the lens of a camera. They each left behind thousands of snapshots—mostly of buildings and landscapes. Like his father, Everett enjoyed traveling to foreign countries first-class, relaxing in the bar of luxury cruise liners. Both men smoked tobacco: most photographs of Hugh, Jr. show him attached to a pipe, and his adult son smoked two to three packs of Kents a day through a long, tapered filter. They were both heavy drinkers, and they enjoyed flirting with women at cocktail parties; the younger Everett treated sexual conquest as a game.

Everett had an ambiguous relationship with his mother: seeking her approval, affection, and warmth—remaining aloof when rejected. Each tried, in their own way, to reach out to the other. In the end, Everett did not know how to recognize her love, much less accept it when it was shyly offered. Sadly, and as a consequence, he did not learn how to proffer unconditional love to others, nor how to reciprocate when it was offered to him. He craved companionship, but remained a loner. He craved happiness, but did not know how to make others happy. He craved love and attention, while failing to recognize the emotional needs of those closest to him.

And there is something else: His whole life, Everett's father, mother, stepmother, grandparents, uncles, and aunts called him "Pudge," because he was chubby. A quarter century after his death, his cousins still referred to him as Pudge.

He *hated* to be called Pudge.

2 Katharine:
the Dark Star

The child, in these decisive first years of life, has the experience of his mother as the fountain of life, an all-enveloping, protective, nourishing power. Mother is food; she is love; she is warmth; she is earth. To be loved by her means to be alive, to be rooted, to be at home.... The longing for this situation as it once existed never ceases completely.

Erich Fromm, 1955.[1]

In 1996, Everett's widow, Nancy, wrote a letter to Mark, touching on his father's rocky relationship with his own mother, or "Mum Mum" as she liked to be called. Katharine suffered from manic depression, and

> Initially, during separation, he lived with his mother, but apparently she was not able to swing it, financially or emotionally...Now you know that all this must have had an emotional impact on a 6–8 yr old. I'm sure it made him wary of getting too close to people emotionally....He wanted to get along with K.K. [his mother] but he was very cool toward her – just normal civilities – no warmth there – something there he couldn't forgive her for.

Whatever the messy details of Everett's parent's break-up, their incompatibility is easy to discern. Military-minded and lawyerly, he loved male camaraderie and the Officer's Club. She was an artist, a bohemian, a woman struggling to carve out a place for herself inside a culture that had only recently accorded her the right to vote. She was unwilling to subvert her career to the needs of her husband and her child. And her art reflected this inner turmoil, she was always writing and rewriting her own story—and her story captures the difficulty of being a self-reliant woman in mid 20th century America.

In 1939, The Banner Press published a book of Katharine's poetry, *Music of Morning*. Typical of popular magazine and newspaper verse of the era, her poems pulse with colorful, disconnected modifiers, important-sounding references to Greek mythology, crazy-mixed-up metaphors, rampant anthropomorphism, and purple-bruised stanzas bristling with "burning swords" and "cold scimitars." She spun off florid poetic phrases, e.g. "Titan-souled," "futilitarian," "silver mirth," "fragile lusts," "caesuras of darkness," "pale anemones stare at the stars," "ghost of scarlet laughter in the rain," and the stunningly oxymoronic: "All the soul's vast measure is / A flowering abyss."

[1] Fromm, E. (1955). 39.

Katharine Kennedy Everett, circa 1935.

One of her better offerings is near-Blakean with bittersweetness:

PRODIGAL

And if he should once again
After long, long pain
See my threshold in the night
Through the bitter rain,

And if he should speak to me,
Standing there apart,
And if he should lay his hand,
Quiet, on his heart,

Saying in a gentle voice
All that was not said,
Shall I not bring forth the wine,
Break at last the bread?

Music of Morning airs Katharine's abandonment issues (by parents, husband, son, self, cruel fate). She oscillates between exhilaration and melancholia, celebrating freedom from her unhappy liaison: "She dreamed of breaking / Unbreakable bars."

"Saint's Season" captures her joyous state:

In the sudden wind, your exile over,
Run through the grass on naked feet,
Take life at last to your breast for lover:
Put on a scarlet dress, my sweet!

"Spinner" looks back at the family she left behind:

Long ago she left us.
She who used to fashion
Lovesong at her spindle:
Now the fire is ashen.

Now the latch is broken,
Quenched the golden candle;
She spun stuff like dreamlight,
That we couldn't handle.

During her writing career of three decades, Katharine's poems evolved from structured love sonnets to free verse about religion and spirituality to paeans to time, space, and gravity. Represented by literary agents based in New York City, "Mrs. Katharine Kennedy" published fiction, non-fiction, and verse in *McCall's, Colliers, Overland, Outlook, Harper's Bazaar, Southwest Review, The New York Times, The Washington Post, Saturday Evening Post, Christian Science Monitor, Catholic World, New Mexico Quarterly*, and some British publications. She occasionally used the pen names, "Kay Everett" and "Penelope Ross."[2]

In addition to writing for fees, she taught writing workshops and advised non-professional authors. She was listed in the 1939 edition of "Who's Who in

[2] KK resume, basement; Readers' Guide to Periodical Literature annual editions.

the East." She was prominent in the affairs of the National League of American Pen Women, the Poetry Society of London, and the Free Lance Writer's Society of Washington D.C. Faded newspaper clippings report that she read poetry in public, and received a literary award for a collection of sonnets, "Armor Against Autumn." In black and white photographs, she appears every bit the literary sophisticate with her dark suit, cloche hat, curly tresses, haunted eyes, and thin, knowing smile.

Over the years, Katharine barely supported herself through a combination of teaching, writing, and working as a secretary, researcher, and editor. She traveled to Europe, Cuba, and Canada, and lived briefly in the United Kingdom. But she always returned to the Washington area, trying to stay in touch with her only child. In 1955, she took a fulltime position as an executive secretary at George Washington University, leaving in 1959. During that period, she earned her third degree from that institution, a bachelors in English. The following years were tough; she moved from job to job, and place to place, hurting for money, but cherishing her infrequent visits with Everett and his young family. She was very proud of his "space-age" career and his "cosmic" top secret clearance. She regularly sent him newspaper clips about rockets, satellites, esoteric science, and, one time, a mathematical proof that the death penalty does not deter murder.

Her final job was as a research assistant to the cultural anthropologist, Edward T. Hall, who had recently published *The Silent Language* (1959), an instantly famous study of how people define social spaces, and communicate unconsciously through body language and subtle cues.

After she died, Nancy remembered,

> She had been such a strong personality it is hard to forget her influence, which I spent most of my time trying to forget or to ignore.... I am not so good at getting Hugh to talk about family matters; yet I'm sure he can't really feel as aloof as he seems.... Just one of the ways in which she revealed her disdain for, or unacceptance of, reality was her almost always mistreatment of material objects. I don't know what she did to things, but they didn't get handled very carefully.

> She seemed to me a truly tragic person. Why should this be the penalty for some creative persons, while others can adjust to [the] world as well as be intensely creative? She herself felt that her unhappy childhood was to blame, though she never outwardly complained to me.... When the marriage fell apart... this only added to the tragic pattern.... Perhaps her personality was one which never could have adjusted to any marriage or to any mundane type of existence.

Social critic

As America warred with Germany, Italy, and Japan, Katharine's poetry took on a political tone. In December 1943, writing in the newsletter of Pearl S. Buck's The East and West Association, Katharine, who was a Christian, penned "Morning on Sinai." In it, we meet "shepards, chanting psalms barbaric" as,

In the East
Light breaks; the Mountain wakes.
Timeless, beyond the curve and flow of Space
Light from the East
Illumes the Mountain's Face.

Politically, Katharine was classically liberal and anti-communist; she, like Buck, preferred Chiang Kai-Chek's nationalist organization to Mao Zedong's communist party. During the post-war years, however, she actively opposed the witch-hunting tactics of Senator Joseph McCarthy and the reactionary antics of the House Un-American Activities Committee.

As the Cold War heated up, a barrage of films, magazine articles, books, and newspaper editorials encouraged patriotic Americans to build fall-out shelters in preparation for atomic invasion by "godless" communists. Not surprisingly, UFO-spotting was all the rage. Katharine walked in fear of annihilation by nuclear war, nuclear accident, or an alien attack from outer space.

In her short story, "Dark Hawks Hear Us," the main character is a reporter, Crosby, who is covering an atomic bomb test near Los Alamos, New Mexico. With the Sangre de Cristo (Blood of Christ) mountains as a back-drop, Crosby, thinks about,

> Men all over the planet digging shelters in the bowels of the earth for shelter against the atomic death they'd managed to manufacture for themselves, and at the same time gawking up to the sky, as though they expected to find an answer there.

Someone asks,

> Do you believe our minds are like islands, subterraneously linked with other island minds, Mr. Crosby? Do you believe, for instance, that your own individual mind is separated only so far as it believes itself separated from a kind of group mind? Do you agree, Mr. Crosby, that minds have the capacity to communicate with other minds independent of space? For instance, Mr. Crosby, in the case, say, of sudden sweeping national enthusiasms or uprisings, couldn't these be instigated by telepathic influences operating in the world group-mind, directing mass emotion and mob action?

Suddenly overwhelmed by an alien group mind communicating in secret code through a United Press International teletype machine, Crosby leaves his earthly body behind, lifeless, and the story ends.

Although attracted by science as a subject for her fiction, Katharine drew most of her material from personal experience. The pitfalls of conventional marriage were a theme she dwelled upon. "The Apartment" is set in Arlington Towers, where Katharine resided when she wrote the story, and where newly-weds Hugh III and Nancy Everett lived briefly. A single woman, Janet, invites a man, Jerome, for dinner and he refuses to leave after eating, taking over her apartment. She is afraid to call the police. She loses sleep and, then, her job. She takes to wandering the streets, hanging out at the zoo, at the Christian Science

Reading Room, returning evening after evening, to find Jerome cooking in the kitchen.

Eventually, she melds with him.

> There was a placid mask on her face, she lifted her head, listening. As though in response to some unseen wire of communication between them, she was aware that Jerome, humming as he prepared dinner in their kitchenette, wished the volume turned up on the TV.

> Obediently she rose and turned the dial. . . . Since their marriage six months ago, there had been little need for words between them, she knew his thoughts as well as he knew hers. . . .

> 'But I die,' she said, aloud.

> Jerome, in the kitchenette preparing dinner said nothing—he was busy dismembering the chicken; tenderly he cut off its wings, popped it in the glass skillet, put on the lid, and turned up the fire.

Another short story, "Pride," is about Marta, who had walked out of a marriage with Rod, leaving behind two small children, Kitten and Tim. Three years later, Marta, now a successful actress, drops in to see the youngsters for the first time since leaving. "Into the room with Marta had blown a bright wind of disaster; the familiar patterns of the place were shattered." (One can visualize Katharine breezing into her ex-husband's home in Bethesda, back from a trip to London, looking for a hug from Pudge.)

> Tim stood with his toy elephant clutched close, looking helpless at his mother. His wet hair was plastered back; there was a clean, shining look about him.

> "I built a house, Mummy," he offered politely. "I built it myself. Out of blocks. Look!"

The children are shunted off to bed. Marta pours a drink, lights a cigarette. Rod asks here why she has bothered to visit the children.

> Her face was pinched; she looked for a moment again as though she were about to cry. But she spoke quietly.

> "I don't expect you to understand, Rod, But if you'd been a bigger man—a man with more vision. If you hadn't been so insufferably self-righteous—"

Rod shouts that she made her choice and must live with it. Marta replies,

> "I never wake up mornings but what I think of Tim. I wonder if he's eating his cereal. If you are attending to his teeth. If he's forgotten his mother—"

Marta flees, driving, "down the hill, zigzagging crazily."
Nancy reflected about her mother-in-law:

> She seems to have been wildly pursuing truth and or beauty ever since [Hugh III went to live with his father], flitting from one answer to another, delving into all philosophies and religions, cults, etc. never being satisfied with any; flitting also from short-term job to job, apartment to different living arrangements, (in the

six years I knew her in Washington she lived in at least that many different places), never finding peace of mind, never finding the warmth and love she must've needed so desperately.

She seemed tragic to me because of her aloneness, but I could not sympathize with her unwillingness to face things in their proper proportion. Her insistence on exaggerating every day events with rosy-rimmed enthusiasm and continually trying to identify herself in any way with others—so as to belong—this insinuating behavior clashed on many occasions with my Yankee bluntness of expression, etc.

But I had no intentions of becoming further involved in her problems. I knew they were more than I could cope with, that no amount of love that I might be able to give, could undue [sic] the wrong that had made her so. (I couldn't even love anyone without any problems.) This realization that I abandoned her, so to speak, is one reason why I am haunted by her still. Yet, I know, I could never have given more....She was a true poet...like the oracle at Delphi, a medium, through which the words of truth and beauty must pass.

In the realm of the Mind

In 1953, *The New York Times* published[3] a breathtakingly abstract poem by Katharine—dedicated to Einstein:

UNIFIED FIELD
(FOR A.E.)

Pure fields of space,
O flowing fields of light,
who shall gauge
these lightning flights
of mind,...only Mind
arching from sun to sun
beyond the blind
groping of scattered minds
who cannot see
the Great Reality
whispers its ultimate Secret
to this mind:
the atom and the universe are One.

In his essay "Religion and Science" Einstein asked why we so easily resort to mystical, anthropomorphic explanations when faced with ontological puzzles.

---

[3] After Katharine died, Everett gave a collection of her published poems about time and space to her alma mater, George Washington University, which republished them in the Winter 1965 issue of its alumni magazine, including "Unified Field."

With primitive man it is above all fear that evokes religious notions—fear of hunger, wild beasts, sickness, death. Since at this stage of existence understanding of causal connections is usually poorly developed, the human mind creates illusory beings more or less analogous to itself on whose wills and actions these fearful happenings depend.

But as mysteries are unraveled by scientific experiment, the rational scientist, wrote Einstein,

feels the futility of human desires and aims and the sublimity and marvelous order which reveal themselves both in nature and the world of thought. Individual existence impresses him as a sort of prison and he wants to experience the universe as a single significant whole.[4]

Katharine, like Einstein, like her son, desired a universe that lent existence meaning as part of a sensible whole. Confronting a limit on understanding, she poetically linked her consciousness to a (God-like) Mind. Indeed, some scientifically serious advocates of the Many Worlds theory have also appealed to a linked-minds concept of a non-physical consciousness in an effort to make sense out of quantum reality.

For his part, Everett explicitly rejected consciousness (or Mind) as a causal element in physics—Katharine, however, was a causal, if unconscious factor in his life. His inability to resolve his ambivalent feelings toward her festered, causing him to distrust humanity for reasons he could not fully explain.

[4] Einstein, A. (1954). 46-47.

3 The Scientist as a Young Man

Peace
by Hugh Everett III[1]

Millions will rejoice on the day of peace,
After this desperate war has ended,
When the murderous cries of battle cease,
And the deep gnarled scars of war have mended.
Then mankind will work to rebuild the earth,
And the draftsmen will plan both night and day,
As earthlings unite in freedoms rebirth;
But the leaders that launched the war must pay.
All those at fault must be taken in hand,
And be punished for their barbarous crimes.
Justice shall reign again through all the land,
And happiness shall once more bless the times.
We must scour the earth of war's harsh sorrow.
So that lasting Peace shall rule the morrow.

Childhood

When Everett was growing up in the 1930s and 40s, Bethesda, Maryland was a racially segregated suburb serving highly educated professionals. The neighborhood provided plenty of parks and recreational space for children, although, even as a youth, Everett was not one for taking nature walks or playing sports. He loved reading science fiction and toying with gadgets and playing practical jokes on his elders.

Nancy wrote in her journal that, all of his life, Everett was an "Army brat." On the one hand, "He could talk with grownups from an early age and loved to embarrass his father in front of 'top brass' and vice versa." On the other hand, "This is a guy who at the tender age of 12 wrote a letter to Albert Einstein, and received a reply! I think his mom—K.K.—may have influenced him [to write that letter]."

---

[1] Written as a class assignment, circa 1940, typed by Katharine (whose pacifist politics probably influenced her son's poem), basement.

Everett's long lost letter to Albert Einstein apparently claimed to have solved the paradox of what happens when an irresistible force meets an immoveable object. Nancy thought he had written it as a "hoax," to see if he could fool the great man. Graciously, Einstein wrote back on June 11, 1943,

> There is no such thing like an irresistible force and immovable body. But there seems to be a very stubborn boy who has forced his way victoriously through strange difficulties created by himself for this purpose.

For his entire life, Everett was irresistibly attracted toward wrangling solutions out of logically immovable paradoxes. He enjoyed pondering looping, recursive statements, such as, "All men from Crete are liars; I am from Crete." And he had fun toying with supposedly logical arguments for the existence of God as a prime mover that fell into infinite regression by failing to answer how there can be a first cause of the universe that does not itself have a cause which requires a cause which requires … and so on.

He looked at the universe as an engineer, searching for its central mechanism. But Katharine tried to cultivate the dreamer in him. His post-apocalyptic "Peace" is remarkable considering his ultimate vocation. More typical was "Checkers":

> In checkers when you get a king
> You'r so happy that you sing.
> Then the other person says, "I quit"
> And you'r so mad that you just sit.

He also wrote a very short play about winning the favor of a girl he liked by saving her from kidnappers. And for a seventh grade class assignment (also typed by Katharine), the future game theorist wrote a two-page short story, "The Bone of Mutt'sburg." The dog-King of Mutt'sburg is preparing to eat a meal of "bone al la mode." A cat from Catsville complains that the dogs are killing all the cows for meat, depriving the cats of milk. The dog-King shrugs off the cat, who snatches the bone al la mode. Enraged, the dog-King sends an army of dogs after the thieving cat, who slyly leads them away from the cow pasture, while other cats steal the cows. In the end, the dog-King proposes to kill only half the cows, if the cats return them all. The warring animals sign a truce, and live in peace happily ever after.

But for unconditional love, Everett turned to his step-mother. He drew a woman and a boy kissing, making the sound: "SMACKO!" It was captioned, "Dear mama Sarah: I love you very much and I would like to kiss you. In fact, I will!"

High school

Shortly after Einstein replied, Everett won a half scholarship to St. John's College, a private military high school in Washington D.C. run by Catholic Christian Brothers. The school motto, "Religio Scientia," means "religion and

Sarah Thrift Everett, circa 1935.

"Pudge" and Sarah, circa 1944.

knowledge." Everett was a life-long atheist, but he did not let that stand in his way as St. John's was well-regarded academically and socially. Cadets were enrolled in the Army Junior Reserve Officers Training Corps and required to take parade drill and weapons training. The United States was at war and the Army JROTC was a conduit into combat.

In the spring of 1944, Everett was rated an overall "perfect" for attendance, general conduct, and grades. A typical report card for his high school career shows a 95 in Mathematics, a 93 in English, and a 90 in Christian Doctrine. But he flunked Military Drill with an amazingly low 45 (70 was failure). The

St. Johns cadet Hugh Everett III, circa 1944.

St John's yearbook for 1946 shows a uniformed, sullen Everett standing at attention with Company G. In a happier photo, he calculates at a blackboard with other members of the Mathematics Club. The caption: "Professor Einstein's successors at practice!"

Never athletic, Everett found a role for himself at student sporting events: Using one of the first commercially available strobe lights, he took photographs of the skirmishes in high school football and basketball games and sold them to local newspapers. He had a keen sense of lighting, once winning an award for a pretty photograph of Tidal Basin cherry trees in full bloom.

His two best friends at St. Johns were also "non-athletes," Ralph Mohr and Fred Wilson. They remember Everett as a "chunky, round-faced" kid, brilliant in math and physics, and also a good pool player. Because of his loudly avowed atheism, he was labeled "the heretic" by devout classmates.

Everett was "embarrassed to death" when Sarah[2] called him "Pudgie" in front of his friends. "He told us *never* to call him that," Mohr recalled.

Mohr and Everett had a lucrative business taking photographs of couples at high school dances and selling orders for prints on the spot. Later, they went on numerous double dates around Washington, including to Rosecroft Raceway, a local horse track. Using a slide rule, Everett developed a system for handicapping horse races, but got bored when he realized that his predictions were exactly the same as the racing form's. After graduation, the friends lost touch. Mohr, a retired mechanical engineer, was pleasantly surprised 60 years later when he picked up a copy of *Scientific American* at his dentist's office and found a profile of his old pal. He called Wilson, a retired Air Force chaplain, and they reminisced, digging out fading photographs of Everett.

"He was a good friend and I enjoyed him tremendously," Mohr said. "How sad that he did not have the rewards which his great intelligence should have obtained for him."[3]

The freshman

Everett graduated with honors from military school in 1948, and transformed into a freshman at nearby Catholic University, where he majored in chemical engineering. The university was owned and operated by the Roman Catholic Church; its 200-acre campus surrounded by an urban community known as "Little Rome," because of the large number of Catholic institutions settled in the shadow of the Basilica of the National Shrine of the Immaculate Conception. Everett's paternal grandparents lived a short walk away on Grant Circle, and he often roomed there, enjoying the friendly, cosmopolitan neighborhood.

He quickly became an academic star, and a campus character. Classmate Louis Painter says, "I remember Everett as easygoing, as a brilliant student. He saved our butts in calculus class, as we had a total klutz as an instructor." Everett

---

[2] His friends thought that Sarah was his birth mother; he never once mentioned Katharine to them.
[3] Mohr, Wilson interviews with author, 2008.

Col. Everett and Pudge, far left.

and Painter ran on a joint ticket for freshman class office—Everett for President, Painter for Vice President. Despite campaigning from the back of a new convertible, they lost.

Painter, a chemical engineer, remembers that, "Everett was much interested in the science of Dianetics," a self-improvement program invented by science fiction writer, L. Ron Hubbard.[4]

In late 1949, Hubbard published his first article on Dianetics in a pulp sci-fi magazine beloved by Everett, *Astounding Science Fiction*. The core idea of Dianetics was that by "auditing" painful memories a person can increase his intelligence while eliminating emotional problems and physical illness. There is no evidence that in his post-student years Everett gave Dianetics or, the shadowed, quasi-religious cult into which it evolved, Scientology, any attention, but, for the rest of his life, he refused to visit doctors, and he believed in the social superiority of uncommonly intelligent beings, such as himself. And like Hubbard, he was a sexual libertine.

Although Catholic University offered expertly taught, non-theologically driven science courses, Everett was required to take two classes in religion: Fundamental Beliefs and Spiritual Foundations of American Life. Nancy wrote, "He drove devout Jesuits to distraction with scientific questioning."[5] Everett, whose academic file records him as "non-Catholic," aced all of his freshman year science and math courses and was handed a "C" in Fundamental Beliefs. But he earned an "A" in Philosophy of Science, which is noteworthy, considering the philosophical waves his many worlds theory is generating.

---

[4] Wilson confirmed that Everett was a Hubbard fan.
[5] Nancy to Mark, 4/26/96.

The case against God

While at Catholic University, Everett reportedly constructed a "logical proof" against the existence of God that caused one of his professors to despair of religious faith. Distressed by causing ontological horror, Everett vowed never to use this argument on a person of faith again; but that did not stop him from tinkering with it, or trying it out on people from time to time, despite his best intentions.[6]

According to a Pentagon colleague, Joseph Clifford, he once applied what he called the "universal existence theorem" to a discussion of the existence of the mythological winged horse, Pegasus. One discussant was a Roman Catholic, so Everett had politely substituted the horse for God. According to Clifford, Everett's logic caused the Catholic to admit that believing in God is purely a matter of faith, and not subject to the rigor of mathematical proof.[7]

Theologians throughout the ages have striven to construct a logical proof of God's existence. The abstract reasoning of St. Anselm of Canterbury's ontological argument in the 11th century is typically expressed: "God is greater than anything that can be conceived. God exists as a conception in the human mind. Existence in reality is greater than existence in the mind. Therefore, God exists in reality." This is sophistry, though. Anselm presupposes that God exists by defining God as existing, so it is not a proof.

In the 1970s, the legendary logician, Kurt Gödel, privately circulated his attempt to improve on St. Anselm's ontological proof; it would have intrigued Everett as Gödel employed "modal" logic—a philosophical method that examines the validity of all possible truths, or contingencies, or "ifs." In his modal argument, Gödel postulated the existence of all possible worlds. He then argued, in essence, that *if* it is possible that God exists, God must exist in some possible world. Therefore, God exists.

A quarter century after Everett's death, in his son's basement, a scrap of paper was found inside a cigar box full of torn ticket stubs and assorted junk. On it, in Everett's handwriting, was scrawled a deliberately fallacious "Improvement on ontological proof," written in the symbolic language of the predicate calculus.

It is, of course, not possible to *prove* that God or winged horses do *not* exist, so Everett's disproof focuses on the fatal flaw in the argument for God's existence that begins with "*If* God exists..."

Mocking the argument as "tautological,"[8] Everett began with the statement that either a proposition is true or its negation is true. He then highlighted the flaw in the standard ontological argument, which treats existence as a property. In order to have a property, an object must already exist. This logic leads to the

---

[6] Reisler, Clifford interviews.
[7] Clifford interview; the test subject was, according to Clifford, Misner, who worked at WSEG one summer.
[8] A tautology is an abstract, self-referential logical statement that does not have to conform to physical reality.

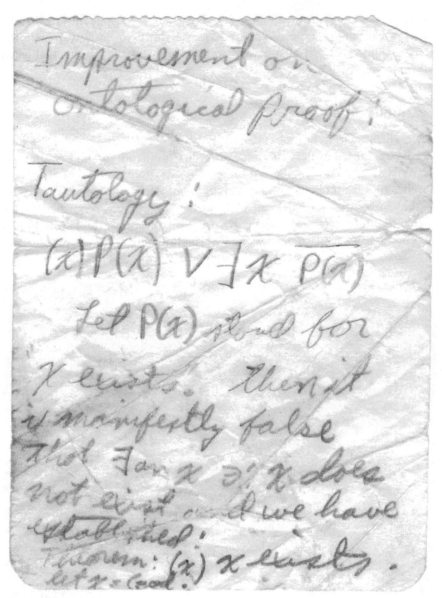

Ontological "proof."

absurd statement: "Everything exists, or there is something that does not exist." As there cannot exist even one thing that does not exist, he concludes, "Everything must exist. In particular, then, if x exists, and x is God, then God exists." His point was that it is not valid to use "if" in an *existence* argument because "if" presupposes existence in arguing for it.

Everett may have had Gödel on his mind when he scrawled his tautology because the many worlds interpretation is sometimes construed by religious people as validating the possibility of God's existence—despite the fact that the many worlds interpretation does not declare all things to exist. At any rate, the scrap debunks that sort of argument and it is probably the critique that felled his professor's faith.[9]

Learning curve

After returning from Europe with a shoebox full of campaign ribbons, Colonel Everett was posted to Headquarters European Command. He and Sarah moved to Wetzlar, Germany. In the fall of 1949, Everett took a year-long leave of absence from Catholic University to join them. He monitored several classes at the University of Heidelberg in Heidelberg, Germany, renowned for its theology, philosophy, and science. On frequent vacations, the family toured the continent's war-blasted cities; the two Hughs taking hundreds of black and

Col. Hugh Everett Jr. (far left) in Sicily, circa 1945.

[9] Thanks to Jeffrey A. Barrett and Simon Saunders for explicating the "existence scrap." Barrett comments, "Everett may have heard this argument from someone else since there are similar reconstructions of the ontological argument that predate Everett. It is quite impressive if he got this himself." Barrett, private communication, 2009.

Military Wives, circa 1950, Sarah Everett third from left.

white photographs of landmarks and landscapes, which were later pasted into albums almost devoid of people.

Ever the prankster, Everett sneaked in and out of East Germany to visit a Spring Fair in Leipzig. Light-hearted rebellion only went so far, though: His first security clearance application, filed three later, related that, "Upon return reported to the S-2 of Wetzlar Military Post and gave a detailed account of what I saw."

It was back to Catholic University for his sophomore year. He shouldered a seven-course load of advanced chemistry and mathematics, racking up a 3.8 grade average. During summer break, he traipsed around central Europe again with his family. In the fall, he signed up for ten courses in engineering, chemistry, and physics. Gung ho for science, he went to summer school, studying electricity and partial differential equations; and moved back in with his father and step-mother.

Colonel Everett now commanded Cameron Station in Alexandria, Virginia, a large Army base that was headquarters to the Defense Logistics Agency, which managed supplies required by the globe-trotting American military.[10]

---

[10] Everett, Jr. cut short his tour of duty in Europe as a post quartermaster in Germany due to a run in with an inspector general. An undercover investigator for the IG found a range of improprieties with contracts and funding for salvage operations under Everett, Jr.'s command. Claiming that the IG's findings were "fly-specking on minutia," Everett asked to be transferred back to the states a year before he was due back because he believed the IG was out to get him and that his inspectors were "hatchets" and "rats." Everett, Jr. was very well-connected at the Pentagon and he was allowed to return to an important position. Letters March–May, 1952, Everett, Jr. to commanding officers.

"SQUAW WORK"

Sarah and Hugh Jr. Photo by Hugh Everett III.

Richly salaried at $12,000, he bought a four-bedroom, colonial-style brick mansion in a hilly, wealthy suburb, Belle Haven, Virginia, a dozen miles south of Washington D.C.

During his senior year at Catholic University, Everett tackled twelve technical courses, including Atomic Physics. His relatively low (for him) 3.6 grade point average that year was the product of a "C" in chemistry lab, and "Bs" in Lectures on Contracts and Spiritual Foundations of American Life, which offset his usual harvest of "As." He and a classmate wrote an unpublished chemical engineering paper, "Fluid Flow Through Randomly Packed Beds of Raschig Rings," which appears to have been expertly typed by Katharine, no mean feat considering the number of equations used. (Raschig rings are small tube cuttings used to pack volatile liquids in industrial settings, and the paper was heavy on statistical analysis.)

In June 1953, Everett graduated with a bachelors degree in chemical engineering, *magna cum laude*; he had also completed the course work for a degree in mathematics. That summer, Everett received the first of many security clearances from the federal government and worked as an associate mathematician doing operations research at the Silver Spring, Maryland field office of the Applied Physics Laboratory run by Johns Hopkins University. He earned a respectable salary of $320 per month.

His specialty was analyzing the dynamics of servomechanisms, i.e. analog devices, such as bomb sights, that rely on "feedback loops" to automatically correct tracking errors. During this time, the lab (working with the Army's Operations Research Office) ran one of the first computer simulations in the history of operations research, a study of the air defense of North America.[11]

It must have been an inspiring few months for a young mathematician with a practical bent: Operations research was an exciting, growing field. It applied "cybernetics" to the study of warfare. An unprecedented amount of government funding was flowing into military research labs designing advanced weaponry systems controlled by computers. And Everett was eager to ride the wave of money and high technology.

"Cybernetics" (based on the Greek word for "steersman") was coined by the polymath Norbert Wiener. Pioneers in cybernetic theory included Wiener; John von Neumann of the Institute for Advanced Study in Princeton; the British computationist, Alan Turing; and Claude Shannon, an engineer at Bell Telephone Laboratories. Cybernetics gave birth to an epochal epigram: "Information is physical."[12]

Defining information as physical removed *thinking* from the exclusive realm of human consciousness, or Mind. As a discipline, cybernetics—the interlocking relationship of information, entropy, and probability—began shaping the development of computer-based technologies while Everett was in elementary school. Cybernetics inspired the invention of the digital computer; it extruded "C3" (command, control, and communication) systems that enabled the waging of hair-trigger nuclear war; it profoundly influenced the structure of corporate advertising and psychological warfare strategies; it affected quantum theory. Everett's innovative work in game theory, quantum mechanics, and software design was a product of the merger of cybernetics and information theory; we shall delve deeply into these matters as his life unfolds.

The best student ever

As graduation from Catholic University approached, Everett scored in the 99th percentile on the advanced physics section of the Graduate Record Examination, a newly invented standardized test administered by Educational Testing Services, of Princeton, New Jersey. Tellingly, he scored below average in social studies, literature, and fine arts. In April, he applied (six weeks past the official deadline) for admission to the graduate school in physics at

---

[11] Ghamari-Tabrizi, S. (2000).
[12] Phrase coined by physicist Rolf W. Landauer of IBM.

Princeton University. As a back-up to Princeton, the tardy applicant also applied to Catholic University's graduate school, but he need not have worried: Princeton's physics faculty, especially its world-class quantum theorist, Eugene Wigner, was eager to accept him on the basis of a portfolio of strong recommendations from his undergraduate professors.

Associate Professor of Mathematics at Catholic University, William W. Boone, wrote,

> This is a once-in-a-lifetime recommendation for I think it most unlikely that I shall ever again encounter a student I can give such complete and unreserved support.

> Everett is by far the best student I have had at Princeton, Rutgers, or Catholic University. Everett has a better knowledge of mathematics than most of the graduate students at Catholic University and probably <u>no</u> graduate student is his equal in native ability. Since his sophomore year he had been taking advanced mathematics courses—combinatorial topology, point set topology, real and complex variables, modern algebra, etc.—many simply as an auditor because of the administrative limitations on the number of hours a student may carry.

> In all of these courses he has given an outstanding performance, showing a thorough and frequently creative mastery of the subject. Everett also has worked widely on sophisticated mathematical texts of his own.... (This year, by the way, he has been taking for credit the game theory course given by Princeton University's Professor [Albert] Tucker at American University.)

For Tucker's game theory course, Everett wrote (and his obliging mother typed) a paper called "A Simplification of the Procedure of Determining the Basic Solutions of Matrix Games." Never published, the paper is in the words of Harold Kuhn, who co-taught Tucker's course in game theory, "a nice piece of juvenelis and might have merited publication at the time it was written."[13]

The introduction to game theory profoundly influenced Everett and he quickly absorbed its utilitarian ethos into his personality.

The Head of the Physics Department at Catholic University observed, in a separate letter of recommendation, "He has a very agreeable personality and excellent manners." Even the local draft board was impressed: "You attained a score of 85 on the Selective Service qualification test, which is very good. Congratulations!" As the last place on Earth that Everett wanted to go was boot camp, that was not good news; the ceasefire that ended the bloody Korean War was not to be signed until July 1953, and other wars and invasions and occupations loomed. Fortunately for him, his student deferment was extended when he was accepted into graduate school. He would not be safe from the draft until reaching age 26, but he planned to go into military operations research after he got his doctorate, which would exempt him from armed service.

---

[13] Kuhn, H. Private communication. 2008; Boone to Taylor, April 17, 1953.

In only its second year of existence, the National Science Foundation granted Everett a stipend to pay his Princeton tuition fees, plus $100 a month for room, board, and expenses. The federally funded science foundation was mandated to sponsor the higher education of a virtual army of promising young mathematicians and physicists demanded by what President Dwight Eisenhower was to label a "military-industrial complex." Eisenhower could have said "military-industrial-academic complex," as many of the nation's top science schools, including Massachusetts Institute for Technology, Stanford University, California Institute of Technology, Johns Hopkins University, and Princeton University, were deeply woven into the institutional fabric of Cold War militarism. The Pentagon depended on elite colleges to train tens of thousands of technicians and theoreticians to staff a spreading public-private network of corporate- and university-sponsored research labs and "think tanks" dedicated to inventing not only the machinery of the Cold War, but also to inventing moral, philosophical, and political rationales justifying it.[14]

Everything is information

Life was good for Everett in 1953: in late summer, scholarship and draft deferment secure, he packed up his belongings, ready to move to Princeton, hoping to leave Pudge behind. And to fire up a sex life.

He boxed up his personal library. It included Wiener's best-selling book, *Cybernetics or Control and Communication in the Animal and the Machine* (1948), which was all the rage among operations researchers. In many ways, Everett's early exposure to Wiener was the touchstone of his intellectual life.

In *Cybernetics*, Wiener defined information as "negentropy," i.e. the opposite of "entropy," which is a measure of the energy in a system. The Second Law of Thermodynamics declares that in a closed system entropy increases over time, leading to heat-death, dispersion, chaos, ever-increasing disorder. Wiener noted that the greater the negentropy, or order in a system: the easier it is to predict specific events, because in a *more* ordered system there are fewer "degrees of freedom," defined as the number of variables, or the number of possible configurations of the system's components. Conversely, in a *less* orderly system, there are generally *more* degrees of freedom and, consequently, less information, and less predictability.

Crediting Claude Shannon and statistician R. A. Fisher with discovering the same idea about the physical nature of what constitutes information, Wiener observed,

[14] As Jessica Wang points out in *American Science in an Age of Anxiety*, "[W]ith the budgetary stresses of the Korean War, the NSF could only attain significant funding by selling Congress on the importance of basic research for defense purposes. The NSF's educational programs, even before the Soviet launch of Sputnik in 1957, owed much to Cold War concerns about the supply of scientific and technical personnel." Wang, J. (1999). 261.

Hugh Everett III enters Princeton, 1953.

The transmission of information is impossible save as a transmission of alternatives.... To cover this aspect of communication engineering, we had to develop a statistical theory of the *amount of information*, in which the unit amount of information was that transmitted as a single decision between equally probable alternatives.... The notion of the amount of information attaches itself very naturally to a classical notion in statistical mechanics: that of *entropy*. Just as the amount of information in a system is a measure of its degree of organization, so the entropy of a system is a measure of its degree of disorganization; and the one is simply the negative of the other.[15]

For Everett, like Wiener, probability was a measure of information, and vice versa. But probability is more than a simple measure: it is also a system of belief. We use the laws of probability to assign proportional values to beliefs about the future. These beliefs about future events are based on having observed the relative frequency of past occurrences of that same type of event. It is thought that the normal precepts of rationality require us to make decisions based on probabilities. Precepts about probability and rationality inform all of Everett's most original work, including his theory of multiple universes, and we shall explore them thoroughly.

During the Second World War, Wiener performed vital operations research, including working on radar and bullet and bomb targeting mechanisms. But he refused to do any more research for the military after Hiroshima and Nagasaki. And he criticized scientists who sold their services to Moloch. In all likelihood, Everett read Wiener's admonition in *Cybernetics*:

> A certain precise mixture of religion, pornography, and pseudo science will sell an illustrated newspaper. A certain blend of wheedling, bribery, and intimidation will induce a young scientist to work on guided missiles or the atomic bomb.[16]

But he did not follow Wiener's example.

His box of books also contained *Patterns of Sexual Behavior*, by C. S. Ford and Frank Beach (1952). This soon-to-be-classic study of sexuality in human and other animal species was probably of less theoretical import to Everett than it was of operational necessity. Post-war, suburban American youth tended to be woefully uninformed when it came to the nitty-gritty of sexual practice; but that was changing for Everett's generation. Alfred Kinsey's bombshell *Report on Sexual Behavior in the Human Male* was published in 1948, followed even more explosively by his *Sexual Behavior in the Human Female* in 1953. Kinsey gave middle-class intellectuals permission to enjoy sex and, in many cases, to do so promiscuously.

During his married years, Everett collected sex manuals, pornographic novels, and hard-core 8mm porn films. At the time of his death, he possessed several well-thumbed copies of *The 120 Days of Sodom or the School of Licentiousness* by the Marquis de Sade. He probably enjoyed reading *Sodom* for its incestuous, ultra-violent, taboo sex, and he espoused a form of de Sade's ultra-libertarian, solipsist philosophy. But his attraction to de Sade was to come much later. In the summer of 1953, unbeknownst to the relatively innocent young scientist as he packed up his books, his future wife, Nancy Gore, was newly arrived in Princeton town.

It was to be two years before they hooked up for life.

---

[15] Wiener, N. (1948). 0–11.
[16] Ibid. 160.

4 Stranger in Paradise

If I stand starry-eyed
That's a danger in paradise
For mortals who stand beside
An angel like you.[1]

Nancy Gore Everett, circa 1957.

[1] Forrest and Wright's "Stranger in Paradise" was one of Nancy's favorite songs.

Nancy Gordon Gore was born at Amherst, Massachusetts on February 13, 1930 to Jane Pollard Gore and her husband, Harold "Kid" Gore, who headed the Department of Physical Education at the University of Massachusetts, Amherst. The Gores operated Camp Najerog ("Gore, Jan" spelled backwards), a 300-acre wilderness camp for children on a Vermont lake. The family was outdoorsy, close-mouthed, politically conservative.

Nancy attended private girl's schools. She spent her summers idyllically: horseback riding, swimming, boating, hiking and singing campfire songs; as an adult, she remained a nature lover, enjoying long walks, bird-watching, gardening.

Her friends found her to be a pleasant, kind, fun-loving adult, if somewhat shy. She had a sense of humor, laughing with a self-conscious giggle, and, in private, she was given to crying jags. Early in life, she developed the habit of writing her streaming thoughts in dairies, unmailed letters, and handy scraps of paper. The intimate record of her life reveals an inwardly tormented soul, struggling to make sense out of romance, sexuality, politics, nuclear war, and her strange, brilliant husband.

In her journals, she described herself as a child trapped in an adult's body. She evaluated herself as if through the eyes of others, and found herself lacking. And she was indecisive, usually waiting for others to act, before following. Although she cherished the idea of personal independence, she was afraid of being without a man; so, she modeled the behavioral conventions of suburban housewifery, clinging to the subservient role that her mother-in-law, Katharine, had so fiercely rejected. The cheerful independence she had so effortlessly enjoyed as a child with the run of a summer camp eluded her adult grasp.

In 1952, Nancy graduated with a B.A. in English literature from Hollins College in Roanoke, Virginia. During her years at the all-female school, Nancy suffered from bouts of exhilaration alternating with existential nausea. She was easily floored. One day she declared her strong affection for a vivacious classmate, and was devastated by the woman's cool response.

> I am never sure of myself, and I am suspicious of people who seem to be, and admire those who really are....For me, any original idea that comes to me (and the creation of something completely new is a miracle to me) is so quick and fleeting that it has disappeared before I can catch hold of it....I know I haven't the capabilities to achieve outstanding feats, but still and all, somewhere there must be a little bit of zip, squeak, life, or whatever, that is there waiting to be awakened, stimulated....I can sum up nothing, but see only each individual part separately, subjectively, and without a clear picture of the whole....I am sitting stagnant in a mist of nowhere and nothing....The only stimulation that seems to move me is to get thrown into a situation where I must either sink or swim.

Nancy's poor scores on the Graduate Record Examination did not reflect her natural intelligence or abilities. Her verbal ability was ranked in the 58th percentile, i.e. 42 percent of the test-takers scored higher. On the literature test, her strong point, she scored in the 74th percentile, but she sank to the 9th percentile for the advanced literature test. She barely registered as having a brain wave in physics, chemistry, biology and social studies, scoring in the 4th percentile overall for quantitative abilities.

After graduation, she moved back in with her parents, and worked as a counselor at Camp Najerog. If she had stayed with the family business, fashioning a career for herself from her love of nature and outdoor activity, her life would have taken a very different course; instead, she took classes in shorthand and typing at a local business school. In April, 1953, she moved to Princeton to start an entry-level job at Educational Testing Services (ETS) (owners of the Scholastic Aptitude Test and the Graduate Record Examinations). The job paid $220 a month, which was enough money to rent a small room, eat well, and entertain herself. Her Princeton diaries chronicle her boredom with clerical work. She failed to impress her supervisors because she often arrived late, she day dreamed, she made careless mistakes. But she had an active social life, attending Episcopal church services and singing in the choir, biking, sailing, playing tennis, going to beer parties, square dancing, and "air-plane spotting."

Cold War culture

Like hundreds of thousands of young people in the early 1950s, Nancy volunteered with the Ground Observer Corps, watching the skies for Russian bombers. Because radar systems could not track airplanes flying under 4,000 feet, volunteer clubs were tasked by civil defense agencies to monitor the night skies, especially in states that bordered oceans. "Highway billboards and movie shorts solicited volunteers. Local chapters hosted talent shows, beauty pageants, square dances, ice cream socials, and bake sales."[2] Even prison inmates and Catholic monks meditating on mountaintops were recruited to watch the heavens for enemy aircraft.

The function of the Corps, "like so much of the macabre apparatus of nuclear war, was primarily ideological: a genuine defense being impossible, a symbolic one was provided instead."[3] Not only was it technologically futile, but conventional sexual morals were, no doubt, tested as legions of young watchers stayed up all night, lying on their backs in the dark, staring at the stars. It certainly kept Nancy up till the wee hours, making her sluggish at work.

Day dreaming of landing a job as an executive secretary in a publishing company, Nancy kept banging her head on shorthand classes, never mastering that skill. Dreaming of travel, she consider joining the Foreign Service (and was rejected after failing an aptitude test). Desiring to write a novel based on her life experiences, she began an outline, but became distracted by a romantic crush and abandoned the project.

Nonetheless, she pondered important issues, including the nature of God, the viability of democracy, the merits of capitalism versus communism, and

---

[2] Ghamari-Tabrizi, S. (2005). 99.
[3] Edwards, P. N. (1996). 90. The Corps program was terminated in 1957.

the omnipresent threat of atomic warfare. Her political thoughts were shaped by Cold War liberalism. She favored granting civil rights to "Negroes," distributing welfare checks to the urban poor, and sending foreign aid to starving masses abroad. Her support for liberal government policies was not particularly rooted in altruistic feeling, but stemmed from the view that alleviating poverty might put a damper on social unrest and revolution; she was for "containing" the spread of communism through police actions by the military.

Like many intellectuals of her generation, Nancy was attracted to the idea, but not the practice of communism. She worried about losing her life in a nuclear conflagration, but she also worried about losing her "individualism" in a communist take-over. As she came of age, the country was plagued by McCarthyism, which aimed to stamp out political dissent. Anti-communist paranoia seeped into every corner of the culture. Liberal intellectuals walked in fear as careers and families were destroyed by accusations of socialist sympathies and secret black listing. Nuclear hegemony was promoted by the political culture of military-industrialism as the guarantor of freedom and democracy against the Communist Menace—embodied by Hollywood as body-snatching aliens and Godzilla.

Not long after Nancy moved to Princeton, she wrote in her diary:

McCarthy's speech tonite. I think he's damn right about communism, but a liar…making it a party issue and blaming Democrats for being party-minded. You certainly cannot blame all the c[ommunist] infiltration on one man, Truman….Its bad when people are motivated by fear…but I think that is why Eisenhower got in so overwhelmingly last year. I for one was scared to vote for a Democrat.

She sought solace in religion:

To me communion is a reminder: a time to remake your vow. Simply to remember why you are living and to remind self to live by the creative spirit that is in all people….I shall keep on trying to do so just as long as I am able to receive assurance that God does exist in life.

Considering the pros and cons of sharing an apartment with Milly, a "party girl," Nancy reflected,

Egads: I hope I don't (should the opportunity occur) choose a husband as blindly as I seem to be choosing a roommate.

A visit to an impoverished section of Trenton,

inspires me to do social work. (But will I ever?) Heel! I think I'll take one of those aptitude tests to see what combines appreciation of the arts, outdoors, and sports for a living no less.

In addition to continuing the torture of learning shorthand, she took courses in Comparative Religion and English at Princeton Adult School. She studied French Conversation and Figure Drawing. She read books by Willa Cather, Edith Wharton, and Theodore Dreiser, but gave up on Henry James, whose writing she found full of "innuendo." Theodore Dreiser, on the other hand,

> was genuinely concerned about the drastic difference in living standards and privileges between 'lower' and 'upper' classes. But I think he was feeling sorry for himself, too, which tainted his philosophy. He turned to communism finally for an answer to the problem, not knowing any better, I believe. I don't think he could see beyond the immediate problem.

Like many young Americans, she thought,

> Communism doesn't look so bad on paper, I s'pose; and maybe some forms of it are workable. Maybe a combo of communism (little 'c' please note) and 'democracy' is the answer to the problem…I believe there are good pts in each and bad pts in each as they are being practiced today.

At the end of February 1954, after a romantic set-back, her mood sank:

> I refuse to recognize the fact that I exist. This makes for one long period of…nothingness…of being a blob of protoplasm, a disgusting form of being.…What a frightening feeling to have that you are not existing. Its like that damn tree that falls down in a forest. There is no one there to hear any sound—was there any noise when the tree fell? It's the same with me now.…My life is one big mass of irrelevant nervous twitches with no pattern holding them together.

One day, she befriended a young Latvian couple, Aria and Voldus, who had immigrated to America after the Second World War, fleeing Soviet repression. Of her new friends, Nancy wrote,

> They of course appreciate American democracy more than I could start to conceive of.…[In the Soviet Union] kids don't have a chance. They can't think for themselves at that age. The Communists therefore start with the youth organizations.…If they didn't join they were suspected.…And we worry about social inequality. God.

She grappled with a particularly Latvian view of post-war politics.

> Another point Voldus brought up which I had never heard concerns Jews. The way he sees it, people who are Jews have been responsible for mass atrocities in Latvia and elsewhere, because they are sheltered by Russia. Russia makes use of them while other countries persecute them. The Jews are evidentially given the power in satellite countries…

> It looks as tho' these particular Jews are completely mercenary—playing ball with the powers that be. I don't know where their religious beliefs fit into this, but I should think it wouldn't be hard to fall back on wealth, money, power as the most important value for keeping them alive after the way they have been persecuted. What do they have to live for—no country. Who can they trust?

> Voldus mentioned before about Jews having all the power in N.Y.C.…but last night he said McCarthy was persecuting Jews in this country, just as affectively

[sic] as done in other countries....But this I'm not taking as a fact yet! It just happens that Rosenberg, jailed spy, is a Jew.[4]

Kennan speaks

Nancy was invigorated by hearing George F. Kennan of the Institute for Advanced Study speak:

> [Went] to hear Kennan tonite, [the] former ambassador to Russia. Very good. Tempted to skip class tomorrow to hear him again. He even help[ed] complete my understanding of the conflict in America betw. the 'American Dream,' tradition etc. vs. realism etc., or always looking at the bright side of nature vs. accepting differences as such and attempting solutions to problems, not ignoring them. The world is not that rosy.

After serving as a State Department official in Moscow during and after the Second World War, Kennan had written a politically influential paper calling for the "containment" of the Soviet Union. Later, hard line American militarists used Kennan's paper to justify increasing the production of nuclear armaments and upping the level of confrontation with the communist states. Ensconced in Princeton, the enraged Kennan loudly objected to such bellicose use of his ideas, saying he had assessed the Soviets as definitely not interested in world domination. He believed that the Stalinist "autocracy" would crumble in the face of internal democratic movements, and that a nuclear arms contest would only prolong the existence of that autocracy. Kennan insisted that his call for "containment" was a call for economic and political isolation, not military encirclement. Allied with J. Robert Oppenheimer, who was his boss at the Institute, he had strongly opposed America's engineering of the hydrogen bomb, seeing it as an impediment to peaceful co-existence with the socialist bloc.

A few days later, Nancy went to hear Kennan speak again:

> He "seemed" very comforting in that he expressed a sober view of situation, but very alarming in that he expects so much of the American people and gv't. He said, in short, that Russia must work out her own problems from within....Any thought of war on either side is ridiculous in that it solves no one's problems....He says [the Russians] are only misguided people, not out to envelope the world (but they are?)...We could only attack in retaliation if we ever need to otherwise no one would be on our side! (Now [I] see why McArthur [was] taken out of Korea so fast)...I can see where [Kennan] could be branded by some to be a 'pinko' or 'subversive' because he is 'against war'—but I am all for him (mostly, tho', it isn't all clear)....

---

[4] On June 19, 1953, Julius and Ethel Rosenberg were electrocuted for passing atomic bomb secrets to the Soviet Union, so either Nancy's use of the present tense is a slip, or she had missed the news of their execution.

He says, in summary… [Those who want war are] the McCarthys who cause fear and dissension, the capitalists who demand war profits, the service leaders, and, I suppose, many people in higher up positions who think in terms of 'civil defense' and nothing else.

The next night, enthralled, she attended a third lecture:

Tonite, Kennan emphasized [the] duty of Americans to be open-minded enough to respect others' opinions, and to accept and give positive suggestions to [other] countries' problems…. [He] treated [the] U.S. just like [a] psycho patient, as being neurotic and provincial, [and] unable to receive stimulus from outside if [we] put up protective barriers!! (immigration laws, military defense fear, witch burnings etc etc!!)

SO—we ALL need positive direction….Marge seemed more interested in snagging interesting looking males at [the] lecture. At least she admits what Jeanne and I don't, I s'pose? But there is a time and a place for almost everything.

But after talking some more with her Latvian friends, Nancy was no longer sure if she agreed with Kennan that socialism was evolving toward democracy.

For these people, death by cobalt bomb is better than [the] life of terror they now endure (which we have no conception of).

Paradise lost and regained

In April 1954, Nancy met a Ukrainian friend of her Latvians, an artist named Marco Stefanovitch Zubaretsky, "a dashing creature from a far-off land," with whom she embarked on a tempestuous affair (losing her virginity). She neglected to write in her journal for four months. In August, she returned,

I've been living in a whirlwind of activity, but…there's no meaning to it….I can't write what I really feel, although I s'pose no one will ever read this.

On a separate piece of paper, tucked into the diary, she penned her thoughts about having sex with Marco, with whom she was squabbling:

My next husband [sic] won't like it, but he will understand what he's got. It meant a lot to me to be a virgin when married; but a life lived with one who is a mature person means more, call me what you will…I was and still am under [Marco's] physical power—seduced, mesmerized or whatever…It does not follow that if a person is taken, conquered physically, and conditioned to being with one person perpetually, that [a] person will not be able to live happily without the other. It does NOT follow that LOVE grows from physical relationship alone….[But] I can't live without [Marco's] physical presence.

Marco moved to nearby Camden, and the relationship alternately cooled and steamed. By December, Nancy's bosses at ETS were fed up with her inattention and mistakes. They asked her to find a new job, telling her that she could *not* use ETS as a job reference.

Stuck in a behavioral pattern that stayed with her for the rest of her life, Nancy rode a roller-coaster of uncontrollable emotion—rising in a moment from the pit of rejection to the height of romantic joy and falling again into despair and ennui superceded by skyrocketing elation. She became desperate to break with Marco:

> I am dabbling beyond my ken. He is of some other world which I can never grasp [the] meaning of.

Eager to change, she gave up cigarette smoking, but it did not help:

> I must, I think I really must leave Marco. But I can't really face the problem sensibly since there is nothing logical about the whole relationship. I'm afraid it will go on like this until he finds someone to marry. Then he will drop me....but in spite of his horribleness I can not yet let him go. Fool.

In September 1954, she typed a long letter to Marco, who was attracted to socialism:

> Don't you know yet that there is no easy way to achieve social justice, if there should be any such thing?...I have a vague idea that [the] United States will certainly not stay the same as it is now forever; that some day it will be socialistic rather than democratic; all well and good if we've ironed the little problems, like loss of individual choice, and a lot of other things that you actually want.

Then, nothing in her diary until Feb 27, 1955:

> I try to understand Marco through Nietzsche in whom he believes as I believe in Christianity. Damn oh damn....It is a world of dog-eat-dog, but if you are passive like I am, and slow, then the other fellow is going to be using me....God, we don't have much to say about our lives, I swear to God we don't...Damn it, I am mad and scared because I am afraid of becoming a whore, prostitute, slut or something. I've never been tempted before, and now, at [the] tender age of 25, I am falling fast....

Impulsively, she slept with another man.

> And then, out of the blue, with no forethought or intent of malice. I've done what I've always been afraid of [doing] since Marco has been fading from the picture. (I hope my grandchildren enjoy this, but better to learn.) Now I'm afraid to go out with any one.

In April, she found a job as an administrative assistant at a community planning firm in Princeton. As the United States rattled its nuclear saber at China for coveting Taiwan, Nancy's world view took an aggressive turn:

> Shall we drop [the] A Bomb on China and get it all over with? Maybe!?! They're just baiting us to see how much they can get away with without starting a war (a real one). Well, the sooner an end is put to it the better!!

Then, she stopped writing in her diary until February 22, 1956, seven months later. On that day, we learn that her heart has been won by a new man, Hugh Everett III, whom she met at a "physicists party" at the graduate school in June.

He had helped her move her furniture into a new apartment. He was taking a class in ballroom dancing with her. "When I met him, he was looking for a typist" [for his dissertation].

She has "exchanged" Marco for Hugh.

> When I used to watch Marco paint, it made me want to do it, and when Hugh is madly writing on his thesis (must finish 1st draft by April 1st), it makes me want to write and [face] myself a little…I am unsatisfied, even unhappy with my existence.

Portrait of Hugh Everett III, circa 1958.

Portrait Nancy Gore Everett, circa 1958.

She reflects:

I must look at human relationships as a game like Hugh is able to do, but [I] can't get out of self. Poo.

And in June 1956:

And NOW that Hugh has gone the way of all Princeton Grad students and has had to seek work in Washington....I [realize that I] have allowed myself to live perfectly happily, obliviously, under Hugh's influence, even tho' I can only admit to myself that we will never be serious [and] marry....I tried to believe that we might someday, but I just don't think we will. But now I feel alone again, as after Marco. You do have to have someone once you have had someone, it seems. That is what I'm afraid of.

Nestled inside the last few pages of Nancy's Princeton diary is a carefully torn newspaper clipping, a few paragraphs from a literary review of a book about Lucrezia Borgia, the 15th century femme fatale, daughter of a Pope, suspected murderess:

> 'Even as a woman,' writes Miss Haslip, she 'retained a curious look of immaturity,' the look of one from whom 'you would not expect a strict moral standpoint or an irrevocable decision,' a person 'with whom everything was soft and pliable, [whose] very nature was fluid, flowing with the tide, accepting with a smile whatever life had to offer.'
>
> In other words, a woman so lacking in definition that she would have attracted no more than passing notice had it not been for the extraordinary background against which she played a part.

And it is in this curious image of the wall-flower cum poisoner, Lucrezia Borgia, that Nancy, who was to became obscured by Everett's shadow, once saw her own reflection.

BOOK 2
GAME WORLD

5 Demigods

There is a tendency to forget that all science is bound up with human culture in general, and that scientific findings, even those which at the moment appear the most advanced and esoteric and difficult to grasp, are meaningless outside their cultural context.

Edwin Schrödinger, 1952[1]

Shortly before the start of the fall term of 1953, Everett arrived in Princeton driving a Buick loaded with an automatic transmission (a luxury item in those days). Eyes of grey, he was five feet, nine inches tall and trim at 155 pounds. His red-brown hair was slightly receding, worn slicked back. Parking at the Graduate College, he lugged his possessions to the top floor of the residence hall, which resembled a gentlemen's club. Everett and his British roommate each had a tiny bedroom; they shared a bath, and a living room with a fireplace. The basement sported bowling alleys, ping pong tables, pool tables, a television, and an ice machine.

Established in 1746, Princeton's neo-Gothic architecture was patterned on the gargoyle-studded buildings at the British universities, Oxford and Cambridge. Cultural traditions were also imported. Black, floor-length gowns were worn to dinner in oak-paneled Proctor Hall under a stain glass window representing the "Light of the World Illuminating the Seven Liberal Arts of Christian Learning." Tipping the housemen, waitresses, kitchen help, and gardeners was forbidden, because, according to the student handbook, that biased the servants toward wealthy students.[2]

The economic and social life of Princeton town revolved around the Ivy League university, the headquarters of Educational Testing Services, a Radio Corporation of America research lab, and the internationally renowned Institute for Advanced Study. Princeton was a major center of revolutionary work in the physical sciences, attracting gown-wearing geniuses to settle in the bucolic town. In the early 1930s, the Institute's star catch, Einstein, described

---

[1] Schrödinger, E. (1952). 110.
[2] Information about the graduate college is from Everett's basement files and interviews with his classmates.

Princeton as, "A quaint, ceremonious village of puny demigods on stilts."[3] But in the neon glare of post-war consumerism, as the academy allied with the national security apparatus and weapons manufacturers, many a demigod exchanged gown for business suit.

Tea time for brainiacs

The Institute for Advanced Study was founded by the millionaire owner of Bamberger's department stores in 1930. It supports theoretical work in history, public policy, mathematics, physics, economics, and information theory. For many years, it was the intellectual home of Einstein and John von Neumann, both of whom profoundly influenced Everett's intellectual development. It was also a fount of operations research. In 1952, the Institute fired up the first of the "Princeton-class" computers, digital machines built to von Neumann's specifications with military funding, and installed at the Los Alamos, Oak Ridge, and Argonne National Laboratories, as well as at the RAND Corporation, an Air Force "think-tank" based in Santa Monica, California.[4]

The Los Alamos computer (acronym: MANIAC) cut its baby teeth crunching numbers for the hydrogen bomb project.[5]

Although the Institute was not formally affiliated with the university, it's woody campus was within walking distance. Its faculty mixed easily and often with their college-based counterparts, attending afternoon teas sponsored by Princeton's mathematics department in Fine Hall. Said Oppenheimer of the teas: "It is where we explain to each other what we don't understand."[6] Chatting informally, professors and students discussed the logic of economics and warfare in ways that rippled out into the world. Everett was a regular at the teas.

In his first year at Princeton, Everett attended weekly game theory seminars at Fine Hall led by Albert Tucker and Harold Kuhn. The two professors also organized a series of formal game theory conferences featuring the illuminati of game theory: von Neumann; Princeton professor of economics Oskar Morgenstern; Massachusetts Institute of Technology (MIT)'s John Forbes Nash; and RAND's star gamer, Lloyd S. Shapley. At one of these conferences, Everett presented a paper on military tactics, "Recursive Games," which is considered a classic in the annals of game theory.

In its early days, game theory was intended to provide rational solutions to problems in economics, sociology, and military planning. But as the shadow of McCarthyism darkened the halls of academia, the game-playing habitués of Fine Hall quickly learned that dissenting from political agendas (however

[3]   Pais, A. (1982). 453. Written in a 1933 letter to Queen Elizabeth of Belgium.
[4]   Institute for Advanced Study. (1954); Edwards, P. N. (1996). 61; Wheeler, J. A. and Ford, K. (1998). 209. RAND is a contraction of "research and development." It was founded by Douglas Aircraft Company a few weeks after the atomic bombing of Hiroshima. It became a separate non-profit firm in 1948 and remains a major player in national security contracting.
[5]   LANL History.
[6]   DeWitt, C. M. and Wheeler, J. A. (1968). x.

irrational) came with a price. Shortly after Everett arrived in town, the Institute's director, Oppenheimer, was publicly disgraced by the board of the Atomic Energy Commission. Oppenheimer had been the scientific leader of the Manhattan Project, but, after Hiroshima, he openly opposed developing a hydrogen bomb. In retaliation, the commission stripped him of his security clearance. The spurious reason given for the discipline was that in his youth, Oppenheimer had leaned to the left ideologically, and was now (although not when these matters were reviewed prior to his hiring by the Manhattan Project), "susceptible to influence" and, therefore, unreliable.[7]

In the wake of political persecution, Oppenheimer kept his job; unlike Princeton assistant professor David Bohm, a stellar quantum theorist, whose academic career was derailed by McCarthyism (and, to a lesser extent, by his renunciation of the prevailing interpretation of quantum mechanics).

Memory lane

During the first semester, Everett made three good friends among his classmates. Charles Misner was his sounding board as he worked up his interpretation of quantum mechanics during their second year in grad school; they remained friends until Everett died a quarter century later. Hale Trotter ended up chairing the mathematics department at Princeton. And Harvey Arnold retired from academia after teaching statistics for many years at Oakland University in Rochester, Michigan. In 2007, the three men, then in their late seventies, got together to reminisce about Everett.[8]

The four classmates met regularly for cocktails before dinner. Everett was fond of cherry herring, a cheap, sugary liquor that packs a wallop. And he had a collection of hard liquor in his room, mostly purloined from his father's liquor cabinet. Arnold, Trotter, and Misner liked traipsing upstairs to Everett's room with a bucket of ice from the basement. He poured them nightcaps from an odd bottle with four spouts, each spout tapping a different liquor.

TROTTER: Everett was a lot of fun. He enjoyed needling people.

MISNER: But always in good humor.

ARNOLD: He was very competitive at whatever it was, if it was a poker game or it was ping pong.

MISNER: But it was always friendly competition.

ARNOLD: But he always wanted to go away the winner and he would make you stay there until he succeeded which didn't usually take that long with me. But if I'd win one he'd say, 'We're playing another game!' And it surprised me after I got to know him that he was as brilliant as he was. It didn't come across until you got close to him. And then you would recognize that this guy would be on top of the world. He was smart in a very broad way. I mean, to go from chemical engineering

---

[7] Bird, K. and Sherwin, M. J. (2005). 541.

[8] Mark Everett interviewed the men during the filming of *Parallel Worlds, Parallel Lives*.

to mathematics to physics and spending most of the time buried in a science fiction book, I mean, this is talent.

The men toasted their departed friend with a glass of sherry.

TROTTER: Everett's father visited Princeton once, briefly. He let it out that within the family Hugh was called 'Pudge.' Hugh was not pleased.

During their third year of grad school, the four shared an off-campus apartment. Everett liked to gorge, especially on steak. That was the year he met Nancy Gore. "She was a delightful, outgoing, pleasant person," Arnold recalled.

Probability is everywhere

Physics graduate students were expected to be self-starters. Professors did not take class attendance, nor issue grades. Students were required to pass their "general examinations" at the end of the second year (equivalent to a Masters degree), and then to write a doctoral thesis.

During his first semester, Everett took a course in electro-magnetism, a seminar in algebra, and Introductory Quantum Mechanics, with Robert H. Dicke, a model of the scientist-businessman. During the Second World War, Dicke worked on radar at MIT's "Rad Lab." In the 1960s, he helped confirm the existence of the microwave background radiation, which is a lingering record of the state of the universe as it existed immediately after the Big Bang. He also did important work testing predictions about gravity made by Einstein's theory of general relativity. Uniting theoretical and applied physics with information theory, he invented and marketed a type of amplifier (called a lock-in amplifier), which pulled coherent messages out of electronic jumbles of white noise.

Under Dicke's tutelage, Everett studied John von Neumann's classic text, *Mathematical Foundations of Quantum Mechanics* (1932), and David Bohm's textbook, *Quantum Theory* (1951). As a student of quantum theory and, also, game theory, (both fields disciplined by the mathematical rigor of von Neumann), Everett was immersed in the mathematics of probability: that mysterious combination of information and belief that obeys formal laws. The science and art of measuring probability is embedded in most decision-making processes: from cooking to playing dice to making love or war. Probability quantifies risk; and probability is the principal tool wielded in game theory, quantum mechanics, and operations research.

The study of how people make decisions, or more bluntly, how people gamble for stakes, began in the mid 17th century in France. Blaise Pascal, Pierre de Fermat, and, later, Pierre-Simon Laplace were among those who discovered the axioms of the probability calculus, a method of adding and subtracting and multiplying ratios that gave birth to statistics, the study of data on behalf of the political state, hence the appellation.

In the latter part of the 19th century, James Clerk Maxwell and Ludwig Boltzman applied probabilistic methods to the microscopic world: constructing a statistical explanation of thermodynamics (i.e. laws that govern the transformation of energy from one form to another). They uncovered a dialectical relationship between entropy[9] and probability, between microscopic systems and macroscopic systems. Much later, in the 1920s, probability was, surprisingly, found at the existential root of quantum physics.

Practitioners of quantum mechanics and game theory (and bookies!) make lists of possible events and rank them with odds. They chart directions in which the future may unfold based upon information—a history of the frequency of past occurrences—and the *belief* that future events are related to past events. There are, however, important differences between how the laws of probability pertain to the "classical"[10] world, composed of macroscopic objects, and the quantum world, composed of elementary entities describable both as objects (particles) and probability waves.

In classical physics, reality is governed by deterministic laws, i.e. classical mechanics. The probability of accurately predicting an outcome in classical mechanics is a function of possessing information about inputs. For example, if we know the mass of our planet and the mass of our star and the distance between them, we can use gravitational formulas set down by Isaac Newton in the 17th century to predict the path of the Earth's orbit around the Sun with close to 100 percent accuracy. In classical physics it was thought that an observer who possessed perfectly accurate information about the initial state of every object in the universe at a given instant, could, in theory, exactly predict the configurations of all future states of the universe. Of course, the universe is far too complex (and relativistic) for us to access that much simultaneous information, but, in theory, the motion of the classical, macroscopic world is *predictable* because its laws are determinist.

In classical mechanics—and game theory—the more information we have with which to calculate predictions, using the "relative frequency" with which certain outcomes have previously occurred, the less *uncertainty* we will have about a possible outcome in the future. Possessing information about the past allows us to hold a *degree of belief* that past relations will continue to prevail in the future. For instance, when heads emerges from a series of coin flips 50 percent of the time, it is reasonable to believe that this ratio will continue into the future. But for any *single* coin flip, we cannot know with any certainty that the coin will be heads, even if the last 100,000 flips have all been tails! All we can reasonably believe is that there is a 50 percent probability that it will be heads (or tails) each time the coin flips. It is often said that probability is a measure of ignorance (and ignorance is the opposite of information). So, the greater the information content (e.g. a record of the relative frequency of heads turning up in the past), the greater the predictive power of the probability calculus.

---

[9] Entropy is a measure of disorder in the energy of a closed system.
[10] "Classical" physics is the mechanics based upon the paradigmatic theories of Newton and Maxwell that ruled science until quantum mechanics and relativity were discovered at the turn of the 20th century, rendering the previous paradigm "classical."

But there is no such thing as possessing (even in theory) perfectly accurate information about the properties of a quantum object at any given moment. Consider an electron: it has no identifiable position until we "measure" or interact with it. And knowing something definite about one property of an electron, such as its exact *position* in space, means that we cannot simultaneously and precisely know its *momentum*.[11]

Uncertainty rules the quantum world: an electron does not have a definite orbit around an atomic nucleus, rather, its path is smeared out inside a range of possible orbits, a "cloud," which we know as a probability distribution.

It appeared to most physicists of Everett's day that the quantum world was fundamentally indeterminist, although Everett was to disagree with that interpretation; his model of the quantum world is thoroughly determinist. He had no particular longing to live in a deterministic universe, he was simply following the logic of the standard equations. But before tackling the quantum question, he cut his probabilistic teeth on game theory, which is classically determinist.

[11] Momentum is mass times velocity; according to the Heisenberg Uncertainty Principle, position and momentum are a unity of opposites. Increasing the definiteness of one quality means that the other quality becomes less definite.

6 Decisions, Decisions— the Theory of Games

Decisions made in the face of uncertainty pervade the life of every individual and organization.... Reasoning is commonly associated with logic, but it is obvious, as many have pointed out, that the implications of what is ordinarily called logic are meager indeed when uncertainty is to be faced.... [L]ogic alone is not a complete guide to life.

Leonard Savage, 1954[1]

For millennia, soldiers have mirrored real life by playing military games. In the 19th century, Prussian generals played *Kriegsspiel*, a game that tested rules for maneuvers with chance occurrences, such as a surprise attack. At Princeton, Everett learned the military application of modern game theory directly from the academics who invented it; he absorbed that dark art into his personality.

The Theory of games was largely systematized by von Neumann, the Hungarian-born polymath. In 1928, he proved an important theorem showing that two players with completely opposed interests always have a rational strategy for playing a "zero-sum" game, defined as an outcome where one player's gain mirrors the other player's loss. In 1944, von Neumann and Morgenstern published *Theory of Games and Economic Behavior*. The massive, highly technical volume was the Book of Genesis for the game theory industry that rose out of the ashes of the Second World War on the back of operations research, defined as the application of statistics to warfare and other complex undertakings. Stating their goal, von Neumann and Morgenstern wrote,

> We wish to find the mathematically complete principles which define 'rational behavior' for the participants in a social economy, and to derive from them the general characteristics of that behavior.[2]

At its foundation, game theory posits the existence of rational players who make decisions based on the relationship between information, probability, and personal preferences, or "utilities." Its efficacy, naturally, hinges on how one defines "rationality." And, in von Neumann's scheme, rationality is equated with

[1] Savage, L. J. (1954). 6–59.
[2] von Neumann, J. and Morgenstern, O. (2004). 31.

the maximization of "expected utility"[3] in a list of ranked preferences for a certain outcome.[4]

Although von Neumann and Morgenstern's work targeted how profit-seeking individuals should act inside a capitalist economy, the mathematical techniques that they developed were quickly applied to optimizing choices in military tactics and strategy, including nuclear war planning.

Game theory calls for assuming that *equally rational* opponents will pick their optimum choices out of a finite set of possible strategies. The "pay-off" values are arrayed in matrices, i.e. intersecting rows and columns displaying the consequences for each player's chosen strategy as dependent upon combining the decisions of the players.

The following definition of game theory was written in 1967 by former RAND[5] economist (and the 2005 Noble Laureate in economics), Thomas Schelling, not long after he played an instrumental role in designing the rationale for carpet bombing North Vietnamese cities during America's war on Vietnam:

> Two or more individuals have choices to make, preferences regarding the outcomes, and some knowledge of the choices available to each other and of each other's preferences. The outcome depends on the choices that both of them make, or all of them if there are more than two. There is no independently 'best' choice that one can make; it depends on what the others do....Game theory is the formal study of the rational, consistent expectations that participants can have about each other's choices.[6]

The only group that benefited from bombing Vietnamese civilians was American munitions contractors, as the brutal campaign was militarily ineffective. On the other hand, from a humanitarian point of view, carpet bombing was preferable to using nuclear weapons! So, within that decision matrix, using conventional explosives was more rational than using hydrogen bombs. The weakness in game theory, of course, is that the decision matrix may be constructed from fundamentally irrational choices (such as the utility assigned to invading and trying to occupy Vietnam in the first place.)

The meaning of zero sum

In a zero-sum contest the winning player gains the same value that his opponent loses, i.e. wins and losses add up to zero. For two-player zero-sum games, von Neumann defined the solution as a "saddle-point," or "mini-max" equilibrium. In this equilibrium, each player tries to minimize his worst

---

[3] Luce, D. R. and Raffia, H. (1957). 50.
[4] To greatly simplify how choices can be quantified: A 30 percent outcome of event A multiplied by a utility of 10 would be 300; a 40 percent outcome of event B multiplied by a utility of 2 would be 140.
[5] RAND played a large role in Everett's career. He was a frequent visitor to its headquarters in Santa Monica, California.
[6] Schelling, T. (1984). 214–15.

possible outcome or, conversely, tries to maximize his best possible outcome. Tic-tac-toe and chess are zero-sum games in which each player has exactly the same information concerning the playing field and defeat is total. As children quickly learn, tic-tac-toe is easily solved: the saddle-point is a draw. In chess, each player has the same information about the state of play, but the game is much more complex than tic-tac-toe and its saddle-point strategy (if it has one) is unknown. Nonetheless, making random moves is not a rational strategy in tic-tac-toe or chess.

But in zero-sum games without a discernible saddle-point, such as poker, where vital information is hidden, it is rational to adopt a "mixed-strategy," e.g. allowing bluffing or chance to enter the fray. Bluffing has perils: double-guessing a double-guessing opponent can lead to an infinite regression. Over the course of a poker game, players consciously and unconsciously absorb information about each other, while attempting to shield information about their own situation. Bluffing on a random basis is more rational than always bluffing.

In the much simpler zero-sum game, matching pennies, each player secretly turns a penny heads or tails up. The rule is that if the pennies match, one player wins, if they do not match, the other wins. The most rational strategy is for both players to randomly select either heads or tails, without trying to outguess the other, because guessing leads to a feedback loop of, "he thinks that I think that he thinks that I think," ad infinitum. Random play in matching pennies (equivalent to flipping coins) ensures that each player will win 50 percent of the time over the long run. So, why would anyone want to play?[7]

It turns out that *non*-zero-sum games with multiple players get closer to reflecting how real life plays out than do two-person zero-sum games. But game theory, in general, diverges from real life by assuming that all players know and agree to the rules of the game. It presumes that players understand the stakes, and can calculate the probabilities, or "expected values," associated with making certain choices. Furthermore, it assumes that the rational players equally desire a particular type of outcome, a quantifiable utility. But as preferences obviously vary from person to person, and group to group, agreements about what is rational smear into subjectivity.

Despite its weakness as a replica of reality, game theory has been usefully applied to retroactively modeling how decision paths emerge in political science, to tracing stages of development in evolutionary biology, and to analyzing the history of competitive forces in capitalist economies. It is not very successful at predicting the future. Nonetheless, von Neumann-type game theory was influential for about two decades as a guide to war-planning for the governments of the United States and the Soviet Union. Some credit the use of game theory with successfully deterring nuclear war during that period, others say it brought the world unconscionably close to destruction.

Everett was its adept.

---

[7] Oddly, taking risks is often more fun than not.

A critique of military game theory

In 1960, one of RAND's top game theorists, Hermann Kahn, wrote a controversial book, *On Thermonuclear War*. Kahn rationalized the fighting of "limited" and "preventative" nuclear wars on the basis of a high expected utility for liquidating communism. Kahn calculated that "winning" a nuclear war was worth the sacrifice of scores of millions of American lives and the genetic mutation and damage to industrial and agrarian infrastructure that would set capitalist society back thousands of years—until it rebounded as a consumer heaven! Some of Kahn's colleagues, including Everett, admired his cold-blooded, cost-benefit approach to "thinking the unthinkable." Others were righteously horrified and attacked both Kahn and game theory, particularly its assumption of rationality.

One particularly vocal Kahn critic was University of Michigan mathematician, Anatol Rapoport, a Russian émigré with sterling credentials as a statistician and a game theorist. In 1964, Rapoport penned *Strategy and Conscience*. In it, he explained how and why (in his opinion) game theory fails as a model for policy-making:

> It is generally assumed in a theory of decision under risk that a rational decision maximizes expected utility. This is a tautology if utilities are so *defined* that the action with the maximum expected utility is always preferred.[8]

It is not always rational to maximize utility. For example, if the preferred utility for players in a war game is to win all battles in the shortest time possible, then the rational choice for winning a battle might be to launch a surprise attack; but using up all of one's resources on that attack might cause another battle (and the war) to be lost. It is more important, therefore, to define the situational context of a game in relation to the particular rules of the game. In order to reflect reality, games should not be treated in isolation from their greater context. Rapoport summed up,

> Therefore, if a normative [ethically prescriptive] theory of risky decisions is not to be vacuous, utilities must be defined independently of the choices made in the context examined. If this is done, the principle of maximizing expected utility becomes an *additional* criterion of rationality.[9]

For instance, if an important but non-contextual utility of a policy choice is a preference for peaceful co-existence, then it would not necessarily be rational for a government to chose a political strategy focused on military victory, as a greater game will be thereby lost. Contrary to the basic assumptions of von-Neumann-style game theory, said Rapoport, true rationality cannot be bound by the limits of a subjectively constructed decision matrix.

He commented,

---

[8] Rapoport, A. (1964). 72.
[9] Ibid. 72.

In the last analysis, then, arguments in support of probabilities assigned to events are pleas to pay attention to some facts more than to others. A change in our attitude brings up a different set of facts as the relevant set and changes our perceived probability of the event in question.[10]

In other words, the assignment of probabilities to possible events is a subjective act; political and ideological agendas bias the choice of utilities in a policy game, thereby defeating the purpose of using game theory as an objective model.

Rapoport struck at the heart of the matter:

> For the most part, decisions depend on the ethical orientations of the decision-makers themselves. The rationales of choices so determined may be obvious to those with similar ethical orientations but may appear to be only rationalizations to others. Therefore, in most contexts, decisions cannot be defended on purely rational grounds. A normative theory of decision which claims to be 'realistic,' i.e., purports to derive its prescripts exclusively from 'objective reality,' is likely to lead to delusion.[11]

Everett's game

Princeton's Harold Kuhn has made many significant contributions to game theory as a scientist and historian. As a young professor of mathematics in the 1950s, Kuhn co-edited two collections of seminal papers on game theory for "Annals of Mathematics Studies." And in 1953 he published one of the field's first textbooks, *Lectures on the Theory of Games*. Kuhn was so lastingly impressed by Everett that he included a paper he wrote during his first year at Princeton in a book of "pioneering" work from the "heroic age" of game theory titled *Classics in Game Theory* (1997). Everett's relatively unknown "Recursive Games" paper appears alongside a famous paper by John Forbes Nash, the co-winner of the 1994 Nobel Prize in Economic Science.

Kuhn recalls,

> I had no social interaction with Everett outside the weekly game theory seminars and teas, but I got to know him very well. In retrospect, I thought of him as a physics grad student slumming over in math. 'Recursive Games' was a beautiful piece of work; it would have been sufficient to get a Ph.D. in math.
>
> The weekly seminars were small and informal, but extremely important in the history of the Cold War. Practically everybody working on game theory in

---

[10] Ibid. 102.

[11] Ibid. 75; In other words, rationality is a (sometimes) quantifiable *quality*. Most human beings would agree that it is not a rational act to cross the street in front of a speeding bus, or to poison the water supply in search of short term profit, or to depend on fossil fuels, etc. But people in power who do obviously irrational things are often compelled to rationalize these actions by falling back on agendized utility values and probability statements. Of course, if you start with an irrational premise, e.g. "nuclear war is a rational option," no amount of utilitarian quantification can, believably, turn it into its opposite. Context is everyting.

the 1950s regularly gathered in Fine Hall, including von Neumann, Nash, Shapley, and Everett. We also held four formal conferences that were mile markers in the history of game theory.[12]

On January 31, 1955, von Neumann chaired a formal game theory conference at Fine Hall. Everett, who was mid-way through his second year as a grad student, and already writing his thesis on quantum mechanics, presented his "Recursive Games" paper on the first day of the conference; among a dozen other presenters were Nash and Shapley.[13] Kuhn describes how Everett's attempt to model the complexity of real life, broadened the scope of game theory:

> There were two competing, but related theories at that time. One, formulated by Shapley was called stochastic games, and the other by Hugh Everett was called recursive games. And the idea of both of them was that in real life we keep making choices and formulating strategies, and we do not necessarily get a pay-off, but, rather, we are led into another game. So instead of having a matrix of pay-off numbers, we have a matrix in which some entries are other games. And these were attempts to make game theory more realistic.[14]

Shapley's paper, "Stochastic Games," defined a type of zero-sum operation with randomized moves where the probabilities relevant to deciding the next move do not depend upon the history of the game. Shapley found a way to design optimal strategies for each player, so that pay-offs could be determined at some point, and the game would end after a finite number of steps. He theorized that his method could be extended to cover games with "infinite sets of alternatives."

Everett picked up where Shapley left off. He defined "recursive" games where the outcome of a single randomized play can either be a pay-off, or another game, allowing for games with infinite moves. The problem, then, was how to identify an acceptable pay-off in the face of the possibility of infinite plays, (a superior pay-off might emerge in the far, far future). Everett found a strategy for making such decisions, and this was an important discovery. His fascination with cybernetics paid off in game theory. He remarked:

> The situation is fully analogous to servomechanism analysis, where the complex behavior of a closed loop servomechanism is analyzed in terms of the (open loop) behavior of its parts [via feedback]. The theory of servomechanisms is concerned solely with the problem of predicting this closed loop behavior from known behavior of the components. An appropriate name for recursive games would be 'games with feedback.'[15]

Everett illustrated the military application of his theory:

[12]   Kuhn interview, Feb. 2008.
[13]   In 1950, Nash invented his Nobel Prize-winning theory of "equilibrium points" in games, which has many socio-economic applications. And RAND's Shapley is most remembered for the "Shapley value," which measures distributions of political power inside groups and coalitions.
[14]   Kuhn interview, Feb 2008.
[15]   Everett, H III. (1957A). 74.

Colonel Blotto commands a desert outpost staffed by three military units, and is charged with the task of capturing the encampment of two units of enemy tribesmen which is located ten miles away.[16]

Under the rules of Blotto's recursive game, daylight raids are forbidden, and night raids require that the attacking force have one more unit than the defending force. Each player must divide his units into attacking and defending forces. Given the constraints and rules of the game, one can see the possibility for an endless loop of excursions and retreats by larger and smaller forces passing each other by in the night, each hoping to overwhelm the enemy, but beset by the necessity of protecting his home base.

In accord with common sense, Everett's innovative probabilistic approach determined that if Blotto is patient, he will eventually win, because he commands a larger force—but winning might take a *very* long time.

---

[16] Ibid. 76.

7 Origin of MAD

The Kodak slogan, 'You push the button, we do the rest,' which since 1889 has helped so much to popularize photography all over the world, is symbolic. It is one of the earliest appeals to the push-button power-feeling; you do nothing, you do not have to know anything, everything is done for you; all you have to do is push the button. Indeed, the taking of snapshots has become one of the most significant expressions of alienated visual perception, of sheer consumption. The 'tourist' with his camera is an outstanding symbol of an alienated relationship to the world. Being constantly occupied with taking pictures, actually he does not see anything at all, except through the intermediary of the camera. The camera sees for him and the outcome of his 'pleasure' trip is a collection of snapshots, which are the substitute for an experience which he could have had, but did not have.

Erich Fromm, 1955[1]

Prisoner's dilemma

As the Cold War intensified, a fear-based equilibrium evolved. It was called mutual assured destruction, widely known as MAD. As a game, it defined the Cold War, and guaranteed Everett's career in operations research.

The winning strategy for a rational player in a zero-sum game is to wipe out the opponent while minimizing the risk of meeting that fate. Short of winning, the optimal strategy is to avoid defeat by forcing a tie. Risk avoidance mediates decision-making; and game theory attempts to quantify risk. But in real life, not all players are equally rational, and their utility values may be diametrically opposed: one might value revenge, another mercy. And not all stakes are zero-sum. For instance, in *non*-zero-sum games *cooperation* is possible between two or more competitive players. The impulse to make bargains is more reflective of the complexity of real life than winner-take-all, or no-risk options. Consider a game in which players can make enforceable agreements, in which they can maximize their own perceived self-interest, and minimize risk, by choosing the best action available to them collectively *given*

---

[1] Fromm, E. (1955). 137.

that the other players are also choosing their best actions: this is the well-known "Nash equilibrium."

Building on the von Neumann-Morgenstern analysis of cooperative (non-zero-sum) games, Nash's "bargaining" theorem shocked Kuhn and Tucker with its mathematical and sociological beauty. Achieving the Nash equilibrium point is synonymous with the result that no rational player would change his initial decision after the results of the game are known. It rewards cooperative bargaining and penalizes unilateralism. It has many applications in market economics and other forms of group dynamics, including geopolitics. It does not, however, solve the problem that game theory relies upon an ideal of rationality and an assumption of shared utility values.

Nash's breakthrough in 1950 was immediately countered by a game that questioned the rationality of searching for his equilibrium points. Developed at RAND, it was dubbed "Prisoner's Dilemma." It questioned the logic of game theory itself.

Prisoner's Dilemma considers the situation of two prisoners accused of jointly committing a crime (which they did commit). Each prisoner is separately questioned by detectives and informed of the stakes. Each is told that if he testifies against his partner, he will be set free, whereas his silent partner will be sentenced to three years in prison. If they testify against each other they will both be incarcerated for two years. These are the "defect" options. But if *both* prisoners refuse to turn state's evidence, then each will spend only one year in prison. This is the "cooperation" option. The shared utility is that each prisoner wants to minimize the risk of doing hard time; both are presumed to be rational (defined as self-interested).

Obviously, if the prisoners could communicate, the best choice would be for both to cut a deal and clam up (cooperate), and each settle for one year in prison. But that would not be the best retroactive choice (Nash equilibrium) in this scenario, because if your partner defects, while you cooperate, you get slammed with three years. Therefore, the least risky and most rational choice is to defect, as betraying your partner will at worst give you two years in prison, and you have a chance of being set free. But if both prisoners act "rationally" and defect, then each prisoner ends up worse off than if both had acted "irrationally" and cooperated.

Prisoner's Dilemma exposes the contradiction of using a rationality construct based on perceived self-interest as a guide to solving conflicts. It demonstrates that even in a situation with a built-in cooperative option that would equally benefit all players, it is not necessarily rational to cooperate. But it is not necessarily rational to defect, either. The game highlights the paradox inherent in the Cold War strategy of assured destruction, which is based on the assumption that no winner will emerge from the ashes of a nuclear war if one or both defect by striking first—and, yet, it is not necessarily rational not to strike first.

Mutual assured destruction was based on terrifying both sides into building large enough arsenals that it would became rational not to use them. Of course, if a comparative advantage was reached by either side, it would be tempting to preemptively wipe out the enemy. Consequently, the hydrogen bomb-wielding players were trapped in a feedback loop of suspicion and weaponeering as it is not rational to expect the enemy not to strike first when he is afraid that you will strike first.

The decision matrix for assured destruction ignored Rapoport's admonition to look outside the box, in this case at the only truly rational utilities: avoiding war through disarmament or, better yet, never having built the first atom bomb![2] Obviously, the big winners in MAD were weapons manufacturers and operations researchers.[3]

It gradually became clear to game theory practitioners that the theory was more valuable for modeling tactical scenarios, and making cost-benefit analyses of weapons systems, than it was useful as a consistently rational guide to deciding whether or not to strike first, or second, in a nuclear stand-off. A remarkable fact that was not always ignored by Cold Warriors on both sides of the superpower divide was that repetitive plays of Prisoner's Dilemma allow the same two players to gain information about each other's decision-making patterns, thereby reducing the risks tied to cooperation, and making multilateral disarmament an agreeably rational option.

Of great note: modern incarnations of game theory in evolutionary biology show that in life's contest Nature tends to reward group altruism—cooperative behavior—more than individual selfishness and competition.[4] However, in matters of military planning (then as now), cooperative options are easily, and unfortunately, conflated with weakness and irrationality and assigned low utility values.

Everett, who was the quintessential Cold War technocrat, subscribed to the prevalent notion among his generation of operations researchers that, "We prepare for war to prevent war." Values that might have questioned the basis for this rationale were banished from the morally closed world of operations research.

Rapoport observed:

---

[2] Although it was not unreasonable to have raced Hitler for an atom bomb during World War Two, after Nazi Germany was defeated by conventional forces in 1944, that reason for pursuing it ceased to exist. Many Manhattan Project scientists urged Truman to refrain from using the new weapon, correctly observing that Japanese militarism was exhausted and at the point of surrender. As they realized, dropping The Bomb on two cities quickly converted a non-zero sum game into zero sum and set a precedent for future conflict.

[3] Operations researchers often claim that MAD *was* rational because it forced the Soviet Union to its knees in 1989, ending the Cold War. This is a dubious argument that will be taken up in subsequent chapters which examine the true risks that were run by both sides. MAD can be viewed as a form of cooperation between the U.S. and the U.S.S.R. military brass and civilian hawks whose careers were mutually advanced by the Cold War.

[4] Wilson, D. S. & Wilson, E. O. (2007). A consistent measure of rationality, it seems, cannot be universally quantified. It should be safe to say, however, that the continued survival of the human species should be a commonly shared utility value amongst genuinely rational humans. As Everett's colleague, George Pugh, pointed out in his 1977 book *The Biological Origin of Human Values*, the imperative of species survival is encoded in our genes and informs our core values.

The strategist defends nightmare visions of the world as a 'realistic' vision, forgetting that any vision of the world is compounded of elements which one has selected for observation. The strategist sees what he has selected to see.[5]

Or as the satirical songwriter, Tom Lehrer, wrote in the early 1960s,

'Once the rockets are up, who cares where they come down? That's not my department,' says Werner von Braun.

---

[5] Rapoport, A. (1964). 125.

8 von Neumann's Legacy

The model nature is quite apparent in the newest theories, as in nuclear physics, and particularly those fields outside of physics such as the Theory of Games, various economic models, etc., where the degree of applicability of the models is still a matter of considerable doubt. However, when a theory is highly successful and becomes firmly established, the model tends to become identified with 'reality' itself, and the model nature of the theory becomes obscured.... It should be clearly recognized that causality is a property of a model, and not a property of experience.

Hugh Everett, III (1956)[1]

In March 1957, Everett, who was by then a heavyweight in the Weapons System Evaluation Group (WSEG) at the Pentagon, attended the Third Conference on Games at Fine Hall. The meeting was dedicated to the memory of von Neumann, who had died a few weeks before at the age of 53 from bone cancer. Dozens of mathematicians and economists flocked to the heady affair, sponsored by the Office of Naval Research. The participants hailed from Ivy League universities, and such military contractors as RAND, General Electric Corporation, and Hughes Aircraft. The roster of attendees was a cross-section of the best minds working for the "military industrial complex," which President Dwight D. Eisenhower was soon to call a threat to democracy.[2]

Olaf Helmer of RAND presented a striking paper, "The Game-Theoretical Approach to Organization Theory." Having learned the lessons of Prisoner's Dilemma, Helmer concluded:

> The theory of games has in my opinion reached a state of near stagnation with regard to its applicability to the real world.... [W]hat it cannot generally do is to predict the behavior, even of rational players with known utility preferences, with regard to their cooperative options, insofar as these depend on the players' attitudes toward their fellow players and toward any behavior patterns which they may have observed in past plays.[3]

A decade before, Wiener had made a similar comment in *Cybernetics*:

---

[1] DeWitt, B. and Graham, N. eds. (1973). 134–136.
[2] Eisenhower's farewell address to the American people, January 17, 1961.
[3] Helmer, O. (1957).

Naturally, von Neumann's picture of the player as a completely intelligent, completely ruthless person is an abstraction and a perversion of the facts. It is rare to find a large number of thoroughly clever and unprincipled persons playing a game together.[4]

But von Neumann-style game theory pointed the way toward the invention of more useful computational techniques for modeling complex situations in real life (including socio-economic and war simulations), and Everett was one of von Neumann's legatees, in more ways than one.

Rich, socially flamboyant, and supernaturally smart in life, von Neumann remained a demigod long after his untimely death. At the Princeton game theory conference, Morgenstern eulogized his recently departed friend and colleague: "There was hardly another contemporary whose mind encompassed so much and who made such a significant contribution to anything he touched."[5]

Gifted on the left side of his brain, von Neumann had been the first to see that the equations of quantum mechanics could be expressed in a highly abstract "Hilbert space" of infinite dimensions. While axiomatizing the new quantum physics in the early 1930s, he postulated a widely accepted solution to the vexing problem of discontinuity in the measurement of quantum objects: "wave function collapse." (As von Neumann lay dying in Princeton, Everett was thinking up a way to debunk that postulate.) In computer science, von Neumann was one of the inventors of "stored programming," i.e. software. The 50-ton digital computers that he designed for weapons research burned through vacuum tubes like tissue paper, crashed every few minutes, and were among the most elegant machines every built.

A death bed convert to Catholicism, von Neumann had long struggled with the mysteries at the heart of physics and socio-economics. Uncertainty was his sworn enemy, whether it appeared in the nature of the electron, or in the nature of society. He evinced no moral uncertainty, however, about designing the implosion method for the plutonium bomb that destroyed Nagasaki in August 1945. And along with his fellow hydrogen bomb-engineers, Edward Teller and John Wheeler, von Neumann was certain that preparing to wage nuclear war was a historical necessity.

Life magazine recalled in his obituary:

> After the Axis had been destroyed, von Neumann urged that the U.S. immediately build even more powerful atomic weapons and use them before the Soviets could develop nuclear weapons of their own. It was not an emotional crusade, von Neumann, like others, had coldly reasoned that the world had grown too small to permit nations to conduct their affairs independently of one another. He held that world government was inevitable—and the sooner the better. But he also believed it could never be established while Soviet Communism dominated half the globe. A famous von Neumann observation at the time: 'With the

---

[4] Wiener, N. (1948). 159.
[5] Morgenstern, O. (1957).

Russians it is not a question of whether but of when. If you say "why not bomb them tomorrow," I say "why not today?" If you say "today at 5 o'clock," I say "why not one o'clock?"[6]

Physicist-historian Kenneth Ford has a more subtle recollection:

I heard von Neumann advocate for the development of the H-bomb on the grounds that it was necessitated by inherent inaccuracy in ICBMs [intercontinental ballistic missiles]. He wanted a multimegaton weapon, he said, not in order to inflict orders of magnitude more damage, but to take out a target if the missile missed it by 5 or 10 miles. Despite his brilliance, he did not at that time envision the pinpoint accuracy of ICBMs that would later be developed.[7]

As a powerful advocate for hydrogen bomb testing, von Neumann told Lewis Lichtenstein Strauss, the head of the Atomic Energy Commission: "The present fear and vague talk regarding the adverse world-wide effects of general radioactive contamination are all geared to the concept that any general damage to life must be excluded." (In his first project at the Pentagon, Everett was to scientifically disprove von Neumann's low assessment of the lethal effects of radioactive fall-out.)

von Neumann remarked of fighting a nuclear war:

Every worthwhile activity has a price, both in terms of certain damage and of potential damage-of risks—and the only relevant question is, whether the price is worth paying....For the U.S. it is.[8]

Super-feisty Hermann Kahn put the "price" worth paying for victory in a nuclear exchange with the Soviets at 60 million American dead.

Writing to Strauss in 1951, von Neumann revealed his Freudian side:

The preliminaries of war are to some extent a mutually self-excitatory process, where the actions of either side stimulate the actions of the other side....As the conflict's 'foreplay' progresses, the original aggression, and its motivation, become increasingly obscured. [I think] the US-USSR conflict will probably lead to an armed 'total' collision, and that a maximum rate of armament is therefore imperative.[9]

Such a ridiculously sexualized, Manichaean comment would be amusing, excepting that during much of the 1950s, the de facto strategy of the Strategic Air Command under General Curtis LeMay was to "preventatively" launch everything in its nuclear arsenal in what Kahn disapprovingly termed a "Wargasm."

As the arms race heated up, von Neumann sold his services as a consultant to the Central Intelligence Agency, International Business Machines, Standard Oil, and Convair (a strategic bomber and ballistic missile manufacturer).[10]

[6] Poundstone, W. (1992). 143.
[7] Ford, private communication, 2009.
[8] Poundstone, W. (1992). 180.
[9] Ibid. 259–260.
[10] Ibid. (1992). 179.

Racing through the revolving door, he headed the "Teapot Committee," a collection of military and scientific experts that, in 1953, recommended going full speed ahead on the production of ballistic missiles capable of carrying nuclear warheads. Eisenhower appointed him to the Atomic Energy Commission, where he oversaw the nation's atomic energy projects. And he helped design SAGE (Semi-Automatic Ground Environment), the top-secret computer system designed to detect a nuclear attack by the Soviets.[11]

Although Everett barely knew von Neumann, he was an aficionado of his work. He aimed to teach digital computers how to play war according to von Neumann. And he devoted his doctoral thesis to a proof that one of the axioms of quantum mechanics (wave function collapse), per von Neumann, was flat out wrong.

So Long, Suckers

One of Everett's favorite recreations was a zero-sum board game called So Long, Suckers, invented in 1950 by Princeton graduate students, including Nash and Shapley. Similar to Prisoner's Dilemma, it was also known as Fuck You, Buddy. The game encourages chip-trading players to cooperate until it becomes more convenient to deceive and murder each other. Everett himself viewed life itself as a winner-takes-all game, and playing mind games was, for him, an acceptable way to optimize his personal utilities—suckers beware.

Kuhn reflects, "Everett had a good deal of ego even as a graduate student. We were all kind of an elite group, though, so it would have been hard for him to play mind games with us." He lost track of the young game theorist after he, "disappeared inside a classified organization [WSEG]."

During his long career in academia and in private operations research, Kuhn did important work in the field of linear programming, which is a method of programming zero-sum, two person games.[12] Linear programming is used to find the optimal solutions to complex problems in economics and operations research, problems that are subject to a large number of constraints. The method can calculate solutions for optimizing the delivery of medical supplies to all sides fighting a civil war, or to budgeting dollars between ballistic missiles, planes, and bombs for maximum punch in a nuclear strike.[13]

Kuhn, now in his eighties, is sensitive to the tendency of some Cold War historians to emphasize the malign applications of game theory and computerized optimization techniques over their more benign uses. Although Kuhn accepted Office of Naval Research research funding, he avoided working directly for the military. He says,

> It is embarrassing when something you do for purely mathematical reasons is turned around and used for dirty applications. I was like a tailor who makes a

[11] Ibid. (1992). 180.
[12] von Neumann, J. & Morgenstern, O. (2004). x.
[13] Kuhn interview, 2008; McDonald, J. "A Theory of Strategy," in von Neumann, J. & Morgenstern, O. (2004). 710.

coat with six arms one day, a coat with three arms the next day, a coat with seven arms the following day, and then a coat with two arms: if somebody puts on the two-armed coat, so be it.[14]

Everett, on the other hand, did not share Kuhn's qualms; he gladly churned out two-armed coats on demand for the war machine.

In 1963, Everett resurfaced in Kuhn's world, making a splash when a professional journal, *Operations Research,* published his "Generalized Lagrange Multiplier Method for Solving Problems of Optimum Allocation of Resources." The "Everett Algorithm" improved upon a centuries-old optimization technique called the Lagrange multiplier method. A creation of the information age, it simplified the solution of hugely complex logistics problems. (Everett discovered the algorithm under the influence of a few beers while visiting Niels Bohr in Copenhagen in 1959.)

The beauty of the algorithm is that it breaks large, intractable optimization problems into smaller ones that can then be solved. Whereas game theory deals with *how* people make interdependent decisions under certain conditions, the Everett Algorithm tells you how to calculate a range of consequences ("prices") for making those decisions in the real world when you expend a specific amount of a resource to overcome a constraint. It can be applied to logistics problems in which there are large numbers of alternative solutions and variously configured obstacles to those solutions. Problems of this type include maximizing the efficiency of assigning nuclear bomb targets, scheduling just-in-time manufacturing runs, allocating bus routes to most efficiently desegregate school systems, or projecting results of funding specific foreign and domestic policies.

The algorithm was the cornerstone of Everett's career in operations research; but it did not drop on him totally out of the blue sky.

Kuhn recounted,

> In 1951, Tucker and I published a paper which extended linear programming as a very useful and important research tool for doing efficient optimization for all kinds of business. Our approach was effectively a Lagrange multiplier approach. Later, Hugh, working at WSEG, had this bright idea about Lagrange multipliers which, in effect, we had previously generalized in our paper. Traditional Lagrange multipliers applied to situations with *equality* constraints. We had *inequality* constraints, and so did Hugh.

Kuhn said that Everett went a step further than he and Tucker; and he enthusiastically quoted from Everett's innovative paper:

> While the use of Lagrange multipliers does not guarantee that a solution will necessarily be found for all problems, it is 'fail-safe' in the sense that any solution found by their use is a true solution. Since the method is so simple compared to other available methods it is often worth trying first, and succeeds in a surprising

[14] Kuhn interview, 2008.

fraction of these cases. They are particularly well suited to the solution of problems of allocating resources among a set of independent activities.

"Everett's Lambda theorem is a lovely idea," Kuhn continued. "It has very little mathematical content, but it has extreme practical application in operations research."[15]

In his path breaking paper, Everett noted that his multiplier method could be applied to

a problem which often arises in military operations research [which] is the optimum allocation of given stocks of several weapons types of differing characteristics to a diverse set of independent targets. For such problems it is often crucial to account for the fact that weapons can be delivered only in integral numbers.[16]

He described the solution matrix clinically:

In this case, the cells are the individual [nuclear bomb] targets. A strategy for a cell is a -tuple of integers, one for each weapon type, representing the number of that type of weapon to allocate to that target. The payoff in a cell is the expected destruction to the target of the given weapon allocation, and the resource functions are simply the numbers of weapons allocated themselves.... The method has been employed in WSEG for several years for solving both production and military allocation problems, and has been quite successful.[17]

Now that we have introduced Everett as a game theorist, optimization expert, and nuclear war planner, it is time to back-track and see what he was doing in quantum mechanics during his second year at Princeton. But, first, we must note that an off-shoot of game theory—decision theory—is now playing a major role in the international debate about the validity of Everett's Many Worlds Interpretation. Philosophers and physicists at University of Oxford and elsewhere are looking toward the utilitarian and probabilistic structures embedded in decision theory as an argument that it is rational to believe that Everett's multiple universes exist.

Whether or not Everett ruminated about decision theory, *per se*, while constructing his interpretation of quantum mechanics is not known, but the structure of his branching universe resembles a decision tree—to which the interplay of information and probability and rationality is fundamental.

---

[15] Kuhn interview, 2008.
[16] Everett, H, III. (1962). 17. Everett's method proceeds by connecting a series of discrete steps, unlike methods based on continuums, which could end up projecting unrealistic unit numbers, such as 7.1984 missiles each carrying 1.666 warheads.
[17] Ibid. 17–18.

BOOK 3
QUANTUM WORLD

9 Quantum Everett

One might still like to ask: 'How does it work? What is the machinery behind the law?' No one has found any machinery behind the law.... We would like to emphasize a very important difference between classical and quantum mechanics.... We can only predict the odds! This would mean, if it were true, that physics has given up on the problem of trying to predict exactly what will happen in a given circumstance. Yes! physics has given up.

Richard Feynman, 1963[1]

Fall, 1954

Beginning his second year at Princeton, Everett took only one class, Methods of Mathematical Physics. With Eugene Wigner, he studied the basic tools of quantum mechanics: vector analysis, Fourier series, and matrix algebra.

Wigner and von Neumann had been close friends and colleagues since their student days in Hungary. Both had worked for the Manhattan Project and both were foreign policy hawks. But Wigner was more deeply troubled by the foundational questions in quantum mechanics than was his friend. In fact, he questioned the reasonableness of von Neumann's "wave collapse" postulate, and he was not happy with the "complementarity" interpretation of quantum mechanics held by Niels Bohr. Ultimately, Wigner forsook a "realist"[2] approach to quantum mechanics, opting for hypostasizing human consciousness—a method soundly rejected by Everett.

Attending lectures was more or less optional for physics graduate students, but if the men[3] goofed off, or proved to be intellectually incapable, they were not asked back for another term. Adhering to a self-starting regimen of lectures, private study, and discussions with teachers and peers, doctoral candidates were expected to pass a "general" examination at the end of the second year. The reward for passing was a Masters degree and a chance to write a dissertation. Given the high standards for admission (only ten percent of

---

[1] Feynman, R. *et al.* (1965). 1–10.
[2] Realism holds that there is an objective, physical reality underlying phenomena.
[3] Princeton enrolled its first female graduate student, Sabra Follett Meserve, as a Ph.D. candidate in Turkish history in 1961. *Princeton Alumni Weekly*, 3/12/03.

applicants were accepted by the physics department), most doctoral students were capable of academic success. Some, like Everett and Misner, were considered by their professors to be unusually intelligent and bound to make important discoveries.

Misner was born in Michigan in 1932. A life-long Roman Catholic, he came to the physics department by way of the University of Notre Dame. His undergraduate background in math and physics was so impressive that he was not required to take classes (although he did take a course in quantum field theory). He spent his first year working on a thorny problem in nuclear physics, (when he could break away from watching the dramatic Army-McCarthy hearings on television). But he took the time to read von Neumann's classic textbook on quantum mechanics in the original German, as did Everett. (German had long been the prevailing language of quantum theory, and von Neumann's textbook was not published in English until 1955.) After passing his general exam in the spring on 1954, Misner spent the summer working at Bell Labs. That fall he briefly "shopped" for a thesis advisor, selecting John Wheeler, and beginning a life-long association with him.

Left to right: Charles Misner, Hale Trotter, Niels Bohr, Hugh Everett III, David Harrison, Princeton, 1954. *Alexandria Gazette*

In the fall of 1954, Niels Bohr was in residence at the Institute for Advanced Study.[4] Inspired by conversations with Misner, Wheeler, Bohr, and Bohr's assistant, Aage Petersen, Everett left game theory behind for the time being. Signing on with Wheeler as his thesis advisor, Everett started to research his dissertation six months before he got his masters degree in the spring of 1955. Soaking up the formalism and philosophy of physics, and in love with paradox *qua* paradox, Everett chose to attack one of the deepest puzzles in quantum mechanics, the "measurement problem." This seemingly intractable paradox was identified during the early, heady days of quantum mechanics in the 1920s. Three decades later it still stymied and stumped and frustrated those who tested their brains against it.[5]

Introducing the measurement problem (a little bit)

The formalism of quantum mechanics works. But why and how it works is a matter of interpretation.

In 1952, one of the founders of quantum theory, Erwin Schrödinger, (he invented the Schrödinger equation), complained about Bohr's "anti-realist" approach to the measurement problem:

> This interpretation is obsolete. There is nothing to recommend it, and it bars the understanding of what is actually going on. It obstinately refuses to take stock of the principle of superposition.[6]

Superposition is a property of waves. In quantum physics, particles are also waves—often described as "probability waves." Probability waves, like water waves, can merge, or superpose. The mathematical description of a probability wave (or quantum "state") is called a wave function, symbolized by the Greek letter psi, ψ. Schrödinger's equation tracks the evolution of ψ through time as it combines (superposes) with other wave functions: ψ_x plus ψ_y equals ψ_z. But if we *measure* (physically interact) with the entity ψ_z it seems to evaporate: leaving us staring at a piece of it, either ψ_x or ψ_y.

This is an extraordinarily counter-intuitive process; and we shall keep returning to it, searching for understanding. For the moment, think of it like this: The wave function, ψ, of an atomic particle contains *all* of the information about the different ways in which that particle could possibly behave. For example, the total wave function for the position of an electron lists all of the positions it could possibly assume in its environment. Each possible position

---

[4] In September, 1954, Everett's hometown newspaper, the *Alexandria Gazette*, ran a photograph of him in a gown, smoking a cigarette, welcoming the "Danish savant," Niels Bohr, to a sojourn in Princeton. During the summer break, Everett had, according to the caption, "[given] a lecture on a special interest of his called 'Game Theory,' a kind of scientific guessing that has military implications."

[5] "'I embarked on [a project] to revise quantum theory,' Nash said.... It was this attempt that Nash would blame... for triggering his mental illness—calling his attempt to resolve the contradictions in quantum theory, on which he had embarked in the summer of 1957, 'possibly overreaching and psychologically destabilizing.'" Nasar, S. (1998). 221.

[6] Schrödinger, E. (1952). 115.

on that list is a wave function. The total wave function for the electron's position superposes (combines) the wave functions for each possible position. Conversely, this means that the total wave function can be decomposed into its constituent pieces, also wave functions. Graphed on a mathematical coordinate system, wave functions can be seen to act like ordinary waves. Where the peaks of wave functions superpose, the probability for certain positions to become definite increases. And where a wave peak meets a wave trough, probabilities cancel.[7]

The superposition principle asserts that, *before measurement*, the wave function for an electron orbiting an atomic nucleus describes the electron existing at all of its possible positions in space and time; it can be envisioned, not as having a definite orbit, but as a cloud of indefiniteness surrounding the atomic nucleus. On a graph, the superposition is represented by a single wave that sums up all of its component pieces.

The *measurement problem* arises when a scientist employs a scientific instrument, such as a Geiger counter, to register the position of a particle. Instead of finding the particle in the multiple positions documented by the superposed wave function, the scientist finds it confined to a single position (as a *result* of measuring it); he would be astonished to find it in more than one position!

This is the mystery: when we measure the position of an atomic particle we record it as existing in a definite place, not in all of the many places it occupies according to its smoothly evolving wave function. The emergence of a single position from the set of all physically possible positions is inexplicable; it creates a logical discontinuity, a gap, a fissure, an interruption in the flow of the Schrödinger equation: it creates a *problem*.

In the early years of quantum theory, leading physicists decided that the Schrödinger equation just does not apply to atomic systems at the moment they come into contact with measuring devices (or our brains). But the equation instantly reapplies to the system subsequent to a measurement or observation, they said. Shortly after Bohr arrived in Princeton, Everett had his own brain storm: he decided to treat the Schrödinger equation as uninterrupted—as correctly describing physical reality as continuous, not broken—and then see what happened. And in doing so, he challenged the most powerful person in quantum physics: Bohr.

Enter Bohr

During Bohr's lifetime, he was lionized as a teacher, a philosopher, and a founder of quantum physics. In the half century since his death, hagiography has given way to more critical analysis. One historian, Mara Beller, portrayed him as caring more about preserving his personal prestige than searching for scientific truth. Other researchers assert that his actual part in developing the

---

[7] Technically, the positive value of the wave function is squared to determine the probability. More about this later.

Young Niels Bohr, date unknown.

Old Bohr, circa 1959.

Copenhagen interpretation of quantum mechanics—which is usually thought to be Bohr's intellectual baby—has been obscured by "mythology."[8] Nonetheless, it is a fact that Bohr decisively shaped *how* quantum physics was and is presented—not only to physicists, but to the world at large—as an exact science troubled by paradoxes that can be dismissed for all practical purposes. And in his thesis Everett equated Bohr's philosophy with the Copenhagen interpretation.[9]

Niels Henrik David Bohr was born in 1885 in Copenhagen to a closely knit, upper middle-class, scientific family. His father, Christian, was a professor of physiology (and later, the rector) at the University of Copenhagen. His brother, Harald, was a famous soccer player and an accomplished mathematician. After receiving his doctoral degree from the University of Copenhagen, Bohr studied in England with the leading atomic physicists of the day, J. J. Thompson and Ernest Rutherford.

In 1913, he claimed his pedestal in the pantheon of physics by explaining why an electron continuously orbiting an atomic nucleus does not radiate energy and fall into the nucleus in a fraction of a second, as was required by conventional electromagnetic theory and Newton's laws of motion. Using a physical constant that Max Planck had introduced in 1900, the "quantum of action," Bohr correctly theorized that electrons orbit in discrete energy levels ("shells") surrounding the nucleus, while seeming to "jump" from one orbit to another—without passing through the space between—as the atom emits or absorbs radiation.

Although Bohr could not explain *why* the elementary particle behaved in this odd, discontinuous manner, his insight helped shatter the "classical" paradigm of physics. Impressed by his achievement, the Danish government appointed him the first professor of theoretical physics in the history of Denmark. He established the Institute for Theoretical Physics in 1921 in Copenhagen with funding from the government and the Carlsberg Foundation. He won a Noble Prize for Physics in 1922 for his work modeling the atom; and his institute quickly became an international locus of research in quantum mechanics.

Pipe-chewing, avuncular, and ruthlessly inquisitive, Bohr was involved in most of the important discoveries in nuclear physics made during the 1920s and early 30s. He mentored the young genius, Werner Heisenberg, as he birthed the uncertainty principle. He drove Schrödinger to distraction with penetrating questions about wave mechanics. He argued endlessly with Einstein about the probabilistic nature of quantum mechanics.

And he addressed the measurement paradox by artificially partitioning the universe between the microscopic world ruled by the Schrödinger equation and the macroscopic world governed by the "classical" laws of motion and electromagnetism.

---

[8] See: Howard, D. (2004). 669–682 and Camilleri, K. (2009). 25–57.
[9] DeWitt, B. and Graham, N. eds. (1973). 110.

In 1927, Bohr declared that all talk of the quantum world must, on pain of incomprehensibility, be couched in the language of classical physics. He modeled the universe philosophically as a unity and struggle of mutually exclusive phenomena and called it "complementarity." And his philosophy was integrated into what eventually became known as the Copenhagen interpretation of quantum mechanics.[10]

Years later, Bohr's assistant, Petersen wrote:

> The word 'reality' is also a word, a word which we must learn to use correctly.... In Bohr's view, the core of the problem of knowledge is in our separation between subject and object.... In physics, if anywhere, we keep ourselves outside the description.... [T]he meaning of a message depends on where the partition between subject and object is placed... The quantum physical regularities responsible for the stability and specific properties of thing can be neither formulated nor explained in the framework of causality. But, as Bohr predicted, these regularities can be encompassed in a frame that is an extension or generalization of the causal description.[11]

As the Second World War broke out in Europe, Bohr partnered with Wheeler to make an important theoretical discovery about nuclear fission that advanced the creation of the first atom bomb. During the early years of the war, he visited Los Alamos and worked on the bomb trigger. But as D-Day approached, ruminating on the likelihood of post-war nuclear proliferation, Bohr obtained a private meeting with British Prime Minister, Winston Churchill and, a few weeks later, with the American president, Franklin D. Roosevelt. He urged them to create an "open world" by sharing the fruits of atomic research with the Russians. Unimpressed by this appeal to reason, Churchill declared that the Dane's hair was excessively long and unruly and suggested, unsuccessfully, that he be detained as a communist sympathizer.[12]

It was not easy to talk with Bohr, especially as he aged. He was hard to hear and legendarily obscure. His complicated sentences hung fire in mid-air while he paused to light and relight his trademark briar. He paced slowly, inexorably in circles, while mumbling almost incomprehensibly and giving, Wheeler said, "the appearance of a man thinking deeply, very deeply, with his deep thoughts struggling to find expression."[13] And although Bohr was a profound thinker, his pronouncements on the interpretation of quantum mechanics were often more oracular than explanatory.

---

[10] Bohr's written accounts of complementarity are famously vague, but, in conjunction with Werner von Heisenberg, Pauli, and others the term "Copenhagen interpretation" emerged in public discourse. By the 1950s, it provided a convenient target for those upset by the status quo (such as Everett or Bohm). As a kind of catch-all interpretation of quantum mechanics it reflects several major ideas, not all of which were overtly agreed to by Bohr. Mara Beller remarks, "The Copenhagen philosophy can thus be seen as a contingent composite of different philosophical strands, the public face on a hidden web of constantly shifting differences among its founders." Beller, M. (1999). 173.

[11] Petersen, A. (1963); see also Bohr, N. (1934). 7–10.

[12] Pais, A. (1991). 502.

[13] Wheeler, J.A. and Ford, K. (1998). 126.

During the post-war years, Bohr, with the prestige and resources of his institute behind him, was the most prominent figure in physics, next to Einstein.
His favor could make a man's career, and his disfavor could stifle it. As he grew
older, he drew his personal circle of physicists ever more protectively about
himself. His long-time assistant and collaborator, Leon Rosenfeld, in particular, became almost fanatical in his devotion to shielding Bohr's complementarity from attack by young Turks like Everett and Bohm.

Bohr was in residence at the Institute for Advanced Study in Princeton for
four months. He enjoyed talking with Einstein, Wigner, von Neumann, Wheeler,
and Oppenheimer, while meeting a new generation of physicists. Harvey
Arnold recalls seeing Bohr, Petersen, and Everett walking the quad at the
Graduate College deep in conversation.

On November 16, Bohr lectured at the Graduate College. Everett and Misner
were there. Bohr's biographer, Abraham Pais, a physicist at the Institute for
Advanced Study, noted in his diary:

> Lecture by Bohr. He thinks that the notion 'quantum theory of measurement' is
> wrongly put.[14]

Aage Petersen summed up Bohr's approach to the measurement problem:

> There is no quantum world. There is only an abstract quantum physical descrip-
> tion. It is wrong to think that the task of physicists is to find out how nature is.
> Physics concerns what we can say about nature.[15]

Bohr's point was both rhetorical and pragmatic. He was not denying that
atomic particles exist. But he believed that we can only experience this world
through the medium of experiment. For instance, when an alpha particle flies
off a radioactive metal to interact with a Geiger counter, that tiny event is
amplified by the machine into an observable, *classical* result: a click, a beep, a
needle pointing to a number on a dial. The unseen quantum event *corresponds*
to a classical event. As Bohr put it,

> Quantum mechanics represents a reformulation of classical mechanics adapted
> to the existence of the quantum of action.[16]

In short, Bohr urged physicists to forgo trying to describe or interpret the
world in purely quantum mechanical terms. We are doomed to be forever ignorant of the inner workings of the quantum world, he insisted, so we must speak
only of what we can know: experimental results. And these results can only be

[14] Pais, A. (1991). 435. In 1956, Bohr noted, "The experimental conditions can be varied in many ways,
but the point is that in each case we must be able to communicate to others what we have done and
what we have learned, and that therefore the functioning of the measuring instruments must be
described within the framework of classical physical ideas. As all measurements thus concern bodies
sufficiently heavy to permit the quantum to be neglected in their description, there is, strictly
speaking, no observational problem in atomic physics." Bohr, N. (1956).
[15] French, A. P. and Kennedy, P. J. eds. (1985). 305. In 1934, Bohr wrote: "In our description of nature
the purpose is not to disclose the real essence of phenomena but only to track down as far as possible
the relations between the multifold aspects of our experience." Bohr, N. (1934) 18.
[16] Bohr, N. and Rosenfeld, L. (1933). 479. In later chapters, we will further explain Bohr's outlook.

expressed in a language of classical physics, as a generalization of the mechanics of the "quantum of action." In other words, out of epistemological necessity, classical physics must explain quantum physics, not the reverse. Everett was of the opposite opinion: he theorized that the universe is fundamentally quantum mechanical. Quantum physics explains classical physics, said Everett, pulling Bohr's theory of knowledge inside out, claiming that the wave function represents physical reality itself, not simply our knowledge of reality.

A slosh or two of sherry

While tape recording their conversation during an alcohol-soaked party at Everett's home a quarter century later, Misner asked his old pal, "How did you get on to, ah, weird quantum mechanics?"

EVERETT: Oh, it was because of you and Aage Petersen. One night at the Graduate College after a slosh or two of sherry, as you might recall. You and Aage were starting to say some ridiculous things about the implications of quantum mechanics and I was having a little fun joshing you and telling you some of the outrageous implications of what you said, and, ah, as we had a little more sherry and got a little more potted in the conversation—don't you remember, Charlie? You were there!'

MISNER: I don't remember that evening actually, but I do remember that Aage Petersen was around—that's entirely possible.

EVERETT: You had too much sherry.... Well, anyway, the whole business started with those discussions, and my impression is I went to Wheeler then later and said, 'Hey, how about this, this is the thing to do…there is an obvious inconsistency in the [quantum] theory.'

MISNER: It is strange that he would be so interested in it—all in all, because it certainly went against the normal tenets of his great master, Bohr.

EVERETT: Well, he still feels that way a little bit.

MISNER: He was preaching this idea that you ought to just look at the equations and if there were the fundamentals of physics, why you followed their conclusions and give them a serious hearing. He was doing that on these solutions of Einstein's equations like wormholes and geons.

EVERETT: I've got to admit that, that is right, and might very well have been totally instrumental in what happened.

Winter, 1954

By Christmas 1954, Everett was hard at work constructing mathematical proofs and logical arguments to show that the "outrageous implications" of quantum mechanics were true. Within the year, he claimed to have solved the measurement problem with his new theory of a universal wave function.

In January 1955, Wheeler evaluated him for the National Science Foundation.

> [Everett is] highly original, originated an apparent paradox in the interpretation of the measurement problem in quantum theory and showed its remarkable difference from other paradoxes in the respect that it deals with amplification processes, encountered in typical measurements, contrary to all other paradoxes so far considered. In discussions of this paradox with graduate students and staff members here at Princeton, and with Niels Bohr, Everett brought to light new features of the problem that make it in and of itself an appropriate subject for an outstanding thesis when further developed. Everett has also done independent calculations on problems in general relativity in association with Misner, each stimulating and providing ideas for the other.... Everett did on his own an outstanding paper—or so I am told by Professor Tucker—on game theory. He really is an original man.[17]

Wheeler was initially enthusiastic about the concept of a universal wave function because he wanted to apply it to a theory of quantum gravity. He was seriously troubled by its consequence: multiple universes. Nonetheless, he was to make a heroic effort to convince Bohr that Everett's idea was useful. Not surprisingly, he failed to obtain Bohr's approval, but he had already changed the history of physics by encouraging Everett to transcend the ontological restrictions imposed by the Copenhagen interpretation. In effect, Everett broke the embargo on talking about the measurement problem by treating the universe as fundamentally quantum mechanical and, therefore, whole.

But before we can describe what it is that Everett did to make his strange theory plausible to Wheeler, we need to delve more deeply into how physicists and philosophers had dealt with the "outrageous implications" of quantum mechanics before Everett arrived on the scene with his bright idea.

It will require learning the basics of quantum theory as Everett learned it, but in ordinary language and without using a single equation!

---

[17] Wheeler to NSF, 1/13/55.

10 More on the Measurement Problem

*Once we have granted that any physical theory is essentially
only a model for the world of experience, we must renounce
all hope of finding anything like 'the correct theory.' There is
nothing which prevents any number of quite distinct models
from being in correspondence with experience (i.e. all
'correct'), and furthermore no way of ever verifying that any
model is completely correct, simply because the totality of all
experience is never accessible to us.*

Hugh Everett III, 1956[1]

Mysterious slits

In a famous experiment in 1807, Thomas Young demonstrated the wave character of light. Shining a single light source through a pair of small holes cut
into a plate onto a screen, Young saw a pattern of bright and dark fringes, rather
than two bright spots. This "interference" pattern was explained by considering
light to be composed of waves, not corpuscles, or tiny particles, as had been
envisioned by Sir Isaac Newton.

The fringes stood out because the peaks and troughs of light waves passing
through the two holes overlapped or superposed like waves of water rippling
out from the epicenters of two stones thrown into a pond. The dark fringes
resulted from peaks and troughs combining to cancel each other and register
as shadow between areas of brightness, while merging peaks registered as
bright lines.

Nowadays the experiment is performed with slits, not holes. The "two-slit"
experiment is important in the history of quantum mechanics because shooting electrons (particles discovered in 1897) at two closely separated slits also
produces an interference pattern. When you beam a lot of electrons at the slits,
an interference pattern builds up showing that the electron beam has wave
properties. Complicating the issue, repeatedly shooting single electrons, one at
a time, at the pair of slits also produces an interference pattern, indicating that
the electron must have flown through both holes simultaneously and inter-
fered with itself. Equally mysterious is the fact that when you repeatedly aim

---

[1] DeWitt, B. and Graham, N. eds. (1973). 134.

and shoot one electron through one hole—after blocking the other hole so that the electron cannot possibly go through it—the interference pattern disappears, as if the electron "senses" *whether or not it is possible* for it to go through both slits at once, or it is limited by nature to passing through one slit.

It seems as if the electron "behaves" sometimes like a wave, and sometimes like a particle, depending on how the experiment is set-up. Of course, the electron is not an animal, so it does not behave. Rather, where it lands—or *might* land—depends on the environment in which it moves. Although the modern experimentalist has no way of knowing in advance exactly where a particular electron will land, using quantum mechanics he can confidently predict the *probability* for it to land at a certain spot. But until it lands and leaves an observable record, the electron, as a probability wave, is in a superposition of positions as it moves through the apparatus toward the screen, where one landing spot out of all those possible is selected by nature, (or so the standard, non-Everettian interpretation goes).[2]

Richard Feynman, a master of scientific summary, remarked of the two slit experiment:

> It is not our ignorance of the internal gears, of the internal complications, that makes nature appear to have probability in it. It seems to be somehow intrinsic. Some one has said it this way—'Nature herself does not know which way the electron is going to go.' A philosopher once said 'It is necessary for the very existence of science that the same conditions always produce the same results.' Well, they do not.

Quantum dawn

In the history of physics, 1905 is known as the Year of Miracles because Einstein published four revolutionary papers, including his theory of special relativity, "On the Electrodynamics of Moving Bodies." A decade later, as the First World War raged, Einstein expanded his model of a relativistic, yet still-classical, still-deterministic universe to include gravity (the theory of general relativity). Abstract, counter-intuitive and, for a few years divorced from robust experimental proof, the basic notions of Einstein's new theories could still be communicated by visual metaphors: by pictures of train passengers experiencing the same event differently when traveling at "relative" velocities; by moving clocks slowing down relative to each other; by space-time as a fabric molding around heavy bodies. Without being able to read the equations of special and general relativity, normal people could still appreciate Einstein's discovery that there are *boundaries* circumscribed by the *speed* of light, that mass and energy *interlock*, that massive objects *curve* the continuum of space-time, that gravity is *geometrical*, and that the *bodily motions* induced by gravitational *attraction*

---

[2] Probability waves move through an infinite-dimensional "configuration" space and are picked out by our limited senses as particles taking definite positions in a three-dimensional space.

or acceleration are indistinguishable. In short, Einstein's relativity theories kept intact causality, determinism, the common sense we use to predict the future.

But Einstein wrote about something else besides relativity theory in 1905: he examined the "photoelectric effect." He asked why shining light on metal causes electrons to zing off the surface. His answer was that as electrons trapped inside the metal substance absorb "quanta" of light energy, they gain energy in proportion to the frequency (the number of vibrations per second) of the light and may escape the metal. The energy came in quantifiable chunks.

Einstein's quantization of light waves used a new mathematical constant, "Planck's constant," signified by "h," discovered in 1900 by Max Planck, a professor of physics at the University of Berlin. Planck's constant (also called the "quantum of action") is a fabulously small number: 33 zeros and a 6 to the right of the decimal when expressed in standard units. Planck unwittingly unleashed a new, non-classical physics by postulating that energy is radiated in discrete bundles, with his new constant h governing the proportion of energy to the frequency of its transmission as a radiating wave. Einstein postulated that not only does radiation exchange energy with matter in quantum units, it also *exists* as quantum units (later called "photons"). Importantly, this meant that systems composed of individual chunks of light energy could now be tracked statistically, i.e. probabilistically.

Despite the excruciatingly high level of technical expertise a person needed to explore in detail the worlds opened up by relativity and quantum theory, the break-through fired up the popular imagination. When relativity was validated in 1919 by the discovery that starlight is deflected by the Sun, the wild-haired, doe-eyed Einstein became a global personality—the cynosure of pure intelligence, blessed with a sense of social responsibility and good humor. He was awarded the Nobel Prize for Physics in 1921 for his work on the photo-electric effect, which, ironically, had set the theoretical stage for undermining the determinist physics that he cherished.

The statistical element

In a burst of group creativity during the 1920s, the founders of quantum mechanics discovered that electrons, as well as photons, are governed by the laws of probability. In fact, all atomic particles behave probabilistically—they simply do not occupy determinate positions in space and time.

The mathematics underpinning quantum mechanics is incredibly precise. Theorists use it to predict outcomes, and experimentalists to test predictions. The ability to make these predictions is based on the collection of data: the key to determining the probability that an electron will show up at particular place when measured is to recreate the same experiment over and over until thousands or millions or billions of measurements on identically "prepared" particles have been logged. Early experimenters performed their tests with small "cloud chambers," we do it now by recording data about particle collisions inside house-sized drift chambers and town-sized particle accelerators.

Quantum mechanics differs from classical physics in this: if you can *exactly* record and reproduce the initial conditions of a classical experiment (dropping a cannonball from a tower, say), then, every time you exactly repeat the set-up of that experiment, the results will be exactly the same. The same is not true in quantum physics: every time you repeat the experiment, tracking the trajectory of an electron, say, the results can be different, within a range of possible difference. Identical initial conditions do not necessarily produce identical results!

Let's stick to measuring position: after logging the data on the relative frequency of the different positions that a series of identically prepared electrons are found at over a large number of measurements, the data is translated into linear algebra, creating a statistical function that obeys wave mechanics: ψ.

ψ represents a quantum state (which can be a superposition). To reiterate: ψ is manufactured from data collected by observing the relative frequencies with which certain positions are repeated in an experiment. ψ sums up experimental results and the wave equation governing it enables us to make predictions. The Schrodinger equation can tell us the probability that we will find an electron at a certain position within a range of possible positions. For instance, we might learn that there is a 40 percent chance that the electron will appear at position X, and a 60 percent chance that it will be found at position Y. But we do not know whether we will find the electron at X or Y, only the probability that it will be at one or the other, within the range of positions already observed. In this case, the range, or "probability distribution," includes only X and Y, but the range could include an infinite number of possible positions stretching from "here" to the end of the universe.[3]

Let's deepen the explanation. The wave function (or "probability amplitude") is not *the* probability, only the seed of a probability. Remarkably (and inexplicably), multiplying a wave function attached to a specific position by itself, squaring it, reveals the probability that the *next* time we measure an identically prepared particle it will be found at that same position.[4] This is the rule discovered by Max Born in 1926: Squaring the positive value of ψ for a property of a particle tells all that we can know about that property. We do not know *if* the electron will be at place X, we only know the probability that it could be at that position, because we have collected enough data to determine the relative frequency with which identically situated electrons have appeared at that position in the past. And that is why quantum mechanics is viewed as statistical and indeterminist.

And, as the founders of quantum mechanics realized, to their great consternation, the micro world of the quantum departs from the macro world of our

---

[3] Feynman, R. (1965). 141.
[4] Technically, it is the modulus of the probability amplitude, a complex number, which is squared. This ensures that the resultant probability will be positive, as complex numbers can be negative. Because these numbers can be negative, wave functions can cancel each other.

experience: Until the position of a particle is measured, *it has no certain position*. Its wave function describes all possible positions, but the precise future of the particle is indeterminable. However, every possible position within ψ is evolved by the Schrödinger equation deterministically, as if it will be at a certain place within a range of places (the probability distribution). And the wave equation does not distinguish between possible positions; each position is treated as equally real by the equation until the measurement interaction occurs and the particle is found at a particular place with a specific probability of reoccurring at that place.[5] Our current understanding of nature offers us no explanation of why it is at *that* place. All it offers is a probability that if we repeat the experiment it will appear at that place again. Nor does nature tell us why we do not find the particle at all of the positions in the probability distribution encoded in ψ.

Hence: the measurement problem.

Indeterminism

The German physicist, Max Born, who introduced the postulate of squaring ψ to obtain a classical probability, wrote:

> One obtains the answer to the question, *not* 'what is the state after the collision' but 'how probable is a given effect of the collision' ... Here the whole problem of determinism arises. From the point of view of our quantum mechanics there exists no quantity which in an individual case causally determines the effect of a collision.... I myself tend to give up determinism in the atomic world.[6]

Born puzzled over how probability related to Schrödinger's equation:

> The motion of particles follows probability laws but the probability itself propagates according to the law of causality.[7]

In other words, a series of single (often different) results emerged from ψ as it evolved causally, deterministically, *linearly* and the particle was measured. This meant that probability was no longer a measure, as in classical physics, of the experimenter's ignorance about a pre-existing condition; indeterminism was, somehow, embedded in Nature's construction of the quantum condition.

In 1927, Heisenberg wrote an epitaph for classical determinism:

> Even in principle we cannot know the present in all detail. For that reason everything observed is a selection from a plenitude of possibilities and a limitation on what is possible in the future. As the statistical character of quantum theory is so closely linked to the inexactness of all perceptions, one might be led to the

---

[5] The mathematical "linearity" of the "unitary" Schrödinger equation conserves probability in the sense that after all of the elements in a superposition are measured, their probabilities total 1 (100 percent). This is a mysterious and deep property of quantum mechanics: chance is not lost, but neither is it chance until it is measured.

[6] Born, M. (1926). 54.

[7] Born, M. (1926B). 804.

presumption that behind the perceived statistical world there still hides a 'real' world in which causality holds. But such speculations seem to us, to say it explicitly, fruitless and senseless. Physics ought to describe only the correlation of observations. One can express the true state of affairs better in this way: ... quantum mechanics establishes the final failure of causality.[8]

Heisenberg's comment was prompted by his discovery of the uncertainty principle, a natural law prohibiting the simultaneous measurement of mutually exclusive quantum properties, such as position and momentum, or energy and time. Everett accounted for the uncertainty principle in his theory, but, contradicting Heisenberg, he viewed quantum mechanics as fundamentally causal and deterministic in the sense that every physically possible event occurs.

Heisenberg made another philosophical comment that has caused much confusion and debate in the years since:

I believe that one can fruitfully formulate the origin of the classical 'orbit' in this way: the 'orbit' comes into being only when we observe it.[9]

Heisenberg, von Neumann, and to a large extent, Bohr, privileged the role of the observer by partitioning the observer from the object observed in a measurement interaction. But both the observer and the object are composed of interacting atomic particles, which are represented by interacting, overlapping, superposing, *entangling* wave functions. So how can they be separate? How could it be possible to stand *outside* the wave function when, according to the Schrödinger equation, the wave function of an object naturally expands to include the wave functions of all the objects with which it interacts, including the observer?

Entanglement

Entanglement is tied to superposition—and it is also inexplicable (albeit describable). Entanglement is the principle that a single ψ can describe the combined state of two or more separate particles. Entangled particles may be spatially separated, yet linked, *correlated*. Consider a pair of interacting electrons that are entangled ("prepared") by a measuring device so that the "spin" of one particle must be "up" if the spin of its partner is "down."[10] After their spins are correlated, the particles fly off very fast and very far in opposite directions. The principle of superposition tells us that until we measure the spin of one of these particles, the composite wave function includes spin up and spin down for each particle. But when we do measure one of the particles, recording spin up, for instance, we then automatically know that its

---

[8] Heisenberg, W. (1927B). 83.
[9] Ibid. 73.
[10] Spin is a purely quantum mechanical property having to do with angular momentum and magnetism. Spin "up" and "down" are convenient notations describing configurations of this property.

entangled partner is spin down, i.e. the second particle is no longer in a superposition of spin states even though we did not directly measure it (only its partner).

Defying the logic of classical physics, what happens to each of these entangled particles *instantaneously* affects its partner, even though they are separated by distances measured in light years.[11] But this phenomenon is not confined to experiment; since the beginning of time particles have been interacting, exchanging energy, entangling, cooking up reality.

Entanglement was at the root of the famous "EPR paradox" posed by Einstein, Nathan Rosen, and Boris Podolsky in 1935 to challenge Bohr's claim that quantum mechanics is a complete description of reality. In brief, EPR (as reformulated by Bohm some years later) pointed out that the instantaneous determination of the spins of a pair of entangled but spatially separated particles would violate special relativity's prohibition against faster than light speed action or "non-locality." The violation was supposed to occur because by measuring the up spin of one particle in the entangled pair, the spin of its (possibly superluminally) separated partner instantaneously became down: Einstein called it "spooky-action-at-a-distance."[12]

As Everett knew, Schrödinger, in 1935, was of the opinion that wave functions exactly model physical reality. The inventor of wave mechanics defined entanglement as,

> When two systems, of which we know the states by their respective representatives [wave functions], enter into temporary physical interaction due to known forces between them, and when after a time of mutual influence the systems separate again, then they can no longer be described in the same way as before, viz., by endowing each of them with a representative of its own.
>
> I would not call that *one* but rather *the* characteristic trait of quantum mechanics. By interaction the two representatives [wave functions] have become entangled....Another way of expressing the peculiar situation is: the best possible knowledge of a whole does not necessarily include the best possible knowledge of all its parts, even though they may be entirely separated.[13]

In quantum mechanics the particles composing the observer continuously "entangle" with each other *and* the particles of the objects under observation as their respective wave functions combine. Everett was to take entanglement very seriously. It is at the heart of his universal wave function, which describes physical reality as a vast superposition of all possible quantum states and all possible worlds. In the end, thought Everett, it is not possible for an observer to

---

[11] Jeffrey Barrett remarks, "On the standard theory, the measurement instantaneously affects the other particle, not by changing the determinate properties of the other particle, but by affecting the composite entangled state, and the two entangled particles only have a well-defined composite state (that is, they do not even have quantum mechanical states to call their own)." Barrett, private communication, July 2009.

[12] Two decades later, John Stewart Bell proved mathematically that quantum mechanics is fundamentally non-local; experiments back him up. More on this later.

[13] Schrödinger, E. (1935). 30.

stand outside a quantum state that necessarily includes himself because the whole universe is entangled with itself.

Schrödinger's jellyfish

It was 1929. Sir Nevill Francis Mott, a British physicist, was perplexed by the results of an experiment showing the trajectory of a particle inside a Wilson cloud chamber, (a sealed container filled with super cooled water vapor). So he wrote a famous paper that made the measurement problem concrete.

Here is the story: In a cloud chamber, alpha particles emitted by the nucleus of a radioactive atom create droplets of observable condensation by interacting with water molecules in the vapor. According to Mott's calculations, the alpha particle wave—think of it as a superposition of possible particle trajectories—spreads in all directions. It is "localized" when it interacts with a water molecule, leaving behind a droplet. A series of such droplets form a track of condensation, depicting the trajectory of the particle. So a spherically spreading wave should leave behind a droplet record that is spherical.

But this is not what happened inside the cloud chamber: only a single track appeared. Somehow, the interaction of the spherical wave with its environment selected only a few interactions to become real. Mott could not explain how or why this happened—why only one element of the superposition became real. But he decided that the spherical wave function of the alpha particle had somehow narrowed or "collapsed" itself, vanishing all possible trajectories but one.[14]

Mott's conclusion bothered Schrödinger, who, like Everett, did not accept the "wave collapse" interpretation. In his thesis, Everett cited a paper in which Schrödinger insisted that experiments such as that described by Mott,

> cannot [be] account[ed] for without taking the wave to be a wave, acting simultaneously throughout the region over which it spreads, not 'perhaps here' or 'perhaps there,' as the probability view would have it.…That would fail to account for the interference phenomena [i.e. the superposition principle].[15]

Wave functions are physically real, Schrödinger insisted, although he could not explain why only one alpha particle track emerged in the cloud chamber.

The problem is that Schrödinger's wave equation evolves all of the possible positions of the spherical alpha wave burped from the nucleus as it interacts sequentially with molecules in the cloud chamber. Until the measurement interaction with the water molecules occurred, the wave equation described the future of each possible state of the particle. It was continuous and causal. The wave equation did not contain a mechanism for selecting out a single track in Mott's cloud chamber to emerge to the exclusion of all other possible tracks, and yet, this happened.

---

[14] Mott, N. (1929). 129–134.
[15] Schrödinger, E. (1952A). 240–242.

In 1935, Schrödinger summarized the problem of measurement:

> Any measurement suspends the law [Schrödinger equation] that otherwise governs continuous time-dependence of the ψ-function and brings about in it a quite different change, not governed by any law but rather dictated by the result of the measurement. But laws of nature differing from the usual ones cannot apply during a measurement, for objectively viewed it is a natural process like any other, and it cannot interrupt the orderly course of natural events. Since it does interrupt that of the ψ-function, the [ψ-function]…can *not* serve, like the classical model, as an experimentally verifiable representation of an objective reality. And yet [it does].[16]

In other words, the emergence of a single alpha particle track from all possible tracks for that alpha particle was inexplicable. Since Mott, several generations of physicists have searched for answers to why and how the classical world of our experience emerges from the multiplicity of possible experiences lurking inside wave functions. It is appropriate to ask, as Schrödinger did, why do we not observe macroscopic objects obeying the Schrödinger equation as their constituent parts must?

For example, when the quantum object named Mott, himself a superposition of possible states, looked into his cloud chamber, why didn't each Mott in the superposition of possible Motts correlate with each possible track of the alpha particle? That was, in fact, Everett's conclusion: Mott did indeed correlate to each element included in the alpha particle wave function and he immediately split into multiple Motts, each Mott seeing and remembering a different, alpha particle track. The sum of single tracks witnessed by the sum of Motts equals the information about possible particle trajectories contained in the spherical wave function, according to Everett.

Following Schrödinger, Everett believed that Born's probability rule was an illusion. Both theorists wanted the real world to reflect the seamless beauty of the *linear* wave equation, which protects the continuous individuality of each and every possibility. But Schrödinger was not prepared to follow Everett to his ultimately unsettling conclusion, one that encompassed many Motts. Schrödinger considered what would happen if we did evolve according to his equation.

In a 1952 lecture at the university of Dublin, he observed:

> Nearly every result [a quantum theorist] pronounces is about the probability of this *or* that… happening—with usually a great many alternatives. The idea that they be not alternatives but *all* really happen simultaneously seems lunatic to him just *impossible*. He thinks that if the laws of nature took *this* form for, let me say, a quarter of an hour, we should find our surroundings rapidly turning into a quagmire, or sort of a featureless jelly or plasma, all contours becoming blurred, we ourselves probably becoming jelly fish. It is strange that he should believe this. For I understand he grants that unobserved nature does behave this way—

---

[16] Schrödinger, E. (1935A). 160.

namely according to the wave equation. The aforesaid alternatives come into play only when we make an observation—which need, of course, not be a scientific observation. Still it would seem that, according to the quantum theorist, nature is prevented from rapid jellification only by our perceiving or observing it. And I wonder that he is not afraid, when he puts a ten-pound note into his drawer in the evening, he might find it dissolved in the morning, because he has not kept watching it.[17]

But he could not accept Bohr's complementary approach to measurement, which he originally characterized in 1935:

> Now while the new theory [Bohr's complementarity] calls the classical model incapable of specifying all details of the *mutual interrelationship of the determining parts* (for which its creators intended it), it nevertheless considers the model suitable for guiding us as to just which measurements can in principle be made on the relevant object. This [is] an unscrupulous proscription against future development…born of distress.[18]

Unable to solve the measurement problem, or to reconcile himself with Bohr's dualism, Schrödinger oscillated between realism (the wave function is physically real), and anti-realism (it is a mathematical ideal). Like Wigner and von Neumann, he eventually developed a "mentalist" interpretation of quantum mechanics based on the intervention of human consciousness in the measuring process.

But Everett found in Schrödinger's pure wave mechanics a platform on which to mount his own realist interpretation, which, he claimed, could be derived directly from the quantum mechanical formalism invented by Schrödinger, without adding new assumptions and postulates. In fact, Everett's model of quantum mechanics was simply based on following Schrödinger's logic to *wherever* it might lead.

To do this, he began by tackling the notion of "wave function collapse," which is the idea that it is acceptable to suspend the smooth and continuous evolution of the quantum world in order to accommodate our perception of single results. And here he confronted von Neumann's most powerful and long lasting contribution to quantum theory: the wave function collapse postulate that vitiated the universality of the Schrödinger equation. The collapse theory was not a solution to the measurement problem, but it effectively swept it under the rug for decades.

---

[17] Schrödinger, E. (1995).19–20; Sean Boocock observes that in modern quantum mechanics (due to decoherence), "the weird jellification that Schrödinger worries about does not come about on large scales simply because the probability of us ever being able to perceive it is too phenomenally low." Private communication, 2009.
[18] Schrödinger, E. (1935A). 154.

Collapse and Complementarity

John von Neumann's *Mathematical Foundations of Quantum Mechanics* was the prevailing textbook on the subject well into the 1950s. In it, von Neumann axiomatized quantum mechanics. He presented "wave function reduction" or the "collapse" postulate as a panacea for the measurement problem. His approach was widely accepted, although it was challenged in the post-war years by some prominent theorists, including Schrödinger, Bohm, and Henry Margenau of Yale University. But it was Everett's frontal attack on von Neumann that allowed critics of his "orthodox" interpretation of quantum mechanics some breathing space to say, yes, there is a problem with the axioms.

Axioms are unproven rules or postulates from which other rules may be derived. One of von Neumann's axioms was that the Schrödinger equation governs how each superposed element in a quantum system dynamically changes over time as it evolves causally, deterministically. Another axiom said that the act of measurement arbitrarily and instantaneously "projects" one element of a superposition into classical reality. This axiom is known the "collapse" postulate; its discontinuity interrupts the continuity of the Schrödinger equation's representation of reality.

Could both axioms be true?

Physical reality told von Neumann that one measurement result emerges, but he could not mathematically derive this result from quantum mechanics. He explained the dilemma:

> In the discussions so far, we have treated the relation of quantum mechanics to the various causal and statistical methods of describing nature. In the course of

[1] Frank, P. (1949). 234.
[2] Zeh, private communication, 2008.

this we found a peculiar dual nature of the quantum mechanical procedure which could not be satisfactorily explained.[3]

The problem he pointed to is that in the objective formalism of quantum mechanics, nature proceeds causally, deterministically, but our perception of nature is subjective: macroscopic reality appears to emerge randomly, non-causally from the microcosm. Deciding in favor of a touchable (if indeterminist) reality, von Neumann continued,

> It is a fundamental requirement of the scientific viewpoint—the so-called principle of psycho-physical parallelism—that it must be possible so to describe the extra-physical process of the subjective perception as if it were in reality in the physical world—i.e., to assign to its parts equivalent physical processes in the objective environment, in ordinary space.[4]

Adopting a dualistic approach, von Neumann decided that if both axioms were correct there are *two* kinds of changes in quantum mechanical states:

- The collapse postulate, which describes "the discontinuous, non-casual and instantaneously acting experiments of measurements," or "arbitrary changes by measurement."
- The Schrödinger equation's "continuous and causal changes in the course of time," which evolve the elements of a superposition as "automatic changes."[5]

It is precisely this dichotomy that Everett was to attack in the first sentence of his thesis when he called for abandoning the collapse rule. There was, however, a rhyme and reason to von Neumann's dualism. He noted that microscopic processes as guided by the Schrödinger equation are, in theory, reversible. But according to the second law of thermodynamics—that entropy increases in a closed system—a macroscopically obtained measurement of a microscopic system is not reversible (for all practical purposes). As the process of change in our macroscopic universe follows an irreversible arrow of time, von Neumann reasoned that, according to psycho-physical parallelism, collapse, as an explanation, overrules the unseen continuity expressed by the wave equation. Only, he could not prove that this was so, so he postulated it. And there were obvious flaws. For example, simply asserting that the wave function of a measured particle is reduced because a particular property of a particle somehow emerges from a superposition of possible results to a single result (without saying why) begs the question of where and at what instant does the collapse of the wave function occur?

Like Bohr, von Neumann believed that, by definition, a valid observation or measurement of a quantum system occurs *outside* the system (Everett claimed to put the measurement process inside the system). But, if the observer and the

---

[3] von Neumann, J. (1932). 619.
[4] Ibid. 620–621.
[5] Jammer, M. (1974). 475.

object observed are microscopically entangled, which von Neumann believed to be the case, where is the *cut*? At what point in space or time does a measurement become externalized, as entanglement with adjacent objects is constantly going on according to the superposition principle? At what precise moment does the wave function reduce into a measurement of a piece of the total system? Does the reduction occur when the Geiger counter needle registers a single measurement? But isn't the needle in a superposition until it is measured by the eyeball of the experimenter? And what about the eyeball and, for that matter, the observer's friend who becomes correlated to the quantum system under observation by the mere fact of entering the laboratory? Did the wave function of the experimenter who is looking at the needle registering the measurement result collapse before or after his friend looked at him? And when does the friend's wave function collapse?

The difficulty of saying where and when the wave function collapses poses the problem of an infinite regression: how can an observer ever get outside the serially entangling system to observe it without being an intrinsic part of it?

Echoing Bohr's arbitrary insertion of an epistemological partition between the quantum and classical realms to achieve externality, von Neumann remarked,

> But in any case, no matter how far we calculate—to the [measuring device], to the retina, or into the brain, at some time we must say: and this is perceived by the observer. That is, we must always divide the world into two parts, the one being the observed system, the other the observer. In the former we can follow up all physical processes (in principle at least) arbitrarily precisely. In the latter, this is meaningless. The boundary between the two is arbitrary to a very large extent.... That this boundary can be pushed arbitrarily deeply into the interior of the body of the actual observer is the content of the principle of the psycho-physical parallelism—but this does not change the fact that in each method of description the boundary must be put somewhere, if the method is not to proceed vacuously, i.e., if a comparison with experiment is to be possible. Indeed, experience only makes statements of this type: an observer has made a certain (subjective) observation; and never any like this: a physical quantity has a certain value.[6]

Unable to mathematically staunch the infinite regression—which threatened to make the whole universe quantum mechanical, thereby making external observation of a classical result impossible—von Neumann (probably influenced by Wigner) arbitrarily declared that the chain of measurement ended with the "abstract ego" of the observer. And that this arbitrary cut was *philosophically* justified by the principle of psycho-physical parallelism.[7]

Collapsing the wave function was a way of making a separate peace with the measurement problem. As once you allowed a superposed quantum system to propagate according to the Schrödinger equation, there was no way to explain why the macroscopic world does not mirror microscopic superpositions.

According to Everett's account, the large does mirror the small.

---

[6] Ibid. 621–622.
[7] Ibid. 623.

Bohr's complementarity

Although Bohr did not dispute that the world is quantum mechanical (if unknowable as such), he called for "neglecting" the atomic constitution of measuring instruments and observers, as to do otherwise would allow for no external ground from which to make a measurement.[8] And although Bohr did not employ von Neumann's wave reduction terminology, his interpretation was analogous to collapsing the wave function. In essence, Bohr made the cut by privileging experimental context and making experimental results the sole criterion for knowledge.[9]

But unlike Wigner and von Neumann, Bohr did not overtly make "consciousness" into a force that extracts material results out of the ether of superposed ideas. Shying away from the heuristic abyss posed by the measurement problem, Bohr acknowledged that classical terms do not completely describe the quantum world, not even in his idealized model of complementarity. Nevertheless, he said, classical terms and concepts are the only guides we have at hand:

> It is decisive to recognize that, however far the phenomena transcend the scope of classical physical explanation, *the account of all evidence must be expressed in classical terms*. The argument is simply that by the word 'experiment' we refer to a situation where we can tell others what we have done and what we have learned and that, therefore, the account of the experimental arrangement and of the results of the observations must be expressed in unambiguous language with suitable application of the terminology of classical physics.[10]

But as the scholar Max Jammer points out, Bohr's interpretation tends to sequester physics from the world it purports to describe:

> Another problematic aspect whose serious implications were only gradually understood was the fact that as long as a quantum mechanical one-body or many-body system does not interact with macroscopic objects, as long as its motion is described by the deterministic Schrödinger time-dependent equation, no events could be considered to take place in the system. Even such an elementary process as the scattering of a particle in a definite direction could not be assumed to occur since this would require a 'reduction of the wave packet' without an interaction with a macroscopic body. In other words, if the whole physical universe were composed only of microphysical entities, as it should be according to atomic theory, it would be a universe of evolving potentialities (time-dependent ψ-functions) but not of real events.[11]

For Bohr, the physical design of an experiment *a priori* divides the effable from the ineffable. Bohr chose not to ask what went on before or after he obtained a measurement result:

---

[8] Beller, M. (1999). 159, 162.
[9] See, for example, Jammer, M. (1974). 98–99, on Bohr's de facto acceptance of wave function reduction when he privileges classicality. For a different reading of Bohr, see Howard, D. (2004).
[10] Bohr, N. (1949). 17-18. Italics added.
[11] Jammer, M. (1974). 474.

Our task is not to penetrate into the essence of things, the meaning of which we don't know anyway, but rather to develop concepts which allow us to talk in a productive way about phenomena in nature.[12]

And there is logic to his method, as measuring instruments are classical and produce classical results for parsing by classical minds:

The experimental conditions can be varied in many ways, but the point is that in each case we must be able to communicate to others what we have done and what we have learned, and that therefore the functioning of the measuring instruments must be described within the framework of classical ideas.[13]

In brief: Bohr declared that although there may be a reality underlying quantum phenomena, we cannot know what the reality *is*. It is accessible to human understanding only through the mediation of experiment and classical concepts. Consequently, generations of physicists were taught that there is no quantum reality independent of experimental result. And that the Schrödinger equation, while incredibly useful as a predictive tool, should not be interpreted literally as a description of reality.

Everett took the opposite view.

Bohr and free will

Although Bohr contended that his interpretation did not posit human consciousness as a causal agent, it easily supported that conclusion, because nothing outside of a measurement registered by a human consciousness counted as real in his interpretation of quantum mechanics. It has been suggested, consequently, that Bohr was influenced by Immanuel Kant, and that Bohr regarded the quantum world as Kant's forever-inaccessible "the thing itself."[14]

In 1968, Bohr's assistant (and Everett's friend), Aage Petersen, claimed that quantum physics "reintroduced Kantian ideas."[15] He elaborated:

The most important feature of the ontological part of Kant's philosophy is his distinction between the world of things as they are in themselves, and the world of phenomena, *or things as they appear to us*. Human reason is a law-giver, but it can only legislate for phenomena, not for things in themselves.... Thus phenomena are conditioned by our way of knowing.... We are members of two worlds. In the world of phenomena, the will is subject to the law of causality. But as a thing in itself it is not subject to that law, and is therefore free... Just as Aristotle was convinced that he had given a complete account of the basic structure of being, Kant held that his critique of pure reason exhausted the subject.[16]

---

[12] Bohr, N. to Hansen, H. P. E. 20 July 1935.
[13] Bohr, N. (1956). 10. A mimeograph of this paper was in Everett's papers.
[14] Beller, M. (1999). 205.
[15] Petersen, A. (1968). 142.
[16] Petersen, A. (1968). 59–61. Italics added.

Bohr's philosophy of complementarity can be viewed as an epistemological framework for holding mutually exclusive opposites: the quantum world is the inaccessible thing itself, the classical world reflects the quantum, bringing it into the realm of reason and knowledge as classically described phenomena. Bohr, in a Kantian echo, wrote that experimental evidence, "exhaust[s] all definable knowledge about the objects concerned."[17] Bohr's resolution of the measurement problem was to declare, congruent with his philosophical basis, that classically defined measurements tell us everything we can know about a quantum system, period.

In part, Bohr's stance was a reaction against the determinism of 19th century classical mechanics that was seen as undermining the agency of free will. Quantum mechanical indeterminism, therefore, had serious religious, philosophical, and sociological implications. No less a scientific light than A.S. Eddington claimed in 1928 that quantum indeterminism liberated free will.[18] And Philipp Frank, an "instrumentalist" who profoundly influenced the development of American philosophical physics (and was admired by Everett), observed in 1954,

> In twentieth century physics, we note clearly that a formulation of the general principles of subatomic physics (quantum theory) is accepted or rejected according to whether we believe that introduction of 'indeterminism' into physics gives comfort to desirable ethical postulates or not. Some educators and politicians have been firmly convinced that the belief in 'free will' is not compatible with Newtonian physics but is compatible with quantum physics.[19]

By the end of his life, Bohr was convinced that his philosophy of complementarity applied to the hard and soft sciences alike and that indeterminism, not determinism, was a basic rule of the universe. In a lecture at the University of Oklahoma on December 13, 1957, a few months after Everett's thesis was published, Bohr pronounced that, "A rigorous deterministic approach leaves no room for the concept of free will."[20]

In Everett's determinist scheme, free will was consigned to the realm of illusion, but it also recognized that we act *as if* we have choices, as if rationality is an option.

A few years later, physicist Frederik J. Belinfante, who was in the process of "reinterpreting" Everett's theory, pointed out, according to Jammer,

> that quantum mechanical indeterminism, may be conceived as harmonizing with the belief in God, who continuously makes his own decisions about the

---

[17] Bohr, N. (1956). 12.
[18] Eddington, A.S. (1929).
[19] Frank, P. (1954). 17; Philosophers were not united in this fear. Sean Boocock points out that the neo-Kantian philospher of science, Ernest Cassirer, wrote in *Determinism and Indeterminism in Physics* (1936) that were quantum mechanics to be indeterminist, free will would be endangered. Ethics would be problematic in a world in which the course of one's actions were subject to probability at all times. One philosopher's free will is another philosopher's ineluctable fate. Boocock, private communication, 2009.
[20] Bohr, N. (1957).

happenings in this world, decisions unpredictable for us, while "if nature would be fully deterministic, one might reason that there would be no task for 'God' in this world."[21]

The determinism inherent in Everett's theory broke decisively with Bohr's view of the universe; although it was not long before free will-loving theists found in his many worlds model room for Divinity: for if everything that is possible occurs, they reasoned, then God must exist.[22] Everett, naturally, viewed that argument as absurd. It is less absurd, however, to ask: What is the nature of probability if everything that is physically possible occurs somewhere in an "Everettian" universe?

---

[21] Jammer, M. (1974). 330, quoting Belifante, F.J. (1970).
[22] Google "Hugh Everett God exists" for treatises using Everett to support Deist theories.

12 The Philosophy of Quantum Mechanics

A new scientific truth does not triumph by convincing its opponents and making them see the light, but rather because its opponents eventually die, and a new generation grows up that is familiar with it.

Max Planck[1]

Copenhagen catch-all

American physicists in the mid 20th century were notoriously unphilosophical about quantum mechanics. Oppenheimer, Wiener, Linus Pauling, John van Vleck, E. U. Condon—leading American physicists—considered von Neumann's "orthodox" postulate of wave function collapse to be a pragmatically acceptable, if mysterious, explanation of experimental results. To the extent that they thought about interpretation, American physicists usually favored the implicit "don't ask, don't tell" attitude of the Copenhagen interpretation because, among other attractions, it left free will intact.[2]

But not all physicists were satisfied with the status quo.

In 1974, Max Jammer, a German-born physicist based at Bar-llan University in Israel, published a comprehensive history, *The Philosophy of Quantum Mechanics*. Chronicling the articulation of interpretive issues in quantum mechanics, Jammer's book accounts for the uneasy marriage between quantum physics and philosophy from 1926 through the early 1970s, concluding with a detailed explication of the many worlds interpretation.[3]

The task of interpretation, said Jammer, is to connect mathematical logic and physical systems. It is not enough for an explanation to be logically consistent; it must explain how the world works by making accurate predictions. And that is why physics concerns many philosophers, even if philosophy does not concern many physicists.

By the 1950s, the identification of Bohr's complementarity with the Copenhagen interpretation had taken on a life of its own (Everett thought them one and the same). But the so-called Copenhagen interpretation was not

---

[1] Planck cited by Kuhn. T. S. (1962). 151.
[2] Cartwright. N. (1987).
[3] Since Jammer wrote his history, much research has been done on the subject of interpretation in quantum mechanics, but his book is a valuable resource for explicating the philosophy of quantum mechanics as it was understood in Everett's day.

completely identical with Bohr's complementarity, nor with von Neumann's postulate of wave function collapse, although many physicists treated it that way. Jammer noted,

> The Copenhagen interpretation is not a single, clear-cut, unambiguously defined set of ideas but rather a common denominator for a variety of related viewpoints. Nor is it necessarily linked with a specific philosophical or ideological position. It can be, and has been, professed by adherents to most diverging philosophical views, ranging from strict subjectivism and pure idealism through neo-Kantianism, critical realism, to positivism, and dialectical materialism.[4]

However,

> In the early 1950s the almost unchallenged monocracy of the Copenhagen school in the philosophy of quantum mechanics began to be disputed in the West. The previous lack of widespread criticism in this field was explained in some quarters as the result of a somewhat dictatorial imposition of what was called 'The Copenhagen dogma' or 'orthodox view.'[5]

Challenging monocracy

In addition to Einstein's well-known skepticism of Bohr's complementarity, Schrödinger agitated against the idea of wave function collapse for decades. And despite the international prestige of Bohr and von Neumann during much of the 20th century, significant challenges to their views were mounted by Americans. Researching his thesis, Everett studied alternatives to the "official" interpretations, while constructing his own model reality.

In 1957, Henry Margenau, a professor of physics and natural philosophy at Yale University, read a preprint of Everett's thesis with pleasure. Subsequently, he sent the young theorist a paper he had recently written. In it, Margenau laid out his own case against wave function collapse, calling it,

> a mathematical fiction ... used persuasively by von Neumann and later by others who were able to derive from this fiction the correct formalism of quantum mechanics, thus adding another example to the vast array of scientific instances in which correct conclusions were deduced from insupportable premises.... Current disbelief in the correctness of the present formulation of quantum mechanics has its source, at least in part, in the grotesque claim of the projection postulate.[6]

Nor did Margenau generally have kind words for Bohr:

> Bohr does not ask science to make a choice—he asks science to resign itself to an eternal dilemma. He wants the scientist to learn to live while impaled on the

---

[4] Jammer, M. (1974). 87.
[5] Ibid. 250.
[6] Margenau to Everett, 4/8/57. Enclosed in the letter was a copy of Margenau, H. (1956) from which the quotation is drawn.

horns of that dilemma, and that is not philosophically healthy advice. [Such dualism] relieves its advocates of the need to bridge a chasm in understanding by declaring that chasm to be unbridgeable and perennial; *it legislates a difficulty into a norm.*[7]

Margenau concluded, as had Everett, that causal relations hold in quantum mechanics and that, "the causal relation is Schrödinger's equation."[8]

Bohm's hidden variables

Bohm broke with the Copenhagen model in the early 1950s, after his misgivings about it were amplified at a meeting with Einstein, who "talked me out of it," encouraging him to focus on a determinist theory.[9] Before he was driven out of Princeton by McCarthyites, Bohm unleashed his "hidden variables" interpretation, which eliminated the necessity of an external observer. Along the way, he shed the superposition principle and the measurement problem, (albeit at the expense of supplementing the quantum mechanical formalism with new variables). Declaring Bohr's interpretation of quantum mechanics to be subjective and "totally inadequate,"[10] Bohm said that his own "non-local" interpretation,

> is based on the simple assumption that the world as a whole is objectively real and that, as far as we can know, it can be correctly regarded as having a precisely describable and analyzable structure of unlimited complexity.[11]

Bohm proposed that an undiscovered physical field permeates the universe. The hidden field guides particles on single trajectories according to the classical laws of motion. Starting from hidden initial conditions, hidden variables determine the visible paths of quantum systems.[12] One problem with his interpretation is that its predictions do not differ from those of the conventional interpretation—and vice versa. It is not falsifiable, because any experiment will give the same results for the new theory as for the old theory. This is also a weakness in Everett's theory—but that does not mean that the Copenhagen approach is better.

Bohr's long time assistant at his institute, Leon Rosenfeld, instantly attacked Bohm's "deterministic" interpretation, not on the grounds of its formal argument, but on the grounds that it did not conform to Bohr's position that quantum mechanics is fundamentally indeterminist.

Everett was influenced by Bohm's work, although he saw no need to add new terms to the Schrödinger equation. In his thesis, Everett opined that hidden

---

[7] Margenau, H. (1950). 422. Italics added.
[8] Margenau, H. (1963).148; Margenau proposed a "latency" interpretation of quantum mechanics: see Jammer, M. (1974). 504–507.
[9] Gell-Mann, M. (1994). 170; Baggot, private communication, 2009.
[10] Bohm, D. (1952). 369.
[11] Ibid. 392.
[12] Jammer, M. (1974). 280.

variable theories, "are indeed possible."[13] Adumbrating an aspect of his own theory, he wrote,

> Bohm succeeds in showing that in all actual cases of measurement the best predictions that can be made are those of the usual [collapse] theory, so that no experiments could ever rule out his interpretation in favor of the ordinary theory.[14]

Nor did Everett care to join in the ballyhoo about God playing or not playing dice:

> The question of determinism or indeterminism in nature is obviously forever undecidable in physics, since for any current deterministic (probabilistic) theory one could always postulate that a refinement of the theory would disclose a probabilistic (deterministic) substructure, and that the current deterministic (probabilistic) theory is to be explained in terms of the refined theory on the basis of large numbers (ignorance of hidden variables).[15]

Everett believed that the best interpretation of quantum mechanics was embedded in the wave equation and that he could tease it out.

Einstein's skepticism

Einstein did not endorse Bohr's complementarity model; he reluctantly favored a "statistical," classically oriented interpretation of quantum mechanics.[16] Although Einstein found von Neumann's projection postulate empirically useful, he did not find it explanatory, and he therefore considered quantum mechanics to be unfinished as a descriptive method. In 1936, he wrote,

> The ψ function does not in any way describe a condition which could be that of a single system: it relates rather to many systems, to 'an ensemble of systems' in the sense of statistical mechanics.[17]

For Einstein, it made no more sense to say that a single particle acts probabilistically than to say that single atom has a temperature. In other words, probability is not in and of itself explanatory, it is a clue that there is a more foundational reality.

In his thesis, Everett cited a recently published essay by Einstein in which he highlighted the implausibility of reducing superpositions governed by the Schrödinger equation into single states by acts of observation. Einstein commented that Schrödinger was "in principle" correct to call for treating the object measured and the measuring apparatus as a combined quantum mechanical system.[18] Referring to the view that reality depends upon acts of observation, Einstein reflected,

---

[13] DeWitt, B. and Graham, N. eds. (1973). 113.
[14] Ibid. 112.
[15] Ibid. 114–115.
[16] Jammer, M. (1974). 440; Einstein, A. (1949). 671.
[17] Quoted in Jammer, M. (1974). 440.
[18] Einstein, A. (1949). 670.

Such an interpretation is certainly by no means absurd from a purely logical standpoint; yet there is hardly likely to be anyone who would be inclined to consider it seriously.[19]

Reiterating that the macroscopic world is "real," Einstein pondered the inadequacy of complementarity for describing the reality of the microscopic world:

But the "macroscopic" and the "microscopic" are so inter-related that it appears impracticable to give up this program in the "microscopic" alone.... To me it must seem a mistake to permit theoretical description to be directly dependent upon acts of empirical assertions, as it seems to me to be intended in Bohr's principle of complementarity, the sharp formulation of which, moreover, I have been unable to achieve despite much effort which I have expended on it.[20]

In his thesis, Everett commented,

Einstein hopes that a theory along the lines of his general relativity, where all of physics is reduced to the geometry of space-time could satisfactorily explain quantum effects. In such a theory a particle is no longer a simple object but possesses an enormous amount of structure (i.e. it is thought of as a region of space-time of high curvature). It is conceivable that the interactions of such 'particles' would depend in a sensitive way upon the details of this structure, which would then play the role of the 'hidden variables'.... [That] possibility cannot be discounted.[21]

Wigner's idealism

Reducing Copenhagen's emphasis on the role of the observer in quantum mechanics to absurdity, Wigner published a broad attack on scientific materialism in 1961, "Remarks on the Mind-Body Question." The paper was the product of years of thought and collegial discussion, so his professor's view that human consciousness rules the quantum world was probably known to Everett prior to its publication.

Citing Descartes' dictum, "Cogito, ergo sum," Wigner dealt with the measurement problem by postulating that,

the content of human consciousness is an ultimate reality [and] the question concerning the existence of almost anything (even the whole external world) is not a very relevant question.[22]

Stating that materialism is incompatible with quantum theory (and all science), Wigner opined that the mind influences the body, but the body does not influence the mind. Defeated, in particular, by an inability to mathematically

---

[19] Ibid. 671.
[20] Ibid. 674.
[21] DeWitt, B. and Graham, N. eds. (1973). 112.
[22] Wigner, E. (1961). 169.

solve the measurement paradox, Wigner decided that the material world must be purely a product of linked human consciousnesses, which are capable of "modif[ing] the usual laws of physics."[23] Consequently, said Wigner, this "may mean that the superposition principle will have to be abandoned."[24] In the interim, he "solved" the measurement problem by postulating that human consciousness (mysteriously, inexplicably) causes wave function collapse—and that we can determine which physical systems are conscious by determining which cause collapse.

Everett preferred to abandon Wigner's idealism and keep the principle of superposition. In his thesis, he parodied an element of Wigner's solipsistic argument, called "Wigner's friend," which he labeled "untenable."

Wigner won the Nobel Prize for Physics in 1963 for contributions to theories of the atomic nucleus (and not for his philosophy of consciousness as determinate of material reality). But his theory of linked consciousnesses continues to be influential. It resurfaced decades later as a "many minds" interpretation of Everett.

Everett's secret

The last chapter of Jammer's book concludes with a lengthy exposition of Everett's "Many Worlds Interpretation."

According to Jammer,

> Everett's theory was first generally ignored, so much so indeed that a recent reviewer referred to it as 'one of the best kept secrets in this century.' *The multiuniverse theory is undoubtedly one of the most daring and most ambitious theories ever constructed in the history of science....* It would also imply that all nonmultiuniverse theories, and that is all the other interpretations described in this book, are logically false.... Although quite a few physicists seem to sympathize, though often with reservations, with the principles of the many-worlds interpretation, it can certainly not claim to have gained wide acceptance.[25]

Jammer covers a large number of interpretations—Copenhagen, orthodox, hidden variables, stochastic, statistical, axiomatic, quantum logical, etc. We have looked at only the interpretive models that Everett commented upon. But he had a deep scientific literature to draw upon, and he was talking to some of the cleverest minds in the physics of his day. He was rather well-informed when he decided that,

> The wave function itself is held to be the fundamental entity, obeying at all times a deterministic wave equation.... The wave theory is definitely tenable and forms, we believe, the simplest complete, self-consistent theory.[26]

[23] Ibid. 179.
[24] Wigner, E. (1963). 338.
[25] Jammer, M. (1974). 509, 517. Italics added.
[26] DeWitt, B. and Graham, N. eds. (1973). 115.

BOOK 4
EVERETT AND WHEELER

13 Wheeler: the Radical Conservative

[There is a] widely held idea that we are distinguished from most periods of history by our greater realism. But to speak of our realism is almost like a paranoid distortion. What realists, who are playing with weapons which may lead to the destruction of all modern civilization, if not of our earth itself! If an individual were found doing just that, he would be locked up immediately, and if he prided himself on his realism, the psychiatrists would consider this an additional and rather serious symptom of a diseased mind.

Erich Fromm, 1955[1]

I personally regard the hydrogen bomb, dreadful though it will be if ever used, as the policeman's stick that enforced the long peace of this era.

John Wheeler, 1998[2]

John Archibald Wheeler was born into a family of librarians in 1911. He died almost 97 years later, after a long, successful career in nuclear and particle physics, general relativity and quantum physics. He was an accomplished teacher, lecturer, and author of technical books, as well as many essays linking science and morality. He was also a enthusiastic designer of nuclear weapons. His obituary in *The New York Times* spoke of these achievements and named his two most prominent students (both of whom pre-deceased him): the Nobel Prize-winning Richard Feynman, much-loved and lionized in life, and Hugh Everett III, who died in relative obscurity.[3]

Wheeler would undoubtedly have been pleased to be associated in death with Feynman, whom he had mentored and collaborated with on important work; but he might have felt less than cheery about having Everett's name carved on his media tombstone. Their quarter century relationship was troubled from the start. As Everett's scientific mentor, Wheeler was forever of two minds about the importance of the many worlds theory. He initially championed it to his own mentor, Bohr, and later publicly disavowed it, but he could never let go of it entirely.

---

[1] Fromm, E. (1955). 170–171.
[2] Wheeler, J. A. and Ford, K. (1998). 270.
[3] *The New York Times*, 4/14/2008.

Cecile DeWitt-Morette remembered conversations between her late husband, Bryce S. DeWitt, and Wheeler regarding the pros and cons of publishing Everett's controversial dissertation in *Reviews of Modern Physics* in 1957:

> I am going to be blunt. Wheeler is a great person, but his total admiration for Bohr was so ingrained in the man, that you do not have any idea what it was like. So, when he first saw the Everett paper, he was actually very uncomfortable because it was questioning Bohr. On the other hand, he wanted to be friendly with everybody. And in case that paper would have an important impact in the future, he did not want to oppose it, but in his heart of hearts he was really uncomfortable with anybody questioning Bohr.[4]

Two years before he passed away at his home in Hightstown, New Jersey, Wheeler said, "How I wish I had kept up the sessions with Everett. The questions he brought up were important. Maybe I did not have my radar operating." When asked why Everett's theory was not well-received at the time of its publication in 1957, Wheeler replied, "Because it made no clear and verifiable predictions." When asked if Everett was disappointed, Wheeler remarked, "He was disappointed, perhaps bitter, at the non-reaction to his theory. I'd love to talk to him today and get his answer to this question. I'd love to have a nice long talk with him today."[5]

John Wheeler with busts of Einstein, Bohr, circa 2003.

[4] DeWitt-Morette interview, 2006.
[5] Kenneth Ford interview of Wheeler on behalf of author.

As an "original" man, Everett had epitomized the clever young scientist that the professor was constantly looking to recruit to work in advanced weaponry research for the government and its defense contractors. But more than that, Wheeler admired Everett's general brilliance; he was disappointed that the young man abandoned physics research after his theory was rejected by Bohr. For nearly a quarter century, Wheeler tried to bring him back into the academic fold, convinced that Everett had significant contributions to make beyond his initial foray.

In 1998, Wheeler published an autobiography, *Geons, Black Holes & Quantum Foam*, co-written with Kenneth Ford, his former student and long time colleague. *Geons* is an informative trip through the nuclear bomb-laden landscape of 20th century physics. And it is a remarkably candid account of Wheeler's fascination with fascism and related methods of social control. The few paragraphs he devotes to Everett, however, do not even begin to tell the whole story of what happened between them. To understand Everett—we must get to know Wheeler.

Nuclear physics comes of age

Growing up in Youngstown, Ohio during the Roaring Twenties, Wheeler enjoyed a Norman Rockwell-type of upbringing; he was happy, patriotic, eager to please. From an early age, he was curious about technology, especially related to things that go bang!, mangling his hand as a teenager while experimenting with dynamite caps. At the age of 15, Wheeler enrolled in John Hopkins University, originally majoring in engineering. But he quickly found his vocation in the brave new world of quantum physics, then centered in Copenhagen and Göttingen, Germany.

In 1933, he received a doctorate in theoretical physics from Johns Hopkins for research on the structure of the helium atom. With a scholarship from the Rockefeller Foundation's National Research Council, he spent a year as a post-doc at New York University doing nuclear physics with a pugnacious Russian émigré, Gregory Breit—an experimentalist and first-rate theoretician. "Regrettably, [Breit] was a heavy smoker, but in those days one put up with it. He also had a habit of snorting every once in a while like a bull about to charge," the non-smoking Wheeler recalled.[6]

The next year, it was off to study the atomic nucleus with Bohr in Copenhagen (supported by more Rockefeller money). At the Institute, Wheeler felt like he had joined an "international family" of physicists, with the smoke-wreathed, pacing Bohr assessing the worth of quantum mechanical conjectures by whether or not they fit his dualist model of reality, complementarity. "Breit taught me new mathematical and calculational techniques. Bohr taught me a new way of looking at the world," said Wheeler.[7] Indeed, he adopted many of his mentor's mannerisms (drawing the line at smoking); particularly his oracular style of speaking about

---

[6] Wheeler, J. A. and Ford, K. (1998). 109.
[7] Ibid. 139.

John Wheeler, 1934, Copenhagen.

physics. Returning to the United States, Wheeler became one of complementarity's strongest advocates, despite feeling (at least, in retrospect) "uneasy" about its explanatory gaps.[8]

At Bohr's institute, Wheeler became friendly with Heisenberg, who, a few years later, ran the Third Reich's atomic bomb research program.

> In our talks, he tried to steer clear of politics. Always circumspect, he neither praised not condemned Hitler. He was a patriot.... I felt sympathy for him. I could understand his commitment to his country. In science, Germany led the world. In culture and arts, it had a centuries-long record of achievement. I was inclined to believe, as he no doubt did, that an immoral dictatorship was a transitory evil, something a great nation could endure without lasting harm. Of course, I was wrong.[9]

Marrying Janette Hegner in 1935, Wheeler accepted an assistant professorship in physics at the University of North Carolina in Chapel Hill. Exploring the atomic nucleus, he developed an important method—called a scattering matrix—for analyzing the probabilities of radioactive events. But North Carolina was not Princeton: the powerhouse of theoretical physics in America.

Fortuitously, in 1938, Wheeler was invited to join the Princeton faculty. Settling into a wood-paneled office in Fine Hall, he groomed like a business man complete with tightly knotted tie and starched cuffs.[10] Radical in his physics, but deeply conservative in his politics, he looked forward to tea-times, when he could commune with similar souls, von Neumann, Wigner, Wolfgang Pauli, Hermann Weyl.[11]

Next door to Fine Hall was Palmer Physical Laboratory: a large building filled with chemical equipment, machine shops for making high-voltage electrical devices, and a laboratory capable of lowering the temperature of atoms until they slowed enough to be observed and controlled. The lab's *pièce de résistance* was a cyclotron that accelerated and collided particles, so that the pattern of their scatter revealed rules of nature.[12]

In January 1939, Wheeler met Bohr's arriving ocean liner in New York Harbor. The great man of physics, accompanied by his loyal assistant, Rosenfeld, bore news of import: Otto Frisch, of Bohr's institute, and his aunt, Lise Meitner, had correctly analyzed the result of a radiation experiment performed by two German chemists. They had concluded that uranium atoms will split when bombarded with neutrons. Excited, Wheeler and Bohr immediately went to work modeling exactly how the heavy atom's nucleus, visualized as a liquid droplet could bisect, or split. Their work led to the fissioning of a rare isotope, uranium 235, and also of a new element, later called plutonium. Atomic fission of these metals set off a chain reaction, unbinding the explosive energy that fuels the stars.

---

[8] Ibid. 124: "I am uneasy [because] I see no bedrock of logic on which quantum mechanics is founded. What is the underlying reason for quantum mechanics? I keep asking myself. It has to flow from something else, and that something else remains to be found."
[9] Wheeler, J. A. and Ford, K. (1998). 140.
[10] Gleick, J. (1992). 93.
[11] Wheeler, J. A. and Ford, K. (1998). 155.
[12] Gleick, J. (1992). 107.

Wheeler and Feynman

Shortly before the Japanese attack on Pearl Harbor, Wheeler's graduate student, Richard Feynman, came to him with the kernel of a grand idea that he called "path integrals." Feynman treated the possible paths of a particle as probability waves (wave functions) that can reinforce or cancel each other. In this model, the actual trajectory taken by a particle is the "path of least action," or the *sum* of all the possible paths it could take—a "sum over histories." After adding and subtracting the wave functions of these superposed, not-yet-real trajectories, the path left standing is the most probable path, the track we would see in a cloud chamber. The paths taken by particles as they scatter off or absorb each other—transforming, thereby, into different kinds of particles—can be expressed graphically by the diagrams that bear Feynman's name. By the end of the 1940s, the splitting and merging patterns revealed by Feynman diagrams allowed physicists to visualize quantum interactions for the first time.

Wheeler often co-authored papers with his students. And he was a sounding board for Feynman during the several years it took for him to work out the substance of his thesis on path integrals. The new theory laid out many of the basic ideas of modern quantum electrodynamics, which applies to particles moving at near light speeds, taking into account relativistic effects, such as different time frames for different velocities. But, in the end, Wheeler was ditched by the smartest student he ever had. As James Gleick reported in his biography of Feynman, *Genius*, "[Feynman] took pains to leave his collaboration with Wheeler decisively behind. He wanted his thesis to be his own."[13] Wheeler later said that working with Feynman was one of the "most satisfying" events in his life.[14] In 1965, Feynman shared the Nobel Prize for Physics with Julian Schwinger and Sin-Itiro Tomonaga for developing quantum electrodynamics.

Wheeler, however, never received a Nobel Prize, (although he was awarded dozens of honors, and was a VIP in the most exclusive and prestigious scientific organizations). Nonetheless, the moral and intellectual support that he tendered to Feynman's and, later, Everett's bright ideas, were major contributions to physics.[15]

War comes to Princeton

Although he was a flexible and creative physicist, Wheeler had an inflexible, "might-is-right" attitude when it came to politics and warfare. Before Pearl Harbor, he was oft-criticized by his more liberal colleagues as a Nazi-sympathizer, which he was, and which he lived to regret. He later explained,

> In the months before Pearl Harbor [while Hitler was crushing Eastern Europe and invading France and Russia], we had a radio in the Fine Hall Tea Room

---

[13] Gleick, J. (1991). 147.
[14] Wheeler/Ford, transcript X, AIP.
[15] Wheeler had many graduate students who made major contributions to physics. And Feynman's work was of a very different nature than Everett's—for one thing, it was experimentally verifiable.

so that we could listen to the war news and discuss the war's progress. I reached the conclusion that a German-dominated Europe might be the best way to assure long-term peace in Europe.... I admired the strength and efficiency of the German state. I cannot claim to have been a naïve young man out of touch with the real world, for I had been avidly interested in history and foreign affairs since my student days. I had learned from German Jewish scientists...what a threat they considered Hitler's Germany to be....I discounted the fears of my friends, believing that no civilized people could translate poisonous rhetoric into inhuman action.... This sympathy did not vanish instantly when America entered the war. It evaporate slowly as I learned more. Even when I was doing everything I could to help defeat Germany, I clung to the belief that people are fundamentally decent everywhere, that German atrocities were as unthinkable as American atrocities....But not until I visited Auschwitz in 1947 was the full horror of German barbarism brought home to me.[16]

On December 7, 1941—despite the existence of military intelligence showing that the attack was imminent—Japanese airplanes sank American warships in Pearl Harbor. The American populace, which had been reluctant to become involved in another world war, instantly mobilized. Fueled by military funding, industry and science supercharged each other. For his part, Wheeler joined the Manhattan project. From 1942–1945, he worked out how to manufacture plutonium as a byproduct of nuclear reaction in concert with the Du Pont Company and the Metallurgical Laboratory at the University of Chicago. Plutonium was the fissionable material at the core of "Fat Man," the atom bomb that annihilated Nagasaki on August 9, 1945.

Purging atomic science

Two weeks before Hiroshima was scheduled to be eradicated, 69 Manhattan Project scientists, including the project's chief physicist, Leo Szilard, signed a petition to President Harry Truman stating that with the recent defeat of Germany, the use of the atomic bomb against Japan could not be justified. The petitioners begged Truman to give Japan an opportunity to surrender before dropping the bomb:

> A nation which sets the precedent of using these newly liberated forces of nature for destructive purposes may have to bear the responsibility of opening the door to an era of devastation on an unimaginable scale.[17]

Neither Wheeler, von Neumann, nor Feynman signed the petition.

And regarding Bohr's call for the sharing of nuclear secrets, Wheeler said, "I lacked the personal passion to push for Bohr's concept of an open world and

---

[16] Wheeler, J. A. and Ford, K. (1998). 42; Wheeler is not on record visiting the ruins of Dresden, Tokyo, Hiroshima, or Nagasaki to bear witness to American barbarism.

[17] U.S. National Archives, Record Group 77, Records of the Chief of Engineers, Manhattan Engineer District, Harrison-Bundy File, folder #76.

did not join in promoting it. I could not convince myself that it was workable in practice."[18]

But other Manhattan Project physicists were so upset by the post-war militarization of science that they formed national organizations to work for disarmament and to protect science from politicization. One such group was the Atomic Scientists of Chicago, which created the famous Doomsday Clock ticking inexorably toward nuclear midnight. Leading atomic physicists, such as Wiener, declined to participate in nuclear weapons research. They gave interviews to the media on the dangers of nuclear armament. They wrote scary books, such as *One World or None,* a collection of essays by Einstein, Oppenheimer, Bohr, Wigner and other scientists dismayed by the violence of the nuclear age they had midwived.

But as McCarthyism and hysterical red-baiting created and fed upon paranoia, formerly outspoken scientists started to keep their heads down, fearing for their jobs. Many were shocked at the callous treatment of the war-hero Oppenheimer by the Atomic Energy Commission, and appalled by Princeton's firing of the radiantly intelligent David Bohm. Members of the Federation of Atomic Scientists, which called for disarmament, were surveilled by informants and FBI agents. Government agents wire-tapped even slightly dissident scientists, harassed them at their jobs, searched their homes and offices for evidence of conspiratorial intent.[19]

Heroically, Everett's game theory mentor, Harold Kuhn, came out in the national media in defense of Bohm, after he was pilloried by the House Un-American Activities Committee. When asked to finger supposedly leftleaning associates, Bohm had taken advantage of his constitutional right to plead the Fifth Amendment, which was the only way to avoid criminal contempt charges. Consequently, he was fired by Princeton, and blacklisted by physics departments in the United States. He moved to Brazil to work. After the United States restricted his passport privileges,[20] he became a Brazilian citizen, eventually moving to England.

Regarding the persecution of Bohm, Wheeler recalled,

> I found it hard to accept Bohm's decision to shield those who adhered to Communist ideology...The university was gauche in its manner of dealing with Bohm, yet I could sympathize with its goal, to preserve its reputation as a center of unbiased scholarly inquiry, not the home of blind loyalty to one ideology or another.[21]

It is estimated that a fifth of the witnesses called before congressional and state "internal security" committees were professors and graduate students, and that half of those were scientists. According to science historian, Jessica Wang,

---

[18] Wheeler, J. A. and Ford, K. (1998). 226.
[19] Wang, J. (1999). 7–9.
[20] Due to his pacifism, physicist-biologist, Linus Pauling, was also forbidden to travel abroad without restrictions, until he won the Nobel Prize in Chemistry in 1954. Wang, J. (1999). 275.
[21] Wheeler, J. A. and Ford, K. (1998). 216.

Most American colleges and universities fired professors, tenured and unten-
ured, who refused to cooperate with HUAC [House Un-American Activities
Committee] and other committees on the ostensible grounds that their lack of
candor [by invoking the Fifth Amendment right guaranteed under the
Constitution] was incompatible with the openness required by scholarly
life....By the 1950s, there was no place to hide....For younger scientists in the
throes of trying to establish stable careers, it was hard to avoid the conclusion
that the best way to protect one's livelihood was to follow the dictates of the Cold
War political consensus and steer clear of controversial political activities
altogether.[22]

Carrots and sticks were used to herd scientists into military work during the
Cold War:

As political elites whose expertise was essential to the military basis for Cold
War foreign policy, scientists had, in theory, more freedom of action than other
groups to defy the politics of anticommunism....[But] Cold War liberalism...denied the importance of popular politics and insisted that all social
problems were amenable to nonideological negotiation through guidance by
responsible elites.[23]

Like many Cold War scientists working for the military, Wheeler and Everett
lived inside a political, cultural, and ethical feedback loop which admitted no
contradictory ideological data; indeed, operations researchers, in particular,
often seemed convinced that nuclear weapons research was an ideologically
neutral act. Despite creating the tools and methods of species-extinction, they
assumed that smart people like themselves held a monopoly on good intentions toward mankind and were, therefore, worthy of respect, responsibility,
and remuneration.

Rising stars

The Cold War turned many American scientists into businessmen. According
to Wheeler,

The rising star of science in the postwar firmament (perhaps it should be called
the stardom of science) affected me personally, as it affected many of my friends.
I received invitations to give talks, write articles for the general public, advise the
government, and sit on boards.[24]

In fact, Wheeler was a military, industrial, and scientific super star.
In 1957, C. Wright Mills, in his (now-classic) study of the post-war "power
elite," observed a sea change in the sociology of science:

Scientific and technological development, once seated in the economy, has
increasingly become part of the military order, which is now the largest single

---

[22] Wang, J. (1999). 272, 283.
[23] Wang, J. (1999). 8, 290.
[24] Wheeler, J. A. and Ford, K. (1998). 161.

supporter and director of scientific research in fact, as large dollar-wise, as all other American research put together. Since World War II, the general direction of pure scientific research has been set by military considerations, its major finances are from military funds, and very few of those engaged in basic scientific research are not working under military direction.... Some universities, in fact, are financial branches of the military establishment.

The top scientific minds, said Mills,

have become deeply involved in the politics of military decisions, and the militarization of political life.... As part of the military ascendancy, there is the felt need of the warlords for theory, the militarization of science, and the present 'demoralization' of the scientist in the service of the warlord.[25]

Wheeler cheerfully served the warlords. He regularly consulted for armament manufacturers, such as Du Pont, General Atomics, General Electric, Consolidated Vultee Aircraft Corporation (Convair), and Lockheed, (consultation rates were typically $200 per day, plus expenses). And for 30 years, he was a trustee of the Battelle Memorial Institute in Columbus, Ohio, a nonprofit military think tank through which many billions in government contracts have flowed. For decades, Battelle has co-managed national laboratories where weapons of many varieties are designed and tested. "I think the effect of the war was to make people appreciate, who were in the field of physics, that you can get money to do physics, you could do it in a big way, and you don't have to be a worm," said Wheeler.[26]

Post-war, Du Pont executives set up a private research fund for Wheeler. The company also paid him a monthly retainer, which got the top-secret-cleared physicist into hot water with his bosses at Los Alamos.

It upset [Norris] Bradbury [director of the Los Alamos National Laboratory] to have me with this connection with Du Pont along with Los Alamos.... I recall his catching me in the corridor one day, drawing me into his office, and there were two or three other of the higher members of the Los Alamos hierarchy there, and he started quizzing me about this relationship with Du Pont—how much time I spent and how much I got paid—which was a little embarrassing for me, and finally insisting that I should give it up, which I did.

[But] my Du Pont friends had a pretty good idea of what I must be doing. Dale Babcock was the member of their staff I was closest to, and I recall one day saying to him, 'It's crazy for us to shoot off our atomic bombs as separate things because they're just match sticks to light the real thing and we should be think of getting on with something like a H bomb.' He was very sympathetic to that, although we certainly didn't talk about any details.[27]

Returning to academia after the war years, Wheeler secured funding from the Office of Naval Research to build and operate a futuristic laboratory at

[25] Mills, C. W. (1956). 216–218.
[26] Aaserud, F. (1995). 41.
[27] Wheeler/Ford, transcript VIII, AIP.

Princeton. He was in charge of experimental research using cosmic rays (barrages of elementary particles arriving from the far reaches of the universe); and he wrote several papers showing how magnetic fields generated by planets affect these particle showers. He participated in a famous conference at Shelter Island, Long Island conference in 1947. Gathered at the Rams Head Inn were the most inventive Americans in quantum physics—Oppenheimer, Hans Bethe, Edward Teller, Enrico Fermi, Isador Rabi, Willis Lamb, Abraham Pais, Julian Schwinger, Feynman—many of them eager to get back to theoretical research after having spent the war years designing such new technologies as radar, microwave devices, analogue computers, and atom bombs.

Cold warrior

Wheeler continued working with Bohr to map the subtleties of atomic structures and, in 1950, he joined forces with his good friend, Edward Teller, to develop the hydrogen bomb. The task was unappealing to many alumni of the Manhattan Project who had seen the ferocious result of their collective brainstorm. But Wheeler had no qualms: "The antagonism I felt from some of my colleagues did nothing to shake my conviction that I was doing the right thing."[28]

Wheeler testified to Congress that the Soviet Union could be ahead of the United States in building a hydrogen bomb. "Our secrecy keeps secret how little we are doing, not how much we are doing." He scoffed at Oppenheimer's warning that the age of nuclear overkill had arrived.[29]

In Wheeler's world, building bigger and better nuclear bombs was a service to truth:

> As a matter of survival the truth about thermonuclear explosions must be known.... How remarkable that today we can...predict in advance the fantastic explosion history of a nuclear device never before seen.... This idea of pinning numbers to ideas...is such a new one that we haven't found out how to pass on the pure delight and wonder of it.[30]

In conjunction with the Atomic Energy Commission, he convinced Princeton to let him set up and operate a hydrogen bomb research laboratory, called Matterhorn B, on land owned by the university a few miles off-campus. For the next several years, Wheeler and his hand-selected team (including several of his students) commuted between Los Alamos and Princeton designing the system for igniting the bomb's hydrogen fuel and calculating its explosive potential.

The first hydrogen bomb, nicknamed "Ivy Mike," was exploded on October 31, 1952 at Eniwetok, a dreamy atoll in the Pacific Ocean. That island paradise was gradually torn to pieces by successive test shots that polluted the

---

[28] Wheeler, J. A. and Ford, K. (1998). 193.
[29] Rhodes, R. (1995). 527–528.
[30] Wheeler, J. A. (1956). 43.

atmosphere of the globe with radioactive fallout. Wheeler watched the first test from afar through dark glasses:

> First a black spot on the horizon, then this opening into brilliance as if the sun had just come into view. Then this getting covered up by a churning mass of clouds. My first reaction was simple relief. It worked. I'm glad I was there. I am ashamed to say that the energy release was about thirty percent more than we figured.[31]

Within a year, the Soviets exploded a similar bomb, and the race for nuclear dominance was on. Wheeler's assessment:

> I am sometimes asked to name the most important peacetime use of nuclear energy. My answer is simple: a nuclear device to keep the peace.[32]

After building the H-bomb, Wheeler consulted for Convair Corporation on the design of the Atlas missile, which was intended to carry nuclear warheads. In 1955, he freelanced for Lockheed Corporation's missile division. But one of his main tasks as a leader in the new industry of militarized operations research was to find and recruit the best scientific minds in academia to design a boutique line of nuclear arms and supersonic delivery systems. Consequently, he was involved in creating several Ivy League-sponsored organizations to do exactly that—steering scientists toward war work and warriors toward scientific solutions.

In the summer of 1958, he convened a group of mostly physicists to address military problems. Out of this effort grew a group of militarist scientists called JASON that, for many years, advised the Secretary of Defense on the feasibility of developing futuristic weapons systems, such as nuclear-powered airplanes and electronic battlefields. Wheeler also worked closely with the Institute for Defense Analyses, which operated JASON[33] and employed Everett in its top secret Weapons Systems Evaluation Group (WSEG) after he left Princeton. And he spent a lot of time, in association with Princeton colleagues, Eugene Wigner and Oskar Morgenstern, lobbying government officials to create an advanced research projects weapons laboratory (separate from the new Advanced Research Projects Agency) that would initially specialize in missile development.

As the Cold War wore on, Wheeler's public role increasingly intersected with his private consulting work. Lobbying for the creation of the missile lab, Wheeler erroneously claimed that the Soviet Union was ahead of the United States in missile R&D and manufacture and deployment. And he angered his colleagues when, just as approval of the missile lab was on the political horizon, he dropped out of the project, saying he preferred to base himself in academia—an action that killed the lab.

Princeton physicist Marvin Goldberger, who was very close to Wheeler while working on the advanced projects lab, wrote to Wigner,

---

[31] Wheeler, JAM. (1998). 225; Wheeler/Ford, transcript VIII, AIP.
[32] Wheeler, J. A. and Ford, K. (1998). 227.
[33] See Finkbeiner, A. (2006).

I must say, however, that my reservations about John's being director, which I'm sure you sensed from our earlier discussions, were reinforced by seeing him in action as a leader. He has many great virtues and his halo is the finest gold. There is however an amorphous quality about him both in his reception of ideas and in his transmission of information to others. I find myself wanting to shake him to make him say something straight out and incisively. I have difficulty putting this into words, but Oskar [Morgenstern] described his own feelings to me in a similar way.[34]

Goldberger's candid assessment sums up an aspect of Wheeler that many of his associates, including Everett, found infuriating: his tendency to play both sides of a game. Of course, in the heyday of game theory, insincerity was often considered to be a tactical virtue. In that light, it was not irrational for Wheeler to recalibrate his political and scientific agendas according to shifts in the balance of power.

In 1959, Wheeler and Morgenstern joined forces with Henry Kissinger, a consultant to the Weapons Systems Evaluation Group, where Everett was employed. Convair Corporation hired the trio of physicist, game theorist, and political scientist to write "A Doctrine for Limited War," which advocated for fighting small nuclear wars. Years later, Wheeler said,

In the end Convair decided not to publish it, because they feared that it would bring down upon them the wrath of people of the kind who had chanted after World War I, chanted about American defense contractors, 'merchants of death.' They didn't want to be called merchants of death.[35]

Kissinger later published an influential book on the doctrine of limited warfare, which allowed for the use of tactical nuclear weapons in the pursuit of "national self interest." Wheeler found Kissinger's book to be "absolutely marvelous." In subsequent years, Wheeler was a super-hawk for invading, bombing, defoliating, and trying to occupy Vietnam. He favored the Kissinger-crafted assassination of President Salvador Allende of Chile in 1973, falsely claiming that Allende was a communist; and he asserted that Chile became more democratic under the brutal rule of the military dictator, Augusto Pinochet.[36] During the Cold War, Wheeler advocated for building the anti-ballistic missile systems and multiple nuclear war head missiles (MIRVs) that kept the arms race pumped up and profitable for the arms manufacturers and war contractors he worked for, especially Battelle Memorial Institute.

Despite his long time devotion to the task of proliferating nuclear weapons, Wheeler was awarded the Niels Bohr International Gold Medal in 1982 for contributions to the peaceful use of atomic energy. (He had helped develop the idea of building nuclear reactors within a containment facility.) Like so many people with access to political power and wealth, Wheeler operated according to a fine-tuned sense of self-interest, *sans* a sense of irony as he flip-flopped

[34] Aaserud, F. (1995). 219.
[35] Wheeler/Ford, transcript IX, AIP.
[36] Wheeler/Ford, transcript XI, AIP.

from one agenda to another. That he could take credit for working for peace while building the engines of war is not a tribute to him, but a sign of the times. Wang observes,

> The Cold War left the United States with an impoverished form of liberalism that viewed policy as driven by process and banished ideals from politics.[37]

It is hard to label Wheeler. He was conservative in his politics and radical in his physics. And he was often of two minds about important matters. Attracted to the "efficiency" of Hitler's national socialism, he eschewed it when Roosevelt efficiently mobilized industry and science against the Axis powers. He made globe-destroying bombs, while remaining a devoted family man. He was a good friend to all, and he had an agenda for all of his friends, whether they knew it or not. He was in love with the idea of truth and beauty, and he cheerfully sold his learning for fees and access to government and corporate power. He vacillated between preaching the absolute truth of the Copenhagen interpretation and praising Everett's anti-Copenhagen theory. In these ways, he was an embodiment of the superposition principle.

[37] Wang, J. (1999). 289.

14 Genesis of Many Worlds

Bohr convinced Heisenberg and most other physicists that quantum mechanics has no meaning in the absence of a classical realm capable of unambiguously recording the results of observations. The mixture of metaphysics with physics, which this notion entailed, led to the almost universal belief that the chief issues of interpretation are epistemological rather than ontological: The quantum realm must be viewed as a kind of ghostly world whose symbols, such as the wave function, represent potentiality rather than reality.

Bryce DeWitt, 1970.[1]

I believe that basing quantum mechanics upon classical physics was a necessary provisional step, but that the time has come...to treat [quantum mechanics] in its own right as a fundamental theory without any dependence on classical physics, and to derive classical physics from it.

Hugh Everett III, 1957.[2]

It was Everett who gave us permission to think about the universe as wholly quantum mechanical.

W. H. Zurek, 2006.[3]

Dice-playing mice

After Ivy Mike exploded, Wheeler returned to full-time teaching at Princeton. He taught a graduate course in general relativity. One spring day, Wheeler and eight of his relativity students took tea with Einstein at his Mercer Street home. They elicited the great man's thoughts on "everything from the nature of electricity and the unified field theory to the expanding universe and his position on quantum theory."[4]

A year later, on April 14, 1954, Einstein gave the last public lecture of his life at the Palmer Laboratory to Wheeler's relativity class. Misner recalled that Everett attended the lecture. That day, Einstein said that quantum mechanics is true, as far as it goes, but he did not see it as fully describing the quantum world:

[1] DeWitt, B. S. (1970).
[2] Everett to Petersen, 5/31/57.
[3] Zurek interview, 2006.
[4] Wheeler, J. A. (1979). 183.

It is difficult to believe that this description is complete. It seems to make the world quite nebulous unless somebody, like a mouse, is looking at it.[5]

(On the cocktail party tape a quarter century later, Everett did not remember attending Einstein's talk, but he had used the mouse metaphor to great effect in his thesis.)

A few months later, Wheeler convened a group of graduate students, including Misner, to explore using Feynman's sum over histories technique as a model for quantizing gravity. This was an extremely ambitious project: it meant trying to figure out how to create wave functions for gravity particles, which were purely theoretical entities (and still are). Misner eventually wrote his doctoral dissertation on that topic.

Misner recalled, "Everyone talking to Wheeler at that time was likely to be encouraged to think about quantum gravity. The question of how to give meaning to the wave function of the universe appears to have played a role in Wheeler's interest in Everett's views of the quantum."[6] Everett did not sign up for Wheeler's relativity class, but he kept a copy of Wheeler's mimeographed "Notes on quantization of gravity and electromagnetism," dated October 15, 1955. The notes are a summary of Wheeler's conversations with students about quantizing gravity. He feared that the equations of general relativity and quantum mechanics were too dissimilar to ever mesh, but he was bound to try. And, as we shall learn, Wheeler considered Everett's theory of a universal wave function to potentially be a key for unlocking the secret of quantum gravity.

Out of the basement

The great find for the history of science in Mark Everett's funky basement was his father's correspondence with prominent physicists and philosophers, and his half-century old handwritten thesis: the penciled first draft, complete with scratched out paragraphs, and pages and pages of private notes about his theory. It is the raw stuff of the many worlds interpretation—and it illuminates how he viewed his own work, particularly his belief on whether or not the branching universes exist.

Fifteen years after the thesis was published, Everett penned a letter (found in the basement) to Max Jammer, who was writing his book on the philosophy of quantum mechanics. Everett told Jammer that "the principle intellectual influences" on the development of his theory were Misner, Petersen, and Wheeler. The latter

> encouraged me to pursue the matter further as a thesis. During the course of this pursuit I would say that perhaps the primary influences were von Neumann's book and the later chapters of Bohm's Introduction to Quantum Mechanics [sic].

---

[5] Tauber, G. E. (1979), 187.
[6] Misner private communication, 4/25/08.

I must answer in all candor the primary motive was, of course, to obtain a thesis. However, I must also admit to a strong secondary motive to resolve what appeared to me to be inherent inconsistencies in the conventional interpretation.

I was of course struck, as many before and also many since, by the apparent paradox raised by the unique role assumed by the measurement process in quantum mechanics as it was conventionally espoused. It seemed to me unnatural that there should be a 'magic' process in which something quite drastic occurred (collapse of the wave function), while in all other times systems were assumed to obey perfectly natural continuous laws.[7]

In 1954, Everett was not alone in his feeling that the collapse postulate was illogical, but he was one of the very few physicists who dared to publicly express deep dissatisfaction with it:

I thought at that time that perhaps the pursuit of this apparent difficulty would lead to a new and different theory which, while resolving the apparent paradoxes, would also lead to new predictions. Unfortunately, as it turned out, the theory which I constructed resolved all the paradoxes and at the same time showed the complete equivalence with respect to any possible experimental test of my theory and that of conventional quantum mechanics. The net result of my theory therefore is simply to give a complete and self-consistent picture (without any particular 'magic' associated with the measurement) that in all practical predictions will of course be identical to the predictions of the conventional formulation.[8]

In other words, Everett had hoped to reinvent quantum mechanics on its own terms and was disappointed that his revolutionary idea was experimentally unproveable, as the only "proof" of it was that quantum mechanics works—a fact which was already known. He saw his interpretation as valuable, nonetheless:

To me, therefore, the real usefulness of this picture or theory of quantum mechanics is simply as an alternative which could be acceptable to those who sense the paradoxes in the conventional formulation, and therefore save much time and effort by those who are also disturbed by the apparent inconsistencies of the conventional model. As you know there have been a large number of attempts to construct different forms of quantum mechanics to overcome these same apparent paradoxes. To me, these other attempts appear highly tortured and unnatural.

I believe that my theory is by far the simplest way out of the dilemma, since it results from what is inherently a simplification of the conventional picture, which arises from dropping one of the basic postulates—the postulate of the discontinuous probabilistic jump in state during the process of measurement—from

[7] Everett to Jammer, 9/19/73.
[8] Ibid.

the remaining very simple theory, only to recover again this very same picture as a deduction of what will appear to be the case to observers.

I therefore believe that my formulation is by far the simplest from an axiomatic point of view. The acceptability, however, clearly is a matter of personal taste.[9]

It is important to note that Everett believed he had deduced from the laws of quantum mechanics the "appearance" of probability to an observer in most of the branches of a branching universe in which everything physically possible happens.[10] In other words, he did not necessarily believe that probability was an objective property of the universe of universes, but that it was a subjective measure of our ignorance about what goes on outside our single branch.

Bohm again

In thinking about the measurement paradox, Everett said he was influenced by Bohm's 1951 textbook, *Quantum Theory*. This highly regarded work explains the basic equations of quantum mechanics in clear language accessible to non-physicists. Bohm employed Bohr's method of treating opposites as complementary processes, but he ignored Bohr's rule that one must not speak about what goes on inside the quantum world. And he spent his last chapter climbing the edifice of the measurement problem:

> At the quantum level of accuracy the entire universe (including, of course, all observers of it) must be regarded as forming a single indivisible unit with every object linked to its surroundings by indivisible and incompletely controllable quanta [i.e. events not controlled by experimental design]. If it were necessary to give all parts of the world a completely quantum-mechanical description, a person trying to apply quantum theory to the process of observation would be faced with an insoluble paradox. This would be so because he would then have to regard himself as something connected inseparably with the rest of the world. On the other hand, the very idea of making an observation implies that what is observed is totally distinct from the person observing it.[11]

Bohm described the contradiction that Everett sought to remedy:

> If the quantum theory is to be able to provide a complete description of everything that can happen in the world, however, it should also be able to describe the process of observation itself in terms of wave functions of the observing apparatus and those of the system under observation [including] the human investigator as he looks at the observing apparatus and learns what the results of the experiment are, this time in terms of the wave functions of the various atoms that make up the investigator, as well as those of the observing apparatus and the system under observation. In other words, the quantum theory could not be

---

[9] Ibid.
[10] Although, there could be "maverick" universes in which the laws of physics as we know them do not apply.
[11] Bohm, D. (1951). 584.

regarded as a complete logical system unless it contained within it a prescription in principle for how all these problems were to be dealt with.[12]

Everett took this to mean that if the object is a superposition of properties, then the observer who is correlating to that object by looking at it will also be in a superposition. A superposition in which each state of the observer is linked to a particular state of the superposed object. In his textbook, Bohm avoided that conclusion by falling back on von Neumann's collapse postulate as an explanation of why macroscopic objects are not superposed. He retreated to Bohr's quantum-classical duality, asserting that, "quantum theory presupposes a classical level," and that,

In order to obtain a means of interpreting the wave function, we must therefore *at the outset* postulate a classical level in terms of which the definite results of a measurement can be realized.[13]

This was necessary, he said, because,

the classically definite aspects of large-scale systems cannot be deduced from the quantum-mechanical relationships of assumed small-scale elements... instead...the nature of what can exist at the nuclear level depends to some extent on the macroscopic environment.[14]

Everett took a different tack. He thought that classicality emerged from the quantum womb, not the reverse. But the basic interpretive questions that Bohm grappled with in the textbook—how single results emerge from superpositions—and his attempt to break out of the Bohr-von Neumann model, must have encouraged Everett, showing him that his idea of a fundamentally quantum universe was not hallucinatory.

---

[12] Ibid. 583.
[13] Ibid. 626.
[14] Ibid. 627–628.

15 Alone in the Room

This problem of getting the interpretation proved to be rather more difficult than just working out the equation.

P. A. M. Dirac, 1977[1]

In the fall of 1954, Everett sat at his desk in his room at the Graduate College. In front of him were sharpened pencils, a yellow legal pad, and a manual on dissertation writing given to him by Wheeler. The guide advised: "In your investigation, think of the subject. In your presentation, think of the reader.... The general tone of one's statements should be cautious, and the strength of the report should come through the nature of the evidence and its logical presentation, rather than through strong personal assertions." Although Wheeler was fond of using provocative language in his own work, he preferred that his students be more circumspect, a bridle at which Everett champed.

During the next year, Everett spent most of his time researching and writing the thesis. In the spring and summer of 1955, he took breaks to take Nancy Gore to parties and football games and ballroom dancing. But back at the desk, he focused on his theme:

> Quantum mechanics is reformulated in a way which eliminates its present dependence on the special treatment of observation of a system by an external observer.[2]

Nothing *too* ambitious: just the reformulation of quantum mechanics.

Because Everett's thesis evolved through multiple versions it had several different titles. To clarify: In his son's basement are the original sheaves of yellow legal paper upon which Everett began writing in pencil during his third semester of graduate school. In this rough draft each chapter evolved through several versions. The equations changed as he refined his mathematical argument; and he repeatedly toyed with images of "splitting" amoebas, cannonballs, and observers. It appears that Wheeler read through some handwritten sections, making suggestions with a red pencil.

Gore, who was an accomplished typist, typed up three mini papers that Everett extracted from his evolving thesis. He gave them to Wheeler as progress reports during the fall term of 1955. They were titled "Probability in Wave

---

[1] Quoted in Pais, A. (1991). 295.
[2] Everett thesis abstract.

Mechanics," "Quantitative Measure of Correlation," and "Objective vs. Subjective Probability," and Wheeler read them carefully, making notations.

The arguments presented in the mini papers explain the main points of his theory. One, that the universe is governed in its entirety by the Schrödinger equation; there is, objectively, no such thing as wave function collapse. Two, that the macroscopic world of our experience emerges from the microscopic world through entanglement. Three, that information theory can generate a probability measure in quantum mechanics without having to postulate the Born rule. Everett originally called his work, "Correlation Interpretation of Quantum Mechanics."

In January 1956, Everett submitted his typed, 137-page dissertation to Wheeler (the "long" thesis), now entitled, "Quantum Mechanics by the Method of the Universal Wave Function."[3] A few months later, after a minor revision, bound copies entitled "Wave Mechanics Without Probability" were distributed to select physicists, including Bohr.

In April, Wheeler wrote to Bohr:

> I would be appreciative of comments by you and Aage Petersen about the work of Everett.... The title itself, "Wave Mechanics Without Probability", like so many of the ideas in it, need further analysis and rephrasing, as I know Everett would be the first to say.

> But I am more concerned with your reaction to the more fundamental question, whether there is any escape from a formalism like Everett's when one wants to deal with a situation where several observers are at work, and wants to include the observers themselves in the system that is to receive mathematical analysis.[4]

In May, Wheeler visited Bohr in Copenhagen and presented the case for Everett's theory. Bohr and his circle vehemently rejected it. Wheeler put Everett's degree on hold pending a drastic revision of the thesis. In June, Everett took a job at the Pentagon doing operations research for the Weapons Systems Evaluation Group. Many months later, in February, 1957, Wheeler and Everett sat down and rewrote the dissertation,[5] excising and condensing three-quarters of it. The final version (the "short" thesis) was retitled, "On the Foundations of Quantum Mechanics," and that is the title of the doctoral thesis that was officially accepted by Princeton on April 15, 1957.

For publication in the July *Reviews of Modern Physics*, it was renamed, at Wheeler's insistence, "'Relative State' Formulation of Quantum Mechanics." Fifteen years later, the *unedited* version (the "long" thesis) was published for the first time, by Princeton University Press, in *The Many Worlds Interpretation of Quantum Mechanics*, edited by Bryces. DeWitt and Neill Graham. Before sending the original manuscript of "Wave Mechanics Without Probability" to DeWitt for publication, Everett made extensive handwritten notes on it,

[3] Wheeler to Dennison, 1/21/56.
[4] Wheeler to Bohr, 4/24/56.
[5] Everett to NSF, Fellowship Report for 1955–56, 6/24/57.

restructuring the section on probability and information theory.[6] He retitled it "The Theory of the Universal Wave Function." Seeking to generate controversy, DeWitt ginned up the phrase "many worlds interpretation" for the book title and it stuck to the theory itself.[7]

Introducing many worlds

In the fall of 1955, Everett outlined for Wheeler the main argument of his interpretation. This unpublished, typed, nine-page work, "Probability in Wave Mechanics," is essentially an abstract, light on mathematical notation, but heavy on metaphor, with numerous descriptions of the observer as "splitting" into multiple copies embarking on different histories in variously branching universes that cannot communicate with each other in any ordinary sense.[8] It is a compressed version of his long thesis.

After defining the measurement problem as the contradiction between the Schrödinger equation and the wave collapse axiom, Everett pointed out that the observer of a quantum system, who is himself a quantum system, is necessarily "correlated" to the object observed. (By "correlated" he meant "entangled.") What happens, asked Everett, when a quantum mechanical observer looks at the quantum mechanical needle (or "pointer") of a meter that correlates with a superposed quantum object? Say, a particle the wave function of which describes a superposition of possible positions.

> Why doesn't our observer see a smeared out needle? The answer is quite simple. He behaves just like the apparatus did. When he looks at the needle (interacts) he himself becomes smeared out, but at the same time correlated to the apparatus, and hence to the system.... The observer himself has split into a number of observers, each of which sees a definite result of the measurement.... As an analogy one can imagine an intelligent amoeba with a good memory. As time progresses the amoeba is constantly splitting, each time the resulting amoebas having the same memories as the parent. Our amoeba does not have a life line, but a life tree. The question of the identity or non identity of two amoebas at a later time is somewhat vague. At any time we can consider two of them, and they will possess common memories up to a point (common parent) after which they will diverge according to their separate lives thereafter.[9]

---

[6] These changes could have been made before DeWitt offered to print the manuscript, but it is not likely that they were made before it was bound and sent to Bohr in 1956 as the notes are very messy and hard to follow, making that section of the manuscript difficult to read. So, although Everett did tell one correspondent in the late 1970s that he had done no more work on the theory after it was first published, it is likely that he was keen to fix what he saw as an inadequate presentation of his argument on the role of information theory in deriving probability from the quantum mechanical formalism, and that he did so before sending the manuscript to DeWitt and Graham for typesetting.
[7] DeWitt, B. S. (2008A).
[8] Although the non-communicating branches are causally separate (physically "orthogonal" in the parlance of physics), there is a sense in which as part of a reality described by a universal wave function they can have some influence on each other through interference effects. David Deutsch asserts that a quantum computer would straddle multiple worlds. Deutsch, D. (1997). 216.
[9] Everett, H III. (1956A). 5.

In this model, each copy of the amoeba is correlated to an element in the wave function of the apparatus, which, in turn, reflects the quantum state of the particle (a superposition): "The apparatus itself 'smears out' and is indefinite, no matter how large or 'classical' it is."[10] Relative to the quantum state of the apparatus, the amoeba is in a superposition of states. After splitting, each copy of the amoeba shares a common history with its immediate ancestor. And after each split, the quantum mechanical amoeba keeps on splitting as it continuously interacts with its changing environment in accord with the Schrödinger equation. Most importantly, the wave function of the smeared amoeba never collapses because each of its possible states becomes concrete in a "branch" of what Everett called a universal wave function. He later used the metaphor of a branching tree to show how the branches were separate, but linked by the past to a common root.

Wheeler wrote in the margin of the paper that the amoeba "analogy seems to be quite capable of misleading readers in a very subtle point. Suggest omission."

Regarding the use of the word, "split," Wheeler noted: "Split? Better words needed. Do first on unconscious object to show ideas more objectively."

Everett observed of his own theory,

> It can lay claim to a certain completeness, since it applies to all systems, of whatever size, and is still capable of explaining the appearance of the macroscopic world. The price, however, is the abandonment of the concept of uniqueness of the observer, with its somewhat disconcerting philosophical implications.[11]

Unlike the conventional collapse interpretation, his non-collapse model, he claimed, explained the emergence of classical objects from microscopic superpositions:[12]

> It is this phenomenon which accounts for the classical appearance of the macroscopic world, the existence of solid bodies, etc. since we ourselves are strongly correlated to our *environment*. Even though it is possible for a macroscopic object to 'smear out',...*we would never be aware of it* due to the fact that the interactions between the object and our senses are so strong that we become correlated to it almost instantly. We now see that the wave mechanical description is really compatible with our ideas about the definiteness on a classical level, due to the existence of strong correlations.[13]

---

[10] Ibid. 4.

[11] Ibid. 8.

[12] See also: "In fact,... whenever any two systems interact some degree of correlation is always produced.... Consider a large number of interacting particles. If we suppose them to be initially independent, then throughout the course of time the position amplitude [i.e. wave function describing all possible positions] of any single particle spreads further and further, approaching uniformity over the whole universe, while at the same time, due to the interactions, strong correlations will be built up, so that we might say that the particles have coalesced to form a solid object." Everett, H III. (1956A). 6.

[13] Everett, H III. (1956A). 6. Italics added.

Everett's analysis of how classical phenomena emerge from the quantum substrate raised questions of the type that were later more successfully addressed by decoherence theory, i.e. the technical description of how a microscopic system starts to behave macroscopically as it irreversibly entangles with its environment.[14] What emerges from decoherence is either a single, macroscopic, classical world, or a "quasi-classical" component of a multiple universe system—depending on one's *interpretive* stance.

Everett explained the ontology of his non-collapse interpretation:

> The physical 'reality' is assumed to be the wave function of the whole universe itself. By properly interpreting the internal correlations in this wave function it is possible to explain the appearance of the macroscopic world to us, as well as the apparent probabilistic aspects.[15]

At the end of the mini paper, Wheeler wrote,

> Have to discuss questions of know-ability of the universal ψ function – and latitude with which we can ever determine it. . . . Question of whether new view has any practical consequence. Also its implications for machinery of the world. Any special simplicity to be expected for the wave fun[ction]? If not, why not? If so, what kind of simplicity? Any explanation then why world doesn't look so simple?

Wheeler's remarks were certainly reasonable, as the universal wave function is not observable. But he viewed the universal wave function as necessary to building a theory of quantum gravity which must apply to the universe as a whole.[16] Wheeler needed a theory that allowed quantum interactions to *include* the observer in a wave function describing the whole universe, as it is not possible for an observer to be *outside* the universe, as would be required by the legislated externality of the observer in both the collapse postulate and the Copenhagen interpretation. Everett's concept of a universal wave function included the observer, along with everything else in a multitude of branching universes. The price was that the observer became many observers, and the universe, many universes.

In September 1955, Wheeler wrote to Everett,

> I am frankly bashful about showing ["Probability in Wave Mechanics"] to Bohr in its present form, valuable and important as I consider it to be, because of parts subject to mystical misinterpretations by too many unskilled readers.[17]

---

[14] For Everett, macroscopic objects are quantum mechanical systems. But Everett's use of "environment" is not necessarily the same as how environment is utilized by decoherence theorists. The role of the environment in interpretations of decoherence will be described in detail in a subsequent chapter as it greatly impacts modern interpretations of Everett's work, and vice versa (i.e., the many worlds theory impacted the development of decoherence theories).

[15] Everett, H III. (1956A). 9.

[16] Wheeler also wrote in a note to Everett about the mini-paper: "Can one generalize your definition of correlation, which is inv't [invariant] so to speak in the scheme of spec[ial] rel[ativity] (against linear transf) [transformation]) so it will be inv't in the sense of general relativity? Probably not except in a very artificial way—but what does this circumstance tell about the meaning of correlation?" Wheeler to Everett, 9/21/55.

[17] Wheeler to Everett, 9/21/55.

Wheeler was more positive about the second mini paper "Quantitative Measure of Correlation." Here, Everett utilized information theory to measure the amount of correlation (entanglement) between two quantum variables in terms of a probability distribution. Everett defined the measure as "learn[ing] something about one variable when [told] the value of the other," which was germane to what he later called the "relative state formulation," i.e. a single quantum state is only describable *relative* to the states with which it correlates (entangles).[18]

In the third mini paper, "Objective vs. Subjective Probability," Everett argued that in the standard interpretation of quantum mechanics a particular probability must be both subjective and objective, and that this dichotomy is "untenable." Everett argued that the Born rule probabilities are not objective. On the contrary, he said, the Born rule is really a measure of the ignorance of the observer:

> A subjective probability refers to an estimate by a particular observer which is based upon incomplete information, and as such is not a property of the system being observed, but only of the information of the observer.

In Everett's developing scheme the Born rule is subjective, because in a branching quantum universe where everything happens there can be no such thing as objective probability. The observer is trapped in a single, classical world and does not have access to all the information in the universal wave function, only to the partial information encoded in his particular branch of it. Therefore, the statistical measure he extracts through experiment is subjective—a measure of his ignorance of the content of the universal wave function.[19]

Hence, what an observer views as the indeterministic collapse of a wave function is not a collapse, but simply a loss of information to him in an otherwise deterministic universe governed by a non-collapsing wave equation. Everett argued that each copy of a branching observer will subjectively experience determinism (everything happens) as indeterminism (chance rules), because each copy accesses only partial information about the total quantum environment.

Philosophical monstrosity

By January 1956, when he turned in his typed thesis to Wheeler, Everett had abandoned the amoeba metaphor, but he did not shy away from painting pictures of superposed, bifurcating observers, cannonballs, and splitting mice. Nor was Everett the least bit bashful about criticizing the prevailing

---

[18] "It is meaningless to ask the absolute state of a subsystem - one can only ask the state relative to a given state of the remainder of the system." DeWitt, B. and Graham, N. eds. (1973). 49.
[19] Probability measurements via the Born rule are physically correct "relative to their information," he wrote, but subjective because "they depend upon the information of the observer." In other words, "The paradox is resolved easily since the outside wave function [the universal wave equation that includes the observer's branch] possesses more information, i.e. phase factors, etc. for the interaction, so that it leads to a causal description." Quotations from typed and handwritten (basement) versions of "Objective vs Subjective Probability."

interpretations of quantum mechanics. He said that the "popular" (von Neumann) interpretation, including its postulate of wave function collapse, was "untenable." Speaking directly of the Copenhagen Interpretation, "developed by Bohr," Everett declared,

> While undoubtedly safe from contradiction, due to its extreme conservatism, it is perhaps overcautious. We do not believe that the primary purpose of theoretical physics is to construct 'safe' theories at severe cost in the applicability of their concepts, which is a sterile occupation, but to make useful models which serve for a time and are replaced as they are outworn.[20]

Lest there be any misunderstanding about the depth of Everett's disenchantment with Bohr, here is what he wrote to Bryce DeWitt in May of 1957.

> The Copenhagen Interpretation is hopelessly incomplete because of its a priori reliance on classical physics (excluding in <u>principle</u> any deduction of classical physics from quantum theory, or any adequate investigation of the measuring process), as well as a philosophical monstrosity with a 'reality' concept for the macroscopic world and denial of the same for the microcosm.[21]

But in a handwritten note in the basement file of his thesis materials called "Random Notes," he wrote: "Complementarity contained in general form in present scheme." In other words, he was not throwing away what he called Bohr's "plausibility arguments to support QM [quantum mechanical] conclusions." His main objection to complementarity was that it precluded the deduction of the classical world from pure wave mechanics. He saw Bohr's partition between the classical and quantum realms as an unnecessary impediment to understanding. So, he held that his own "scheme" included (and improved upon) Bohr's dualistic model by *explaining* how the classical world is contained within the quantum realm.

In the long thesis, Everett concluded,

> Our theory in a certain sense bridges the positions of Einstein and Bohr, since the complete theory is quite objective and deterministic...and yet on the subjective level...it is probabilistic in the *strong sense* that there is no way for observers to make any predictions better than the limitations imposed by the uncertainty principle.[22]

He added,

> The constructs of classical physics are just as much fictions of our own minds as those of any other theory; we simply have a great deal more confidence in them.[23]

---

[20] DeWitt, B. and Graham, N. eds. (1973). 111.
[21] Everett to DeWitt, 5/31/57.
[22] DeWitt, B. and Graham, N. eds. (1973). 111. To deal with quantum uncertainty, Everett suggested treating microscopic systems at a "coarse" level of observation, where approximations override the problem that the position and momentum of a particle cannot be simultaneously and precisely measured.
[23] Ibid. 134.

The picture of smeared observers, and the biting critique of Bohr, were features in the long thesis of January 1956. Those features were excised from the final dissertation as approved by Wheeler and published in 1957. Decades later, in an unpublished referee report, DeWitt commented,[24]

> I know that John Wheeler admires brevity and probably urged Everett to try and 'sum up in a nutshell' the essential points of his new interpretation of quantum mechanics. It is also possible that Wheeler was reluctant to support a more blatant statement because it would mean setting himself into direct opposition to his hero, Niels Bohr.

> What is sure is that Wheeler long ago abandoned his support for Everett. What is equally sure is that if the Urwerk [the original, unedited 137 page thesis that DeWitt published in 1973] had been published [in 1957], Everett would not have been ignored for so long.

[24] DeWitt, B. S. (2008A). 4.

16   Tour of Many Worlds

I can safely say that nobody understands quantum mechanics....Do not keep saying to yourself, if you can possibly avoid it, 'But how can it be like that?' because you will get 'down the drain,' into a blind alley from which nobody has yet escaped. Nobody knows how it can be like that.

Richard Feynman, 1965[1]

Now that the main ideas of the many worlds interpretation have been introduced, we take an informational tour of the long thesis. Leaving out the mathematics, we explain how Everett argued for his main idea in ordinary language that does not require any special training to follow. We talk about the influences of Wiener, von Neumann, Shannon, Schrödinger, Einstein, Bohr, Bohm, and others on the development of the theory. We go where no one has ever gone before, using handwritten drafts and research notes discovered in the basement that illuminate Everett's inner thoughts.

He was not writing for lay people; he presupposed a professional knowledge of quantum mechanics for his readers. So, it is safe to say that this tour for the general reader does not fully explicate the difficult theory (even *with* the mathematics it is not fully explained!). Rather, we focus on the role played by "information as physical" and Everett's struggle with language as he left the realm of the intuitive.

Some physicists and philosophers of science are attracted to the simplicity of the many worlds interpretation, i.e. that everything that is physically possible happens inside a non-collapsing wave function. Some agree with Everett and DeWitt that the branching worlds are physically real,[2] some treat them as useful idealizations. For convenience, we employ a term that Everett did not use—"multiverse"—to describe the sum of branching universes described by the universal wave function, real or not.[3]

---

[1] Feynman, R. (1965). 123.
[2] Everett repeatedly wrote that the branches are "equally 'real.'" On the reasonable premise that he considered at least one branch to be "real," then, for him, all are "real."
[3] "Multiverse" has several meanings in modern physics (see *Epilogue*), but we use it here as denoting the quantum mechanical superposition of all physically possible events as describable by a universal wave function. Energy is conserved in this model of branching universes.

No collapse!

Inside one of the basement files, there is an undated scrap of paper upon which Everett penciled:

> Theory of UWF [universal wave function] is above all a precise, clear, unmystical way to understand just those assertions (rules of thumb of engineers) which are made by Bohr and called 'complementarity' without having a wide, gaping vacuum by denying the very possibility of ever understanding functioning of class[ical] descrip[tion] of meas[uring] app[aratus]! ... Theory acts as a whole to explain phenomena ... [There is] no necessity that elements behave 'independently' [of Schrödinger equation].

It is clear from this scrap and his much more substantial writings that Everett started with the assumption that the universe is completely quantum mechanical. Psyching himself up to overthrow the prevailing interpretations, he conceived of inventing a theory as a competitive game[4] and noted to himself:

> Activate – PUTTING OTHER FELLOW ON DEFENSIVE – If we have a rule that keeps us from using QM [quantum mechanics] on a system that contains an observer
>
> or rather put other fellow on spot – he has to invent some alternative law – Complete history of the universe wherein all systems are physical systems subject to same laws – There should be no mystical sets of observations separate from theory.

The "rule" was the externalization of the observer by Bohr's ontological partition coupled to von Neumann's paradoxical collapse postulate. In best military practice, Everett immediately went on the offensive, declaring at the outset that the rule prevents us from understanding quantum mechanics. He declared his intention to solve the measurement problem by ignoring the rule, and treating the Schrödinger equation as universally valid.

A wave function, said Everett, "objectively characterizes the physical system," provided that the system is "isolated," i.e. considered as not interacting with an external system. His big idea was that as the only truly isolated system is the universe, no subsystem of the universe is forever isolated: every subsystem, every object, can be described as existing relative to the remainder of the universe.

Saying this was one thing, proving it was another.

Building the case for a universal wave function, Everett commenced by explaining the measurement problem as the contradiction between the Schrödinger equation and the collapse postulate. He carefully noted that he intended for his theory to conform with von Neumann's "principle of psycho-physical parallelism," which requires that a scientific formalism must be able to describe a real, physical world. And he set out to show why, in a multiverse

---

[4] Also in his notes, he wrote: "Information: definition – plausibility hypothetical games – entropy = -I." He was clearly thinking about defining probability as purely subjective in terms of "utility," which accords with his game theory background.

in which all physically possible events occur, we experience only one of those events at a time on our particular branch of the universal wave function.

As was and is commonly accepted, the Schrödinger equation documents the continuous, causal evolution of quantum systems as they change through time. The seeds of possibilities are superposed within the wave equation. This phenomenon can be tested by experiment, i.e. the Schrödinger equation accurately computes the set of positions (or momentums, energies, etc.) at which an electron can be found. We use the collapse postulate and the Born rule to assign a probability measure to each of these positions. But according to Everett, the Schrödinger equation ruthlessly evolves everything in a huge number of universes through stages of causal change regardless of how static or disconnected things may appear to be to humans—and the collapse postulate and the Born rule are but useful illusions generated by human ignorance.

In the handwritten draft, Everett explained how observers correlate or entangle with macroscopic objects that, like the microsystems of which they are composed, also exist in superpositions[5]:

> Even though the wave function for such an object may be spread over a large region, interaction with an observer will immediately <u>correlate</u> him to the object, so that he will perceive the object in a definite position, that is, after the interaction there will be a superposition of states, each of which will contain the definite object in a definite position, and an observer who perceives this.

And, like the amoeba, each copy of the splitting observer embarks upon its own future as "it" splits again and again like a branching tree, mirroring the superpositions with which it incessantly entangles. And all copies of the observer can be traced backwards through time to common ancestors, as the foliating branches at the top of a tree can all be traced to a common trunk. It is not the universe that splits, per se, but the observer, and in doing so he correlates with a causally connecting branch of the multiverse within the global superposition described (in theory) by the universal wave function.

Collapse, on the other hand, lops off all branches but one.

The role of consciousness

In his introduction, Everett examined the claim (later concretized by Wigner in a famous paper[6]) that human consciousness intervenes to collapse the wave

---

[5] "Emphasize basic, inescapable fact that state functions for macroscopic objects <u>do not</u> generally describe single, definite configuration, but only superposition of such configurations." Handwritten notation, "Random Notes on QM thesis."

[6] Wigner, E. (1961). In a footnote to the section of this paper on the measurement problem that leads directly to the "Wigner's friend" argument, Wigner commented, "The contents of this section should be part of the standard material in courses on quantum mechanics. They are given here because it may be helpful to recall them [and] because the writer is well aware of the fact that most courses in quantum mechanics do not take up the subject here discussed." It is not a stretch of the imagination, therefore, to suppose that Wigner *did* talk about the measurement problem and some version of his "friend" argument in the course that he taught to Everett at Princeton.

function, discarding all possible futures but one. Everett thought it absurd to say that the consciousness of a single observer can collapse the wave function of a quantum system that includes other observers. Where would the "real" collapse occur then? *Which* conscious being is responsible for the collapse? To escape this problem of infinite regression, Wigner claimed that consciousness is *non-physical* and, therefore, not subject to the laws of quantum mechanics[7] (von Neumann agreed with Wigner's metaphysical interpretation).[8]

Disparaging the idea of consciousness as a determinant force in nature, Everett concluded,

> It is now clear that the interpretation of quantum mechanics [i.e., von Neumann's collapse postulate] with which we began is untenable if we are to consider a universe containing more than one observer.[9]

According to Everett, human consciousness plays no super-causal role in the microscopic realm, for, if it did, then the universe would be a creation of human consciousness (but whose?) He considered consciousness to be nothing more or less than a quantum system itself: the physical state of a brain with no claim to special powers. And he was keenly aware that the "observer" need not be human, although it had to be physical. The only requirement for scientific observation, he said, is that an event leave an impression or create a record in the environment, and that does not require human agency. A computer creates electronic and paper records; for that matter, a fission track fossilized in a mineral is a record. Following the necessity of explaining quantum arguments to human beings from the point of view of human observers, Everett focused on the branching memory state of the observer, e.g. the physical recording in the observer's brain of the click made by a Geiger counter registering the presence of a particle. For instance, an observer retains a memory of a Geiger counter click at precisely 10:43 PM, but one of his copies recalls a click a second later, and so on. It is the observer who splits, and, in Everett's scheme, memory states are *evidence* that a split occurred because they record only *one* result. But, and this is the important part: in nature, objects split regardless of the participation of humans.

Alternative interpretations

To buttress his argument, Everett listed several alternative ways of dealing with the paradoxical rule of perpetual externality.

One, he called the "solipsist" position: there is only one observer in the universe. With no one to tell him otherwise, the solipsist is content with the collapse theory because it privileges his observational power: the world exists *because he sees it* and his memory contains the only record of it. Everett rejected

---

[7] Wigner, E. (1961). 168; Wigner, E. (1963). 338.
[8] Jammer, M. (1974). 480–482.
[9] DeWitt, B. and Graham, N. eds. (1973). 6.

solipsism, because the universe contains more than one recording device (observer).

Another alternative was to simply declare that quantum mechanics does not apply to macroscopic observers (including the "servomechanisms" with which Everett cheerfully proposed to replaced scientists). He considered that notion to be absurd, as macroscopic objects are obviously composed of microscopic systems.

Bohm's hidden variables interpretation tempted Everett, but it got rid of wave function collapse by adding terms to the Schrödinger equation, which Everett was not prepared to do. Unlike Bohm, Everett was committed to treating the universe as a purely quantum playing field in which all of the various properties of particles exist in superpositions of "equally real" possibilities described by a universal wave function that bypasses the need for a collapse theory.[10]

So, the best alternative, said Everett, was his own: "pure wave mechanics." It required *abandoning* collapse as non-physical, as scientifically convenient, but ultimately nonsensical and explainable as an illusion. It required the entire universe to obey the Schrödinger equation at all times: All objects, from electrons to elephants, exist in superpositions that are constantly entangling with other objects in their environments. Each object continuously "splits" into copies of itself *relative* to the states of the objects with which it entangles (and which are entangling with it!). And each copy embarks upon a uniquely branching history inside a foliating multiverse encompassing all physically possible correlations and governed by a universal wave equation.

Everett noted that Schrödinger held the view that his wave equation was universally valid,[11] (although he did not go as far as Everett in arguing for that view, and, later, backed away from it, largely because he feared jellyfishication). But Everett admitted that it was incumbent upon him to explain *why* do we not witness macroscopic superpositions—why collapse *appears* to obtain. (In experiments since Everett's day, "mesoscopic" quantum systems have been witnessed in superpositions, but in 1954 that technology was far-off.)

He summarized his task:

> This theory can be called the theory of the 'universal wave function,' since all of physics is presumed to follow from this function alone. There remains, however,

---

[10] In his handwritten notes: "Bring out the view that it is not proper to consider apparatus as converting pure state into mixture (one or the other of which is regarded as 'really existing'), since [there is] always the possibility of developing interference properties between the elements of the superposition, all of which must therefore be considered equally valid, or 'real.'…All elements of resulting superpositions must, to avoid error, be considered equally valid, or 'real', since the view that 'really' one or the other of them has somehow been realized, to the exclusion of the remainder, can lead to contradictions.…In terms of the theory, all branches are regarded as equally 'real' since the fundamental entity is Ψ itself. In interpreting the theory one should not think of one outcome being selected out of the many possibilities, but of all outcomes existing simultaneously, each with a corresponding observer who perceives that outcome." Source: "Random notes on QM thesis." A "pure" state can be a superposition; a "mixture" is a collection of non-superposed states.
[11] Schrödinger, E. (1952).

the question of whether or not such a theory can be put into correspondence with our experience.[12]

Probability, information, and correlation

Everett was deeply influenced by how Wiener, von Neumann, and Shannon treated information itself as physical and probabilistic. Weiner conceptualized negentropy (information) as a measure: "Entropy and information are negatives of one another. Information measures order and entropy measures disorder."[13] For Weiner, the gathering of information reduced uncertainty and was, therefore, subject to the rules of probability.

Preparing to write his thesis, Everett studied von Neumann's influential paper, "The General and Logical Theory of Automata."[14] Here, von Neumann argued that the human brain is analogous to a machine, that living organisms are analogous to automata—part digital, part analogue—and that neurons are analogous to vacuum tubes, and that information obeys the laws of physics, especially thermodynamics. The article influenced Everett on several levels, including in his description of human observers of quantum events as information processing machines, as equivalent to "servomechanisms."

Inspired by these pioneers of the information age, Everett strove, in his first chapter, "Probability, Information, and Correlation," to develop a probability measure for a single quantum mechanical event emerging from the multiplicity of all physically possible quantum events *without* using the Born postulate. Using (classical) information theory, he claimed to have deduced an explanation for the *apparent* existence of probability in a quantum universe in which everything happens. He defined information as, "the negative of the entropy of a probability distribution as defined by Shannon."[15]

For Everett, as for Shannon, information was subject to statistical analysis. For Everett, all information was physically encapsulated in the universal wave function: "This is the position proposed in the present thesis, in which the wave function itself is *held* to be *the* fundamental entity, obeying at all times a deterministic wave equation."[16]

Wheeler later wrote in his companion article to Everett's published dissertation:

> Every attempt to ascribe probabilities to observables [properties of particles] is as out of place in the relative state formalism as it would be in any kind of

---

[12] DeWitt, B. and Graham, N. eds. (1973). 9.

[13] "Speech to Academy 1946," quoted in Conway, F. and Siegelman, J. (2005). 164.

[14] von Neumann, J. (1951). Everett also studied the classic paper, "Can a Mechanical Chess Player Outplay its Designer?" Ashby, W. R. (1952). Ashby's paper, he noted, was a "good article on production of machines capable of surpassing designer—info theory—natural selection—non-deterministic machines, etc."

[15] DeWitt, B. and Graham, N. eds. (1973). 15.

[16] Ibid. 115; compare to handwritten draft thesis, version #2, 6: "The theory is thus completely determinate and regards the wave function itself as the fundamental physical quantity."

quantum physics to ascribe coordinate and momentum to a particle at the same time. The word 'probability' implies the notion of observation from outside with equipment that will be described typically in classical terms. Neither these classical terms, nor observation from outside, nor a priori probability considerations come into the foundations of the relative state form of quantum theory.[17]

The trick was to deduce a probability measure from the point of view of an observer *inside* a branch of the universal wave function. This was problematic because the mathematical space ("Hilbert" space) in which quantum mechanics is calculated by the Schrödinger equation does not allow for a probability measure, as such. That is why wave function collapse (also called the "projection" postulate) *projects* the description of a quantum system into a different mathematical space, one in which probability can be measured classically ("phase" space). Because we can calculate a measure of probability in "our" branch, Everett needed to show how such a phenomenon could occur, without postulating the Born rule, which he was trying to avoid, because it is tied umbilically to the collapse postulate. He claimed to show that a mathematical equivalent to the Born rule logically emerges from the quantum formalism without being assumed a priori.[18]

Instead of collapsing the wave function to find a probability (the Born rule method), Everett strove to measure the amount of information available after an event happened (e.g. an observer or cannonball split) and to equate the result to a probability statement. He put it thus:

> The information is essentially a measure of the sharpness of a probability distribution.[19]

In Everett's information-theoretic model, every measured event has a probability weight (information). And that weight is proportional to the sum of the weights of the events that split off (into separate universes) from the original event.[20]

---

[17] Wheeler. J. A. (1957). 152.

[18] He used the "Lebesque" measure that Wiener had used previously to put a measure on an infinite set of trajectories. Wiener had demonstrated a close connection between the Lebesque measure and probability theory. Heims, S. J. (1980). 63.

[19] DeWitt, B. and Graham, N. eds. (1973). 16.

[20] Quantum measurements enable predictions of future events that can be recorded. Think about a particle that could be found at 10 different positions, A...J. Say that a measurement gives us a probability of D recurring as 0.7 or 70 percent. The Schrödinger equation does not deal in ratios. And if we believe in the Everett interpretation, we could say that each event, A...J, will occur in some universe with a probability of 1 (100 percent). So why, then, do we get probabilities out of measurements in our universe? What Everett seemed to be saying was that although there is not necessarily a determinate number of universes post-branching, probability weights can be assigned to *sets* of results. In this model, information and probability are conserved and the Born rule (or its equivalent) serves to *weight* the number of universes in which the event (D) has occured (a weight of 0.7) over a larger set of post-measurement branching universes that encompasses all results A...J, and that the weights of all such results sum to unity (100 percent). So, the fact that we experience one result is not contradicted by multiple observers experiencing all possible results. Finding the particle at D is recorded in 70 percent of the set of the post-measurement branching worlds (which can be uncountably infinite).

Everett's information-theoretic measure, conveniently, happens to be mathematically equivalent to the Born rule.[21]

To recapitulate, in Everett's highly technical argument, he claimed to be able to obtain a classical probability measure over branching events *without* resorting to collapse that was the equivalent of using the Born rule *with* collapse. This explained, he thought, why an observer stuck in a single branch subjectively experiences probability in a multiverse in which all physically possible events occur. But his critics, and many of his supporters, note that developing an abstract mathematical equivalence is not the same thing as deriving a measure that can be experimentally shown to fit the actual dynamics of physical reality.[22] His claim to have derived a probability measure from the formalism of quantum mechanics is often seen as unproven and it is the major theoretical weakness in his model. And yet it holds water.

Everett remained convinced until the day he died that he had succeeded in deriving probability from the formalism of quantum mechanics and that he had, therefore, solved the measurement problem: he was puzzled as to why many physicists he respected failed to agree with him. He blamed their incomprehension, at least partly, on the excision of the chapter on probability and information from the first published version of his thesis (the short thesis). In the 1973 publication of the long thesis, the missing chapter was restored.

Relative states

Deepening his argument in the next chapter, "Quantum Mechanics," Everett translated his information theoretic model into a purely quantum mechanical model. Explaining the meaning of correlating information in quantum mechanics, Everett wrote: "If we say that X and Y are correlated, what we intuitively mean is that *one learns something about one variable when he is told the value of the other.*"[23] For instance, if you know that the spin of one of two entangled electrons is "up," then you may be confident that, simultaneously, the

---

[21] In 1962, Everett explained (somewhat enigmatically) to a conference of prominent physicists meeting at Xavier University in Ohio, that, "Since my states are constantly branching, I must insist that the measure on a state originally is equal to the sum of the measures on the separate branches after a branching process." Xavier transcript, TUES AM, 21.

[22] Everett was not the first nor the last scientist to try and extract the Born rule from the formal mathematics of quantum mechanics: In 1957, Andrew Gleason of Harvard University published a highly abstract proof that the Born rule emerges from quantum logic. And many other attempts have been made. But Everett and Gleason added certain formal conditions to the quantum logic, seemingly natural conditions that were necessary to support their conclusions that the Born rule emerges from the operation of the Schrödinger equation. Philosopher Jeffrey Barrett remarks, "Deriving a probability measure from the formalism plus a set of stipulated conditions falls short of justifying the use of the measure to assign probabilities to physical events. Neither Everett's nor Gleason's arguments are quite circular. Their arguments just do what they do—namely, they show that if one adds a few conditions to the quantum formalism, then there is just one probability measure that satisfies the conditions. One might argue that the necessary conditions are natural, but one still needs to add them to the theory in order to derive anything." Private communication, Barrett, 2009. See Gleason, Andrew M. (1957).

[23] DeWitt, B. and Graham, N. eds. (1973). 17.

spin of the other electron is "down," even if the two electrons are separated by 100 million light years.

As touched upon previously, *entanglement* is a very formal concept, which may be described as the experimentally verified fact that when two particles interact (exchange energy), they then share a quantum state described by a *composite* wave function: they are correlated, linked, entangled. Everett argued that entangled *subsystems* (such as X and Y in a composite system Z),

> do *not* possess states independent of the states of the remainder of the system....It is meaningless to ask the absolute state of a subsystem—one can only ask the state *relative* to a given state of the remainder of the system.[24]

After a measurement interaction, every element of the superposition of the measured object is entangled with a copy of the observer. Each copy exists *relative* to the state of the object observed and, also, to the state of the remainder of the branching universe with which both are correlated. And this process is going on everywhere:

> So far as the complete theory is concerned all elements of the superposition exist simultaneously, and the entire process is quite continuous.[25]

Everett evoked the image of a cannonball correlated to an electron in a superposition of spin states.

> Suppose, for example, that we coupled a spin measuring device to a cannonball, so that if the spin is up the cannonball will be shifted one foot to the left, while if the spin is down it will be shifted an equal distance to the right. If we now perform a measurement with this arrangement upon a particle whose spin is in a superposition of up and down, then the resulting total state will also be a superposition of two states, one in which the cannonball is to the left, and one in which it is to the right. There is no definite position for our macroscopic cannonball![26]

It is all about entanglement: when you know something about the electron, you know something about the cannonball and vice versa.

Everett was aware that,

> This behavior seems to be quite at variance with our observations, since macroscopic objects always appear to us to have definite positions.[27]

And indeed the cannonball does have a definite position in his model: exactly one position in each of the two universes that corresponds to the state of a cannonball relative to each possible spin state of the electron. Naturally,

[24] Ibid. 43. In a handwritten notation: "Relativity Principal [sic]: In any composite system there is, in general, no state for a subsystem, but only a relative state, relative to some (arbitrary) specification of the state of the remainder. Sort of parallel to usual Relativity principle since denies <u>absolute</u> significance to subsystem state." Source: "Random Notes on QM thesis."
[25] Ibid. 60.
[26] Ibid. 146.
[27] Ibid. 61.

the cannonball could be simultaneously entangling with systems more complex than a single electron, and, therefore, automatically splitting into an uncountably infinite number of cannonballs, each in a different (non-communicating) branch of the multiverse.

Completely breaking with Bohr and von Neumann, Everett treated measurement as "a natural process within the theory of pure wave mechanics." Eliminating any special role of an observer, he defined a measurement as "simply a special case of interaction between physical systems—an interaction which has the property of *correlating* a quantity in one subsystem with a quantity in another."[28]

To split or not to split

"Split" appears thrice in the long dissertation; only once as a description of the observer state.[29] But the word that made Wheeler so nervous appears numerous times in Everett's notes, the handwritten draft, the mini papers, and in an important footnote added at the last minute to printer galleys of the short thesis. Limiting the use of the offensive verb did not eliminate the concept of splitting, which was embedded in Everett's formalism.

Near the end of the long thesis, Everett referred to a comment made by Einstein at the Palmer Lab lecture in 1954 regarding the role of the observer in wave function collapse. Everett wrote,

He put his feeling colorfully by stating that he could not believe that a mouse could bring about drastic changes in the universe simply by looking at it.[30]

Everett explained, using italics,

However, from the standpoint of our theory, *it is not so much the system which is affected by an observation as the observer, who becomes correlated to the system*....From the present viewpoint all elements of the superposition are equally 'real.' Only the observer state has changed, so as to become correlated with the state of the near system and hence naturally with that of the remote system also. The mouse does not affect the universe—only the mouse is affected [by the universe].[31]

---

[28] Ibid. 53; In a handwritten note: "The 'quantum-jumps' exist in our theory as relative phenomena— the states of an object system relative to chosen observer states show this effect, while the absolute states change quite causally." Source: "Footnotes."

[29] "As soon as the observation is performed, the composite state is split into a superposition for which each element describes a different object-system state and an observer with (different) knowledge of it. Only the totality of these observer states, with their diverse knowledge, contains complete information about the original object-system state - but there is no possible communication between the observers described by these separate states. Any single observer can therefore possess knowledge only of the relative state function (relative to his state) of any systems, which is in any case all that is of any importance to him." Ibid. 98–99.

[30] Ibid. 116.

[31] Ibid. 116–117; see handwritten note from "Random notes on QM thesis": "Note, the mouse does not affect the universe, only the universe affects the mouse!".

But, given that multiple copies of the observer emerge: *who* is the observer? Everett noted (in a much-debated footnote):

> At this point we encounter a language difficulty. Whereas before the observation we had a single observer state afterwards there were a number of different states for the observer, all occurring in a superposition. Each of these separate states is a state for an observer, so that we can speak of the different observers described by the different states. On the other hand, the same physical system is involved, and from this viewpoint it is the *same* observer, which is in different states for the different elements of the superposition (i.e. has had different experiences in the separate elements of the superposition). In this situation we shall use the singular when we wish to emphasize that a single physical system is involved, and the plural when we wish to emphasize the different experiences for the separate elements of the superposition.[32]

That this statement needed to be made makes sense when considering how to trace the transtemporal—or transbranching—identity of an observer or object. In a sense, the splitting observers share an identity because they stem from a common ancestor, but they also embark on different fates in different universes. They experience different life-spans, dissimilar events (such as a nuclear war, perhaps), and at some point they are no longer the "same" person, even though they share certain memory records. As with the amoeba, Everett was not suggesting that the splitting of observers was not physical.[33]

In a handwritten note, he commented,

> There is no question about which of the final observers corresponds to the initial one, since each of them possess the total memory of the first. (Which amoeba is the original one?) The successive memory sequences of an observer do not form a linear array, but a planar graph (tree): That is, the <u>trajectory</u> of an observer forms not a line but such a <u>tree</u>.[34]

And he made a diagram of the trajectory of an observer (or more precisely, the memory records of an observer).

Many years later, DeWitt recalled:

> For [Everett], whether we (i.e. the universe and all that is in it) have an independent existence or are merely solutions of some super differential equation is irrelevant. If there is an isomorphism [direct correspondence] between one and the other they are interchangeable.... Under an isomorphism between formalism and the 'real' world, if something exists in the formalism then it 'exists' in the 'real'

---

[32]   Ibid. 68.

[33]   Some proponents of "many minds" interpretations of Everett see the footnote as substantiating a claim that Everett did not view the splits as physically real, but, in my opinion, this argument fails in the face of such statements as: "The theory is thus completely determinate and regards the wave function itself as the fundamental physical quantity." Handwritten draft thesis, version #2, 6.

[34]   Everett explained further: "We see that the predictions of an observer have fundamental limitations. These limitations arise, however, not from the fact that there is no unique correspondence from initial to final system states, but because there is no unique observer state for an initial state." Source: "Random Notes on QM thesis."

Thus, if we wish, we can regard observation as a process which converts a single observer with memory sequence [A, B ... C] into a number of observers with memory sequences [A, B, ... C, D₂]

There is no question about which of the final observers corresponds to the initial one, since each of them possess the total memory of the first. (which amoeba is the original one?)

The successive memory sequences of an observer then do not form a linear array, but a planar graph (tree):

$$[A] < \begin{matrix} [A, B_1] = \begin{matrix} [A, B_1, C_1] \\ [A B_1, C_2] \end{matrix} \\ [A, B_2] < \begin{matrix} [AB_2, C_1] \\ [A B_2, C_2] \end{matrix} \\ [A, B_2] = \end{matrix}$$

That is, the "trajectory" of an observer forms not a line but such a tree.

Branching diagram.

world. Does this make Everett a realist? In my opinion the views of both Everett and myself lie somewhere between realism and Platonic idealism. We both believe in the 'reality' of the many worlds but we also believe that ultimately the abstract idea, theory, wave function, or ideal form behind it all is the true reality.[35]

---

[35] DeWitt, B. S. (2008A). 5.

Back to information

Everett was confident that he had formally demonstrated that,

> We have thus seen how pure wave mechanics, without any initial probability asser-
> tions, can lead to these notions on a subjective level, as appearances to observers.[36]

Subjectivity occurs because the wave function for each branching universe obviously contains less information than the amount of information embedded in the universal wave function (which is inaccessible to macroscopic observers consigned to single branches). So, the splitting process can be considered overall as entropic, as an objective loss of information to each copy of an observer, as a "decoupling" of one world from the set of all physically possible worlds—even though we one-branchers experience this process as an increase in information (a probability statement). The apparent increase is tied to the fact that after making an experimental measurement, we gain information about the chance that a repetition of the measurement will produce a particular result.

According to Everett, the experimental validity of the Born rule for obtaining a probability measure can be explained by information theory as an *appearance* generated by the ignorance of an observer trapped in a single history. It is worth remembering that the Born rule is itself a postulate, so, at the worst, Everett implicitly, if unknowingly, embedded it in his formal argument even as he claimed to be deriving it. Until the day he died, however, he firmly believed that he had shown why an observer would subjectively experience wave collapse, for all practical purposes.

The preferred basis problem

For Everett, macroscopic objects and, indeed, entire universes coalesce out of the quantum microworld by connecting in a way that is historically consistent with what is physically possible: "All laws are correlation laws."

He explained causality as an *effect* of correlation:

> Causality is a property of a model, and not a property of the world of experience.[37]

In other words, A does not *cause* B, A is *correlated* to B in as many branching universes as it takes to exhaust all physically possible correlations. Take heart: you did not necessarily lose your loved ones in some dark universe.[38]

But there is a related issue identified by both Everett's supporters and critics as a major problem for the theory: It does not seem to provide a precise

---

[36] DeWitt, B. and Graham, N. eds. (1973). 78.

[37] Ibid. 137.

[38] In the handwritten draft, Everett wrote that "classical mechanics is an approximate law regarding the correlations in such systems. Can be seen most easily from Feynman point of view which shows that one classical configuration leads to nearly the corresponding result at later time since this case has <u>largest amplitude</u> over the histories." "Random Notes on QM thesis."

method for determining how the constantly splitting amoebas and cannon-balls and brains keep their individual histories causal and consistent (corre-lated), so that the past does not contradict the present and turn us all into jellyfish. How does each universe keep track of its own history so that physical consistency—causality—is preserved as pieces of each universe split off into different universes? On what physical basis does one branch separate itself from the other branches? Along what specific fault lines of position or momentum or spin or time does a superposed object split? Why would an object split on a position "basis," as opposed to a momentum basis?[39] Why are some branches more stable or robust and enduring than others? How can we be sure that the "present" correlates to the "past" even when there are records? How do we know that finding the fossilized footprint of a dinosaur means that there once were dinosaurs? For that matter, why do we not see dinosaurs shopping on Fifth Avenue? And what about worlds where it is probable that the laws of probability do not apply?

This is called the preferred basis problem; it is suggested by some theorists that it is resolved by decoherence theory; others say it is not resolved. At any rate, in the Everett model the question of preferred basis is intimately tied to the problem of defining probability in a multiverse where everything happens.[40] Everett seems to have had a blind spot on this question; he did not consider preferred basis to be a problem. He seems to have assumed that it was sufficient that the observer's selection of an experimental apparatus deter-mined the preferred basis (i.e. whatever physical properties the machine was rigged to measure). But his critics note that making preferred basis dependent upon an experimental set-up does not answer the burning question of how preferred basis is determined in nature independently of machines and observers! Therefore, "the preferred basis problem is just the original meas-urement problem in another guise."[41]

This was a serious flaw in his formal argument, but, as we shall learn in a subsequent chapter, many modern theorists believe it is solved.

---

[39] Technically, "basis" refers to a "vector" in Hilbert space, which holds an infinite number of vectors, i.e. directions and magnitudes of change.

[40] Henry Stapp poetically asks how the Schrödinger equation can pick our particular history when it must treat them all as viable: "The essential point is that if the universe has been evolving since the big bang in accordance with the Schrödinger equation, then it must by now be an amorphous structure in which every device is a smeared-out cloud of a continuum of different possibilities. Indeed, the planet Earth would not have a well-defined location, nor would the rivers and oceans, nor the cities built on their banks. Due to the uncertainty principle, each particle would have a tendency to spread out. Thus various particles with various momenta would have been able to combine and condense in myriads of ways into bound structures, including measuring devices, whose centers, orientations, and fine details would necessarily be smeared out over continua of possibilities.... But the normal rules for extracting well-defined probabilities from a quantum state require the specification, or singling out, of a discrete set of [separate] subspaces, one for each set of alternative possible experientially distinguishable observations." Stapp is a proponent of Von Neumann's wave collapse postulate and links it to a concept of human consciousness as causal. He criticizes theorists who "claim that decoherence completely resolves" the preferred basis problem; and they criticize him for idealism. Stapp, H. (2002).

[41] Barrett (1999), 176; see 173–179 for an informative discussion of preferred basis problem in Everett.

Bird's eye view

Linking entropy, information, and probability in his long thesis, Everett showed that the universal wave function is intrinsically reversible (time can flow backwards, broken eggs reverse trajectories to reunite yolk and shell). But for us, the motion through time *appears* to be irreversible, said Everett:

> Macroscopically irreversible phenomena are common to both classical and quantum mechanics, since they arise from our incomplete information concerning a system, not from any intrinsic behavior of the system.[42]

And in the handwritten draft:

> Thus the apparent irreversibility of natural processes is understood also as a subjective phenomena, relative to observers who lose information in an essential manner, still within a determinate framework which is overall reversible (in which total information is conserved).

So, even though the universal wave function allows information to be transferred to a scientist who does not stand outside the system observed, his perspective is limited:

> There are, therefore, fundamental restrictions to the knowledge that an observer can obtain about the state of the universe....Any single observer can therefore possess knowledge only of relative state function (relative to his state) of any systems, which is in any case all that is of any importance to him.[43]

Physicist Max Tegmark uses the metaphor of a "birds-eye" view when conceptualizing the totality of Everett's universal wave function as a superposition of all the separate branches.[44] He reserves the "frogs-eye" point of view for a continually branching observer with incomplete information.

In the Everett model, probability only has meaning for frog-like observers stuck inside one branch. They cannot see the whole picture, i.e. the multiverse in which everything physically possible happens. Everett equates the unavailability of information to each branching observer as the *appearance* of probability, as a measure of ignorance. For a frog, the determinist universe appears to be indeterminist. But the probability measure generated by the conceptual tool of wave function collapse is, nonetheless, informative: squaring the wave function works!

From the "bird's eye view," an omniscient observer accesses all of the information included in the universal wave function. Overall, Everett takes the bird's eye view and tries to account for what the frogs see.[45]

[42] DeWitt, B. and Graham, N. eds. (1973). 99.
[43] Ibid. 99.
[44] Tegmark, M. (2008). 10.
[45] In a handwritten note to himself about the global superposition, Everett wrote, "(Emphasize that non-interference of mixtures of combined system holds for operators on a subsystem, not on total.)"

In the last few paragraphs of the long thesis, Everett paid his dues to Wheeler's quantum gravity project by commenting on the application of the universal wave function to general relativity. He boldly asserted that by eliminating quantum jumps—i.e. wave function collapse—his relative state theory allowed the field equations of general relativity to be "satisfied everywhere and every-when."[46] And it is this application that was to later catch the eye of cosmologists and power up the many worlds interpretation of quantum mechanics.

[46] DeWitt, B. and Graham, N. eds. (1973). 119.

17 The Battle with Copenhagen, Part I

Bohr's principle of complementarity is the most revolutionary scientific concept of this century and the heart of his fifty year search for the full significance of the quantum idea.

John Wheeler, 1957[1]

In the beginning, Wheeler was Everett's champion. The publication of his thesis in *Reviews of Modern Physics* was accompanied by his advisor's glowing assessment:

> It is difficult to make clear how decisively the 'relative state' formulation drops classical concepts. One's initial unhappiness at this step can be matched but few times in history: when Newton described gravity by anything so preposterous as action at a distance; when Maxwell described anything as natural as action at a distance in terms as unnatural as field theory; when Einstein denied a privileged character to any coordinate system, and the whole foundations of physical measurement at first sight seemed to collapse.... No escape seems possible from this relative state formulation.... [It] does demand a totally new view of the foundational character of physics.[2]

But the paper printed in *Reviews of Modern Physics* was drastically abridged from the original thesis submitted to Wheeler in January 1956. Everett was displeased by the final product. He never published another word on quantum mechanics.

Wheeler's initial support for Everett was born of a scientific agenda: quantizing gravity. And for this project, Everett's formulation of a universal wave function was useful, provided that its baggage—a non-denumerable infinity of branching worlds—could be, somehow, lightened. But Wheeler was not going to sign off on the thesis until Bohr had a chance to weigh in. And after Bohr rejected it, Wheeler insisted that Everett cut and rewrite it on pain of losing his degree. Under the professor's close supervision, three-quarters of the original was excised or condensed. One of the first sentences to go was Everett's dismissal of the Copenhagen interpretation as "developed by Bohr," because it

---

[1] "No Fugitive and Cloistered Virtue," speech delivered on October 24, 1957 in Washington D.C. ceremony presenting Bohr with Atoms for Peace Award. Wheeler. J. A. (1957A).
[2] DeWitt, B. and Graham, N. eds. (1973). 152: DeWitt later told Wheeler: "It always amused me to read in your assessment of Everett's theory how highly you praised Bohr, when the whole purpose of the theory was to undermine the stand which he had for so long taken!" DeWitt to Wheeler, 4/20/67.

treated the wave function as "merely a mathematical artifice." The basic critique of von Neumann's collapse postulate remained intact, but metaphors of splitting cannonballs and observers vanished, leaving the concept of multiple universes in the formalism, but not in the language. And the entire chapter on information and probability disappeared, along with most of its argument.

When the thesis was revamped to Wheeler's satisfaction, he waxed most eloquently, publicly comparing his student's work to the achievements of Newton, Maxwell, and Einstein. Not everybody agreed with him, to say the least (especially not Bohr and his circle in Copenhagen). But as the many worlds interpretation gradually gained credence and popularity over the next few decades, Wheeler ceased advocating for, and eventually disparaged it.

American Bohr

Misner explains Wheeler's dilemma:

> John Wheeler got along with everybody. But in Hugh's case, Wheeler had a very difficult time applying his usual tactics because he couldn't just encourage Hugh to follow his ideas and present them as powerfully as possible since they ran contrary to Bohr's ideas. And Wheeler regarded Bohr as his most important mentor.
>
> He really adored Bohr and Bohr was a marvelous person. He had ways Wheeler somewhat copied of trying to encourage people and being very careful about getting the best possible people but then giving them a fairly free hand.
>
> Wheeler's theme for the previous several years had been something he called radical conservatism which means, and he was applying it to gravity, that if you've got a theory that you think is sound, you shouldn't just use it to calculate the decimal point about things you understand, you should look at the theory and see if it's telling you about anything you haven't thought of. And that's where he got these ideas of wormholes which evolved into black holes later and geons which were ways of thinking about really curved space time.
>
> He would say, 'Let's push the equations and see if they don't open up a new world.' So Hugh had this dynamic conservatism of just taking the equations and looking to see what they would say. So here was Wheeler with Hugh following his own principles of physics, namely following the equations, and Bohr was his deepest emotional attachment in all of physics. He was a man he really respected and bowed to.
>
> So he was really torn and I think he kept trying to play both sides of that tension by trying to get Hugh to tone down the thesis so it wouldn't be quite so needling to people, and then writing a comment on it himself to publish along side of it to try and smooth things over a bit.[3]

In classical antiquity, a "sibyl" was a woman who uttered prophecies of the gods: an oracle. Among Everett's basement papers is a copy of "A Septet of

---

[3] Misner interview, 2007.

Sibyls," seven essays on the nature of truth written by Wheeler for *American Scientist* in 1956.[4] The article was inscribed: "Hugh Everett from J. A. W." It must have given Everett a chill to read the article, in which Wheeler treated Bohr as a living god.

Wheeler's rhetorical habit was to juxtapose opposites, and he took this model directly from Bohr's philosophy as he summed it up in *Septet*:

> Complementarity...represents...the most revolutionary philosophical conception of our day.... We say, 'The use of certain concepts in the description of nature automatically excludes the use of other concepts, which however in another connection are equally necessary for the description of phenomenon.'[5]

He lauded Bohr's method of unifying old and new physics:

> His judgment and courage, his daring conservatism, carried him to wonderful conclusions.... That human truth is defined, not by text books, but by the battles of men and ideas that have brought us where we are, is a lesson appreciated least of all by those we call cranks and nuts, knowing well that we are all in some measure cranks and nuts.... What a wonderful sorter-out of ideas is the principle that new ideas must correspond to old ones, must include them, but must transcend them![6]

That statement highlights Wheeler's problem: he was not satisfied with the collapse postulate, nor Bohr's partition between the quantum and classical realms—both rules stood in the way of quantizing gravity.[7] Yet, it was not a wise career move to question Bohr on such a fundamental issue as the usefulness of complementarity. So, he set out to convince Bohr that Everett's theory was not a refutation of complementarity, but a *refinement* of it.[8]

The battle is joined

Mid-May 1956 found Wheeler in Copenhagen discussing Everett's work with Bohr and Petersen (Wheeler was teaching that semester at University of Leiden, Netherlands).[9] Bohr had a copy of Everett's long thesis, which contained overt criticisms of the Copenhagen interpretation. Wheeler was hoping that Bohr

---

[4] Wheeler, J. A. (1956).
[5] Ibid. 49.
[6] Ibid. 48–49.
[7] Quantizing gravity (a cosmological project) requires that the universe as a whole be (in theory) measurable. But it is impossible to take a measurement from outside the universe.
[8] And he was not in error. Despite his rhetoric against Bohr, in the basement file "Random notes on QM thesis" Everett wrote: "Complementarity contained in general form in present scheme." But Bohr believed complementarity was universal, and not subsumable into a more comprehensive scheme.
[9] He brought with him there Misner (working on wormholes), Joseph Weber (working on gravitational waves), and was visited there by Tullio Regge (working on the stability of what Wheeler later named "black holes"), all projects impacting the Chapel Hill conference the next spring. Misner private communication, 2009.

and his circle would concentrate on understanding Everett's formalism, and not be too put-off by the loaded language.[10]

He wrote to Everett,

> After my arrival the three of us had three long and strong discussions about [your thesis]....Stating conclusions briefly, your beautiful wave function formalism of course remains unshaken; but all three of us feel that the real issue is the words that are to be attached to the quantities of the formalism. We feel that complete misinterpretation of what physics is about will result unless the words that go with the formalism are drastically revised.[11]

Wheeler urged Everett to struggle it out in Copenhagen directly with Bohr. And he warned Everett that he would not schedule his final exam "until this whole issue of words is straightened out."

A few hours later, Wheeler wrote Everett another letter, this time enclosing a copy of the notes he had taken of his meetings with Bohr and Petersen. He penned,

> Much of what is said in objection to your work is irrelevant. Much is relevant: The difficulty of expressing in everyday words the goings on in a mathematical scheme that is about as far removed as it could be from the everyday description; the contradictions and misunderstandings that will arise; the very very heavy burden and responsibility you have to state everything in such a way that these misunderstandings can't arise.
>
> This appallingly difficult job I feel you (among the very few in this world) have the ability in thinking and in writing to accomplish, but it is going to take a lot of <u>time</u>, a lot of heavy <u>arguments</u> with a practical tough minded man like Bohr, and a lot of <u>writing</u> and rewriting. The combination of qualities, to accept corrections in a humble spirit, but to insist on the soundness of certain fundamental principles, is one that is rare but indispensable; and you have it. But it won't do much good unless you go and fight with the greatest fighter.
>
> Frankly, I feel about 2 more months of nearly solid day by day argument are needed to get the bugs out of the words, not out of the formalism.

Wheeler offered to pay half of Everett's steamship fare to Denmark and said Bohr would cover the rest.[12]

According to Wheeler's notes, Petersen said that Everett should have paid more attention to "Bohr's way," which did not recognize a problem with measurement.

---

[10] A paper trail detailing the stages of a fierce struggle between Everett and Wheeler and members of Bohr's inner circle emerged, not only from the basement archive, but also at the Niels Bohr Archive in Copenhagen, the American Philosophical Society in Philadelphia, and at the American Institute of Physics in College Park, Maryland. Some of these records were unearthed by Professor Olival Freire Jr. of the Universidade Federal da Bahia, Salvador, Brazil, Anja Jacobsen of the Niels Bohr Archive, Stefano Osnaghi of the Centre de Recherche en Epistémologie Appliquée, Ecole Polytechnique, Paris, France and Fabio Freitas of the Instituto de Física, Universidade Federal da Bahia, Salvador, Brazil. With assistance from the author, a study of these records, "The Origin of the Everttian Heresy," appeared in *Studies in History and Philosophy of Modern Physics*, 2009.

[11] Wheeler to Everett-1, 5/22/56.

[12] Wheeler to Everett-2, 5/22/56.

He said that the measuring apparatus simply could not be included in the wave function of the object observed: "Silly to say apparatus has a wave function."

Everett scribbled several caustic remarks on his copy of Wheeler's notes. Next to a comment made by Petersen that wave functions do not make sense outside of viewing experimental results, Everett wrote: "Nonsense!"

Then Petersen argued that formalisms must be constructed on a foundation of communicable meaning:

> Math can never be used in physics until [we] have words. [We] aren't comparing [our]selves with servomechanisms. What [we] mean by physics is what can be expressed unambiguously in ordinary language.

He asserted that experimental results have no meaning in Everett's scheme; and that, "He talks of correlation but can never build that up by ψ [wave functions]." Everett penciled next to this statement, "Obviously hasn't completed reading of thesis! It does just that."

Also in May, Alexander Stern, an American physicist-engineer in residence at Bohr's institute, wrote to Wheeler saying he had just given a seminar on Everett's "erudite, but inconclusive and indefinite paper," and that Bohr had opened the discussion. Reflecting the tenor of that discussion, Stern commented that Everett,

> lack[s] an adequate understanding of the measuring process. Everett does not seem to appreciate the FUNDAMENTALLY irreversible character and the FINALITY of a macroscopic measurement....It is an INDEFINABLE interaction."[13]

Stern flatly stated that wave collapse *does not* contradict the Schrödinger equation—as if the construct of wave collapse was included in the complementary model. Without bothering to comment on Everett's technical arguments, he asserted that Everett's "preconceived model of the universe" was "untenable." And he objected to the claim that observers can be replaced by servo-mechanisms. In response to a suggestion made by Wheeler (in a previous letter to Stern) that physics needs to develop mathematical models *akin to game theory*, "that will include the observer," Stern pointed out that behavioral decisions made in games are inherently subjective, whereas measurement in physics is objective. But it was the determinism of Everett's theory that was grounds enough for dismissing it without further ado. Stern said that by "exclud[ing] probability from wave mechanics [he fails to] understand the concept of 'observer.'" For Stern, as for Bohr, the macroscopic observer is by definition external to the microscopic quantum system observed. Therefore, complained Stern: "Wave mechanics without probability excludes physicists."

Blinded by his assumption that measurement requires an external observer, Stern failed to appreciate that Everett was *totally* eliminating the role of external observation so crucial to Bohr's interpretation. He concluded:

---

[13] Stern to Wheeler, 5/20/56.

If Everett's universal wave equation demands a universal observer, an idealized observer, then this becomes a matter of theology ... The subjective aspect of physics, which some scholars and philosophers have claimed to detect but have not understood, has its origin in the fact that physics must make contact with reality which is, after all, the way the world appears to us, and can be understood by us.

Like many of Everett's critics, past and present, Stern was troubled by Everett's treatment of the Born probabilities as non-objective features of the universe:

I do not follow him when he claims that ... one can view the accepted probabilistic interpretation of quantum theory as representing the subjective appearances of observers.

For Stern, indeterminism *was* quantum mechanics; he was not prepared to accord any validity to a theory that challenged the status quo, simply on the basis that it was not the status quo.[14]

Within Everett's basement archive are two pages of "comments" on Stern's letter:

Technically, 'observer' can be applied to any physical system capable of changing its state to a new state with some fairly permanent characteristics which depend upon the object system (with which it interacts), i.e., any measuring apparatus could be called observer ...

Stern's remarks about [my] misunderstanding of [the] fundamental irreversibility of [the] measurement process indicate rather clearly that he has had insufficient time to read the entire work. Several rereadings on his part seem to be called for. Also, Stern is quite guilty in these remarks of begging the question – one of the fundamental motivations of the paper is the question of how can it be that [many] measurements are 'irreversible,' the answer to which is contained in my theory, but is a serious lacuna in the other theory.

Wheeler had seriously miscalculated: Bohr and his circle were not even remotely open to such a radical idea as a universal wave function. In a lengthy and conciliatory letter, Wheeler replied to Stern:

---

[14] Stern is an interesting character; for starters, he was horrified by the increasingly industrial and social nature of scientific research. For decades, he had earned his living as a civil engineer in Brooklyn, New York, while pursuing his avocation: quantum physics. From the 1930s to the early 1960s, he contributed thoughtful articles on quantum foundations to *Physics Today* and a wide range of papers to scientific, technical and philosophical journals. He was a firm believer that science should not be professionalized: "The growing socialization of science involves serious dangers [to freedom of scientific thought]. One must be alert and guard against scientific research degenerating into rubber, oil, textile, military research.... The pure science of physics ... may disappear. The desire to get at the nature of things would give place to the desire to make 'better things.' Thus, the age of scientific enlightenment and culture may be succeeded by an age of technology, where comfort replaces culture." Stern, A. (1945). Stern agitated against government control of science, which, he wrote, must remain 'an intellectual activity—its very nature is not practical.' Stern, A. (1944). As a frequent visitor to Bohr's institute, Stern did not question the Dane's ideological authority, and he actively promoted use of the philosophy of complementarity. But he encouraged physicists to keep an open mind, quoting the American humorist Artemus Ward, 'It is not what people don't know that makes them ignorant, it is what they do know that isn't so.' Stern, A. (1953).

I fully recognize that there are many places in Everett's presentation that are open to heavy objection, and still more that are subject to misinterpretation....I would not have imposed upon my friends the burden of analyzing Everett's ideas...if I did not feel that the concept of 'universal wave function' offers an illuminating and satisfactory way to present the content of quantum theory. I do not in any way question the self consistency and correctness of the present quantum mechanical formalism when I say this.

On the contrary, I have vigorously supported and expect to support in the future the current and inescapable approach to the measurement problem. To be sure, Everett may have felt some questions on this point in the past, but I do not. Moreover, I think I may say that this very fine and able and independently thinking young man has gradually come to accept the present approach to the measurement problem as correct and self consistent, despite a few traces that remain in the present thesis draft of a past dubious attitude.

So, to avoid any possible misunderstanding, let me say that Everett's thesis is not meant to <u>question</u> the present approach to the measurement problem, but to accept and <u>generalize</u> it.[15]

Wheeler went on to clarify that in Everett's theory *there is no such thing as an external observer*. Mounting a spirited defense, he equated Everett's universe to the deterministic universe of Laplace, with the caveat that,

No one seriously believed that it would be a practical possibility ever to know at one moment the position and velocity of every particle, but it was convenient to postulate that these quantities nevertheless had well-defined values.[16]

Likewise with Everett's universal wave function, which, in theory, contains the coordinates of every particle in the universe.

Wheeler insisted that the logical consequences of the theory were, for the time being, confined to a mathematical "model for our world," which, in his opinion, had not yet been shown to physically correspond to the real world. The question of the reality or unreality of that correspondence was what Wheeler was urging Everett and Bohr to struggle out, in accord with one of Wheeler's favorite sayings, "The kind of physics that occurs does not adjust itself to the available words; the words evolve in accordance with the kind of physics that goes on."[17]

Wheeler copied Everett on his reply to Stern, attaching another pedagogical warning,

I have no escape from one sad but important conclusion: that your thesis must receive heavy revision of words and discussion, very little of mathematics, before I can rightfully take the responsibility to recommend it for acceptance.... I feel that your work is most interesting and am sure that it will receive discussion of a scope comparable to that which has attended Bohm's publications. But in your

---

[15] Wheeler to Stern, 5/25/56.
[16] Ibid.
[17] Ibid.

case I must ask that the bugs be got out and the sources of misunderstanding be clarified <u>before</u> the job is made public, not <u>afterwards</u>.

I hope you will realize that I mean this as what is called here your 'promoter,' and one actively interested in your reputation and promising future.[18]

Continuing his flurry of letter writing, Wheeler wrote to Allen Shenstone, chairman of the physics department at Princeton,

I think [Everett's] very original ideas are going to receive wide discussion.... Since the strongest present opposition to some parts of it comes from Bohr, I feel that acceptance in the Danish Academy would be the best public proof of having passed the necessary tests.[19]

Wheeler reported to the National Science Foundation:

I have to say that I am more and more impressed by Everett's originality, mathematical ability, talent for expressing himself clearly orally and in writing, and self-sustaining research drive. His thesis...develops an idea quite original with him, to treat the problem of observation in quantum mechanics by including the observer himself within the framework or the mathematical formalism. Then the equations themselves become completely deterministic and the unavoidable indeterminism of all quantum systems comes in on a much more subtle level. The mathematics has many new features.... [but the] interpretation and the words that go with the formalism to show how to use it and what it means, make a much more difficult problem....

I feel Everett's very original work is destined to become widely known.... I think of Everett as an example of a National Science Fellowship at its best. The man has an independent status, with no boss to tell him what to do. He is working side by side with the faculty in the pursuit of learning. He is an independent scholar in his own right.[20]

And he mailed Bohr a copy of "Septet of Sibyls," enclosed with a note telling his mentor that he had urged Everett, "to discuss the issues with you directly and arrive at a set of words to describe his formalism."[21]

The situation was delicately balanced. Wheeler was proceeding full steam preparing for the upcoming Chapel Hill conference on quantum mechanics and general relativity, and he wanted to highlight the universal wave function. But if Bohr was adamantly against it, he would have trouble showcasing it. Unfortunately, Everett was getting ready to start working at the Pentagon, and he was not eager to rewrite his thesis.

As May wound down, Petersen sent Bohr's copy of Everett's dissertation back to Wheeler with a note suggesting that Bohr would soon write to him "about his attitude regarding the epistemological situation in quantum physics and especially about the status of the observers in the complementary mode of

[18] Wheeler to Everett, 5/25/56.
[19] Wheeler to Shenstone, 5/28/56.
[20] Wheeler to Dees, 5/24/56.
[21] Wheeler to Bohr, 5/24/56.

description."[22] Obviously, Bohr and Petersen were more concerned with correcting Wheeler's deviation than in becoming convinced by him that Everett was correct, or, worse, that Bohr's philosophy of physics was wrong.

Wheeler and Everett conferred by telephone in late May, and Everett agreed to visit Bohr with the caveat that he had to return to Washington to start work at the Weapons Systems Evaluation Group by June 15. Wheeler cabled Bohr with this news. Petersen cabled back that a "longer stay is desirable." He also wrote to Everett explaining that, "It would be very helpful to us if, as a background to your criticism, you gave a thorough treatment of the attitude behind the complementary mode of description and as clearly as possible stated the points where you think this approach is insufficient."[23]

Referring to Petersen's request that he bone up on complementarity, Everett replied: "I have not done this yet, but while I am doing it you might do the same for my work." He said he'd like to visit Copenhagen in a few months, and that he was enclosing another copy of his thesis, since,

> Judging from Stern's letter to Wheeler ... there has not been a copy in Copenhagen long enough for anyone to have read it thoroughly, a situation which this copy may rectify. I believe that a number of misunderstandings will evaporate when it has been read more carefully (say 2 or 3 times).[24]

Everett was not to get to Copenhagen for three years. And when he did, the meeting with Bohr was, as he recalled at the cocktail party, "that was a hell of a—doomed from the beginning."[25]

---

[22] Petersen to Wheeler, 5/26/56.
[23] Petersen to Everett, cable and letter, 5/28/56.
[24] Everett to Petersen, circa June 1956.
[25] Cocktail party tape, 1977.

18 The Battle with Copenhagen, Part II

> Physicists are to some extent sleepwalkers, who try to avoid
> [interpretive] issues and are accustomed to concentrate on
> concrete problems. But it is exactly these questions of principle
> which nevertheless interest nonphysicists and all who wish to
> understand what modern physics says about the analysis of
> the act of observation itself.... The heart of the matter is the
> difficulty of separating the object and the observer.
> Fritz London, Edmond Bauer, 1939.[1]

In the summer of 1956, Everett took a job with the top secret Weapons Systems Evaluation Group at the Pentagon. Gore became pregnant and they married. Shortly after the January 1957 conference on gravitation at Chapel Hill (which Everett did not attend), he got together with Wheeler to edit and condense the dissertation,[2] making it, in the professor's phrase, "javelin proof."[3]

In March, preprints of the truncated dissertation and Wheeler's supporting article were sent to a score of prominent physicists, including Bohr, Schrödinger, Oppenheimer, Rosenfeld, and Wiener. A few weeks later, Bohr dropped Wheeler a note,

> I have not found time to write to you and Everett about the papers you kindly
> sent me. It appears that the argumentation contains some confusion as regards
> the observational problem and... Aage Petersen will write to Everett about our
> discussions.[4]

Days later, Petersen wrote to Everett,

> As you can imagine, the papers have given rise to much discussion at the
> Institute....I think that most of us here look differently upon the problem and
> don't feel those difficulties in quantum mechanics which your paper sets out to
> remove. Accordingly, we cannot agree with you and Wheeler that the relative
> state formulation entails a further clarification of the foundations of quantum
> mechanics.[5]

---

[1] London, F. & Bauer, E. (1939). 219–220.
[2] Everett to NSF, Fellowship Report for 1955–56, 6/24/57.
[3] Wheeler to Everett, 9/17/56.
[4] Bohr to Wheeler, 4/12/57.
[5] Petersen to Everett, 4/24/57.

Petersen was adamant: there was no point in talking about a measurement problem because it is irrelevant:

> There can on this view be no special observational problem in quantum mechanics – in accordance with the fact that the very idea of observation belongs to the frame of classical concepts.... There is no arbitrary distinction between the use of classical concepts and the formalism since the large mass of the apparatus compared with that of the individual atomic objects permits the neglect of quantum effects which is demanded for the account of the experimental arrangement....

> Of course, I am aware that from the point of view of your model-philosophy most of these remarks are besides the point. However, to my mind this philosophy is not suitable for approaching the measuring problem. I would not like to make it a universal principle that ordinary language is indispensable for definition or communication of physical experience, but for the elucidation of the measuring problems hitherto met with in physics the correspondence approach has been quite successful.[6]

So now we have the curious situation in which Wheeler and Everett had cut nearly three-quarters of the original paper. The chapter on information theory was gone. (Stern had thought this chapter to be the "best in the book.") Much of the colorful language that Everett used to bring his theory alive in "ordinary" terms was excised, as was his criticism of Bohr. The edit did clarify the significance of applying a universal wave function to gravitation. In fact, the revised dissertation was reframed as, "the task of quantizing general relativity," which had not been Everett's primary goal. But fearful for his doctorate, Everett allowed Wheeler to basically dictate what was to remain intact of his original thesis, while removing all mention of splitting. And now Bohr, through Petersen, was complaining that the math was not explained in terms of ordinary language, the language of classical physics, Copenhagen-speak.

Everett replied to Petersen, angrily,

> Lest the discussion of my paper die completely, let me add some fuel to the fire with .... criticisms of the 'Copenhagen interpretation.' First of all, the particular difficulties with quantum mechanics that are discussed in my paper have mostly to do with the more common (at least in this country) form of quantum theory, as expressed for example by von Neumann, and not so much with the Bohr (Copenhagen) interpretation. The Bohr interpretation is to me even more unsatisfactory [with its] strange duality of adhering to a 'reality' concept for macroscopic physics and denying the same for the microcosm.

> Now I do not think you can dismiss my viewpoint as simply a misunderstanding of Bohr's position.... *I believe that basing quantum mechanics upon classical physics was a necessary provisional step, but that the time has come...to treat [quantum mechanics] in its own right as a fundamental theory without any dependence on classical physics, and to derive classical physics from it....* We should

---

[6] Ibid.

no longer regard quantum mechanics as a mere appendage to classical physics tacked on to cover annoying discrepancies in the behavior of microscopic systems.

Let me mention a few more irritating features of the Copenhagen Interpretation. You talk of the massiveness of macro systems allowing one to neglect further quantum effects (in discussions of breaking the measuring chain), but never give any justification for this flatly asserted dogma. [And] there is nowhere to be found any consistent explanation for this 'irreversibility' of the measuring process. It is again certainly not implied by wave mechanics, nor classical mechanics either. Another independent postulate?[7]

In April 1957, H. J. Groenewold of Natuurkundig Laboratorium der Rijks-Universiteit te Groningen wrote a long critique of the edited thesis ("relative states" preprint) in which he "profoundly disagree[d]" with its premise and conclusion. In his letter to Everett and Wheeler, Groenewold said that in the summer of 1956 he had "borrowed" a copy of "Wave Mechanics Without Probability," and that the preprint of "'Relative State' Formulation of Quantum Mechanics" was "much improved." Believing that this preprint was an *abstract* of a likewise improved long thesis, he asked to read that, too![8]

Groenewold wrote,

I fully sympathize with the idea of describing the measuring process on purely physical systems without including living observers. So the 'measuring chain' has to be cut off. But it is extremely fundamental that the [cut] off is made after the measuring result has been recorded [in a] permanent way, so that it no longer can be essentially changed if it is observed on its turn.... This recording has to be more or less irreversible and can only take place in a macrophysical (recording) system.

Everett penciled in the margin,

Nonsense. Whole idea not to cut off till after final observ[ation] Q[uantum] M[echanics] says it effected just like microsystem. Whence this magic irrevers[ibility]?

Groenewold continued:

Because all observable quantities may ultimately be expressed in statistical relations between measuring results and the latter are represented by essentially macrophysical recordings, the former ones may ultimately be expressed in macrophysical language.

In the margin Everett scribbled,

Epistemologically garbage. Lack of understanding of the nature of physical theory. Why base concept of reality on classical macrophysical realms?

---

[7] Everett to Petersen, 5/31/57. Italics added.
[8] Groenewold to Everett, 4/11/57.

When Groenewold complained that Everett's theory could not avoid introducing the "cat"[9] and Einstein–Podolsky–Rosen (EPR) paradoxes, Everett exclaimed, "Didn't even read my paper... the paradoxes [are] more easily explained than usual." In a subsequent letter, Groenewold insinuated that Wheeler and Everett had "abandoned the idea of interaction at a distance." And, perhaps, they had—as Everett believed he had accounted for what he described as the "fictitious" EPR paradox.[10]

Not all of the reactions to Everett's and Wheeler's preprints were negative. Henry Margenau, wrote, "The problem with which you deal has irritated many minds. I, for one, find your disposal quite acceptable."[11]

Norbert Wiener weighed in: "The inclusion of the observer as an intrinsic part of the observed system is absolutely sound." But Wiener remarked that Everett was wrong to introduce a classical probability measure in the mathematical space of quantum mechanics. He concluded, "Your paper should be published, but more as comments on the present intellectual situation than as a definitive result."[12] Everett was disappointed that Wiener had seriously misconstrued his derivation of a probability measure. He wrote him a letter setting him straight about that error.[13]

But he did not reply substantively to Wiener's complaint that, "I do not find an adequate discussion of what it means to say that a certain fact or a certain group of facts is actually realized."[14] This, of course, is a common complaint about the many worlds theory. On his copy of Wiener's letter, Everett handwrote a detailed reply (which he did not send to Wiener). It is worth quoting:

> In <u>theory</u> the universal state function is <u>the</u> realized fact. In superposition after measurement all elements actually realized. I am fully aware that this question of 'actualization' is a serious difficulty for convent. Q.M. and is in fact one of the

---

[9] The "Schrödinger's cat" example of macroscopic superposition appears in Schrödinger, E. (1935A). Everett's solution to the cat paradox was that it was supple in one universe, stiff in another.

[10] Everett, H III. (1957) in DeWitt, B. and Graham, N. eds. (1973). 149; Everett thought his theory made the EPR paradox irrelevant as pairs of non-locally entangled particles would correlate their spin states as "spin up-spin down" in one universe, and "spin down-spin up" in another universe regardless of the speed of information transfer. Jeffrey Barrett brings up the point that this explanation may violate relativity, as "The question 'when does the universe split' cannot have an inertial frame independent answer as required by relativity." Barrett, private communication, July 2009.

[11] Margenau to Everett, 4/8/57.

[12] Wiener to Wheeler and Everett, 4/9/57. It is worth noting that in 1950, Wiener observed of physics, "One interesting change that has taken place is that in a probabilistic world we no longer deal with quantities and statements which concern a specific, real universe as a whole but ask instead questions which may find their answers in a large number of similar universes. Thus chance has been admitted, not merely as a mathematical tool for physics, but as part of its warp and weft." Wiener, N. (1950).11.

[13] In a handwritten comment on a copy of Wiener's letter, Misner wrote: "[Everett] uses a true Lebesque measure. It is a meas. on sequences of outcomes of observations. This seems to be a more reasonable place to want a probability than in Hilbert space, where there can be none." And Everett scrawled on another copy of the letter, "I do not need Leb. measure in Hilbert space. Whole problem neatly avoided by my treatment. My measure on trajectories, i.e. sup[erposition] of orthog. states, not entire H[ilbert] space." Everett wrote to Wiener, "I would like to correct any impression that my theory requires a Lebesque measure on Hilbert space. The only measure which I introduced was a measure on the [trajectories of] orthogonal states which are superposed to form another state... and not a measure of Hilbert space itself, the difficulties of which I am fully aware."

[14] Wiener to Wheeler and Everett, 4/9/57.

main motives for present formulation. No problem in present form, however. [N]o such statements ever made in theory like 'case A is actually realized,' except *relative* to some other state. All possibilities 'actually realized,' with corresp. observer states.[15]

In May 1957, Everett wrote a critical letter to E. T. Jaynes, a physicist at Stanford University who was pioneering the use of von Neumann-Shannon-type information theory in physics. Jaynes had just published an article in *Physical Review*, "Information Theory and Statistical Mechanics,"[16] on the relationship of entropy, information and probability. Jaynes wrote a long response[17] to Everett's critique of his paper saying that he thought that his own approach to statistics was "equivalent" to how Everett claimed to derive probability in his theory:

> You claim that my theory is only a special case of your theory, with one particular information measure. I can, with equal justice claim that your theory is a special case of mine.

Jaynes went on to explain where he saw the equivalence, noting:

> The strange thing about the information principle is that the difficulties are not mathematical, but conceptual. The mathematics is very elementary, but there is the greatest difficulty in finding the proper words to convey its meaning.

He thought it was a shame that information theory was not being used widely in quantum physics, saying,

> I think the reason for it is that the subject appears so sensational at first; one has the impression that one is getting something for nothing.

It turns out that Everett's theory had a very positive influence on the subsequent use of information theory in quantum physics.[18]

Everett, the philosopher

In May 1957, after more than a year of battling unsuccessfully with Wheeler and Bohr's circle, Everett sent his and Wheeler's preprints to Professor Philipp Frank, a philosopher of science at Harvard who had recently edited a collection of essays on "operationalism" and related interpretational approaches to

---

[15] Everett, H III to Wiener, 5/31/57. In July 1957, Wheeler wrote to Everett saying he and Wiener agreed that Everett needed to explain his probability measure better, so he asked if he would, "write up a 7-typewritten page draft of a possible note on the subject for submission to him and me with the idea it would go in Phys. Rev. after it had been modified to meet all objections? Perhaps there is more you'd like to add – but is there any issue outstanding that's more central to the future discussion that will occur on your paper? Like basketball so here I believe the way to win is to keep every player covered!" Wheeler to Everett, H III, 7/23/57. There is no record of Everett complying with Wheeler's request.
[16] Jaynes, E. T. (1957).
[17] Jaynes to Everett, H III, 6/17/57. Found in the basement was Jaynes' letter, two papers by Jaynes with Everett's comments in the margins, and Everett's handwritten draft of a letter to Jaynes critiquing his papers.
[18] See Zurek, W. H. ed. (1990).

physics.[19] Frank had been a member of the "Vienna Circle" of logical empiricists in the 1920s. During the early and mid 20th century, he was a highly regarded physicist-philosopher.[20] His widely read book, *Modern Science and it's Philosophy*, was referenced by Wheeler in his *Reviews of Modern Physics* article praising Everett as a revolutionary thinker.

Frank observed,

> Almost every new physical theory has to face the commonplace accusation that it stands in contradiction to everyday experience or, as it is sometimes put, that it contradicts common sense.[21] ... The special mechanism by which social powers bring about a tendency to accept or reject a certain theory depends upon the structure of the society within which the scientist operates. It may vary from a mild influence on the scientist by friendly reviews in political or educational dailies to promotion of his book as a best seller, to ostracism as an author and as a person, to loss of his job, or, under some social circumstances, even to imprisonment, torture, and execution."[22]

Distressed by the rejection of his theory by people he viewed as having agendas, Everett wrote to Frank.

> I have received several of your works on the philosophy of science. I have found them extremely stimulating and valuable. I find that you have expressed a viewpoint which is very nearly identical with the one I have developed independently over the last few years, concerning the nature of physical reality.[23]

Everett described his theory to Frank as,

> a completely abstract mathematical model which is ultimately put into correspondence with experience.... It has the interesting feature, however, that this correspondence can be made only by invoking the theory itself to predict our experience – the world picture presented by the basic mathematical theory being entirely alien to our usual conception of 'reality.' The treatment of observation itself in the theory is absolutely necessary. If one will only swallow the world picture implied by the theory, one has, I believe, the simplest, most complete framework for the interpretation of quantum mechanics today.[24]

The proof of my theory, said Everett, is that the world appears as it appears. Experience itself is the verification of the theory, because it is impossible to sense self-splitting, or the existence of multiple universes.

---

[19] Frank, P. (1954).
[20] "In a famous paper written in 1907, [Frank] made the original suggestion that the law of causality is a convention. How do we know when an experiment has been repeated 'under the same conditions?' Frank argued that there is no exact method except to find out whether it yields the same result. Hence he concluded that the law of causality is not a statement about observable physical facts but is a definition of the expression 'under the same conditions.'" Whitrow, G. T. and Bondi, H. (1954). 275.
[21] Frank, P. (1949). 144.
[22] Frank, P. (1954). 12–13.
[23] Everett to Frank, 5/31/57.
[24] Ibid.

Concluding his letter to Frank, Everett offered to send him copies of various criticisms of his theory, suggesting he might be interested in reading them for psychological and sociological reasons (not as physics). Clearly, Everett believed that Copenhagen's reaction to his theory had little to do with the validity of his argument.

And he was particularly struck by Frank's essay on Nicolaus Copernicus, a 16th century scientist whose heliocentric theory was not fully recognized as true until Isaac Newton substantiated it a century after it was first proposed. As an example of scientific rigidity toward the counter-intuitive, Frank had cited the example of Francis Bacon—who had rejected the Copernican view because it did not accord to common sense. Frank elaborated,

> Looking at the historical record, we notice that the requirement of compatibility with common sense and the rejection of 'unnatural theories' have been advocated with a highly emotional undertone, and it is reasonable to raise the question: What was the source of heat in those fights against new and absurd theories? Surveying those battles, we easily find one common feature, the apprehension that a disagreement with common sense may deprive scientific theories of their value as incentives for a desirable human behavior. In other words, by becoming incompatible with common sense, scientific theories lose their fitness to support desirable attitudes in the domain of ethics, politics, and religion.[25]

Frank wrote back warmly to Everett, saying he was "attracted" to his idea of a non-collapsing wave function, because he disliked the "traditional treatment of 'measurement' in quantum theory according to which it seems as if 'measurement' would be a type of fact which is essentially different from other physical facts."[26]

Clearly influenced by reading Frank's essay on Copernicus, Everett corresponded with Bryce DeWitt, who was guest editing the issue of *Reviews of Modern Physics* in which his edited thesis was slated to appear, alongside a collection of papers from the Chapel Hill conference. DeWitt had written to Wheeler that Everett's paper was "valuable" and "beautifully constructed." He said,

> Everett's removal of the 'external' observer may be viewed as analogous to Einstein's denial of the existence of any privileged inertial frame.

But:

> The trajectory of the memory configuration of a real physical observer...does not branch. I can testify to this from personal introspection, as can you. I simply do not branch.[27]

Everett rejoined in a letter to DeWitt that the same sort of objection was raised by Copernicus' critics: When he asserted that the earth revolved around

[25] Frank, P. (1954). 9.
[26] Frank to Everett, H III, 8/3/57.
[27] DeWitt to Wheeler, 5/7/57.

the sun, they said that was impossible because they could not feel it move. Everett poked DeWitt: "I can't resist asking: Do you feel the motion of the earth?" He then remarked, "It is impossible to do full justice to the subject in so brief an article as the one you read."

DeWitt recalled years later, "His reference to the anti-Copernicans left me with nothing to say but 'Touché!'" DeWitt did not read the unexpurgated thesis until the early 1970s, but he said he put Everett's paper in *Reviews of Modern Physics* because,

> Although Everett had not been a conference participant and I had never met him, his paper was accompanied by (1) a strong letter from John Wheeler and (2) a paper by Wheeler assessing Everett's ideas. Since Wheeler had been a very active conference participant and since Everett's paper seemed to be relevant to the themes of the conference, I agreed to include it.[28]

Regarding the editing of Everett's paper, DeWitt remarked, "I asked [Wheeler] why the original article, I mean the [Urwerk], wasn't ever published. Wheeler said, 'Because I sat down with Everett and told him what to say.'" Dewitt said, "The funny thing is, you have to read the *Reviews of Modern Physics* article very carefully, as I did, to see what's really there. Whereas in the Urwerk it's quite well spelled out, to me."[29]

In the end, after the rebellious, anti-Bohr comments in the original work were excised, along with much of the explanatory language, and much of the formal argument, Everett's dissertation was accepted by Wheeler. He passed his oral exam with a "very good," the second highest rating, and became Dr. Everett.

One of his classmates, Chuck Rockman, congratulated him on finally having his thesis posted for reading in the physics department. Rockman noted,

> Incidentally, did you know that there was a rumor here that there were no faculty members willing to be second and third readers on it? On checking, this was scotched by Charlie [Misner] who claimed it to be a sort of ploy by Wheeler who wanted you to keep rewriting until it was in shape to convince the world. How do you figure the odds on that?[30]

In mid-April 1957, Wheeler added a memo to Everett's student file:

> This work is almost completely original with Mr. Everett both as to the formulation of the problem and its solution. It is too early to assess its final contribution to physics, but there is a distinct possibility that Everett's work may be a significant contribution to our understanding of the foundations of quantum theory.[31]

[28] DeWitt, B. S. (2008A).
[29] DeWitt-Morette interview by Kenneth Ford, 2/28/95.
[30] Rockman to Everett, 3/2/57.
[31] Dissertation acceptance memo, Everett Princeton student file. 4/15/57. Mudd.

When Everett's paper appeared in the July *Reviews of Modern Physics*, it included "splitting" in a footnote! He had inserted it when he proofed the galleys:[32]

> In reply to a preprint of this article some correspondents have raised the question of the 'transition from possible to actual,' arguing that in 'reality' there is—as our experience testifies—no such splitting of observer states, so that only one branch can ever actually exist. Since this point may occur to other readers the following is offered in explanation.
>
> The whole issue of the transition from 'possible' to 'actual' is taken care of in the theory in a very simple way—there is no such transition, nor is any such transition necessary for the theory to be in accord with our experience. From the viewpoint of the theory all elements of a superposition (all 'branches') are 'actual,' none any more 'real' than the rest. It is unnecessary to suppose that all but one are somehow destroyed, since all the separate elements of a superposition individually obey the wave equation with complete indifference to the presence or absence ('actuality' or not) of any other elements. This total lack of effect of one branch on another also implies that no observer will ever be aware of any 'splitting' process.[33]

Touché.

---

[32] According to a telegram and a letter in the basement archive, in late July, Everett told the RMP editor that his original corrections of the proof had been "lost in the mails." With days, maybe hours to spare before the deadline, Everett gave the editor a second copy, with the "split" footnote inserted in ink. Due to time constraints, it is almost certain that Wheeler did not know the footnote had been added.

[33] Everett, H III. (1957) in DeWitt, B. and Graham, N. eds. (1973). 146–147.

19 The Chapel Hill Affair

The one story I remember Hugh telling us was that he once said that Wheeler was a really, really wonderful thesis advisor because you'd go in to talk to him about what you were thinking about and he'd say 'Gee, that's a great idea. That's wonderful.' And then he said there was this one day that he happened to be walking down the hall and the door to Wheeler's office was open and there was a guy in there who was some obvious nut talking about some completely crazy thing and Wheeler was saying, 'Gee, that's a great idea. You ought to develop that.'

Harvey Arnold, 2007[1]

In January 1957, the efforts of Wheeler's relativity study group were showcased at a conference on quantizing gravity held at the University of North Carolina in Chapel Hill. This conference was the first public trashing of Everett's work—by no lesser a light than Feynman. But before discussing the conference, we explore why Wheeler was so attracted to Everett's theory that he was willing to risk Bohr's displeasure.

It was all about quantum gravity.

The unification problem

In 1918, shortly after he invented the general theory of relativity, Einstein remarked that the most important problem in physics was uniting special and general relativity with quantum mechanics "in a single logical system."[2] Nearly a century later, this goal continues to elude seekers, but not for lack of trying.

Special relativity broke with Newtonian physics by demonstrating that there is no such thing as absolute space, nor absolute time, nor absolute simultaneity of events in the universe. It equated mass and energy; it established the invariance of the speed of light; it wove together three dimensions of space, and one of time, into a new concept: "space-time."

General relativity revealed mass and gravity as dance partners shaping the geometry of space-time. The force of gravity is infinitesimal compared to the

[1] Arnold interview, 2007.
[2] "Principle of Research," in Einstein, A. (1954). 222.

electro-magnetic force that influences photons and electrons. But when transmitted between large objects, as waves or particles, gravity becomes powerfully attractive.

By the mid 1950s, the developers of quantum electrodynamics—notably Dirac, Feynman, Tomonaga, Schwinger, Hans Bethe, and Freeman Dyson—had pummeled quantum mechanics into rough accord with Einstein's theory of special relativity. A theory of quantum gravity, however, was proving to be more elusive. Basically, that was because the equations of general relativity deal with the classical motions of macroscopic objects, whereas quantum mechanics roams the probabilistic microcosm. The mathematical coordinate systems we use to describe motion inside these two realms do not match up. Relativity cannot predict the motion of particles through time, and quantum mechanics cannot predict the effects of gravitational attraction.[3] This is a problem because in the real world, gravity and the quantum of action interlace.

It was this theoretical disunity that Wheeler was determined to fix. He figured that since Feynman's path integral method meshed quantum mechanics with special relativity, it might also work with general relativity.

Quantizing gravity at Chapel Hill

The six-day "Conference on the Role of Gravitation in Physics" was organized by Wheeler, and two professors at the University of North Carolina, Bryce DeWitt and Cecile DeWitt-Morette, (who were married to each other).[4] Three dozen physicists from academia, industry, and the military attended, including Feynman, whose work on quantum electrodynamics was a focus of the historic meeting.

Decades later, DeWitt explained, "Most of you can have no idea how hostile the physics community was, in those days, to persons who studied general relativity. It was worse than the hostility emanating from some quarters today towards the string-theory community." He said that the editor-in-chief of a leading physics journal had decided in the mid 1950s to no longer accept "papers on gravitation or any other fundamental theory," but that Wheeler dissuaded him from "behind the scenes."[5]

The conference was dominated by the work of Wheeler and his relativity students. In fact, he wrote or co-wrote six of the 23 conference papers, including his positive assessment of Everett's theory. Everett did not attend, but, at

---

[3] Whereas gravity acts locally, prohibiting action at a distance, quantum entanglement acts non-locally. And the space-time of general relativity is "curved," whereas quantum mechanics demands a "flat" coordinate system. In general relativity, the universe divides into many different observational frames of reference. What you see depends on how fast you are traveling relative to other frames. Not so in quantum mechanics, which applies a single frame of reference to the whole universe.

[4] Other organizers included Freeman Dyson of the Institute of Advanced Study, Frederick Belinfante of Purdue University, and Peter Bergmann of Syracuse University. It was funded by the National Science Foundation, U.S. Air Force, U.S. Army, and the International Union of Pure and Applied Physics.

[5] DeWitt, B. S. (2008).

Wheeler's insistence, his edited dissertation was included in the official proceedings of the conference, published six months later in *Reviews of Modern Physics* in a special section edited by Bryce Dewitt.

A major theme of the conference was the possibility of quantizing gravity by considering it as a "field" and not a particle. (Fields are energetically more extensive than particles, which can be considered as the "quanta" of fields.) Misner's dissertation on quantizing gravity was published in the conference proceedings.[6] At the conference, he remarked that,

> One aim of unified field theory has always been the notion that fields are more fundamental than particles, and that it should be possible to construct all particles from the purely geometrical representation of the field.[7]

Treating particles as fields, Misner, in his dissertation, suggested a way to think about summing over field histories à la Feynman until the ordinary probabilities of quantum mechanics jelled.[8] He recalled,

> My Feynman path integral approach to quantum gravity is mostly considered an attempt to calculate the operations necessary to *evolve the wave function of the universe forward in time.* A rigid adherent of the Bohr observer-driven collapse of the wave function would have anathematized any attempt to evolve a wave function which served no observer. Thus the awareness that Hugh's alternative view of quantum mechanics existed left me free to think about formulating the dynamics of quantum gravity.[9]

Several different approaches to quantizing gravity were presented at Chapel Hill, but the hottest debate centered around how wave functions might be assigned to yet-to-be-discovered gravity particles (gravitons), and how the superposition principle and the idea of wave function collapse might be treated in a unified field theory. Microscopic gravitons would, presumably, exist in quantum superpositions while, at the same time, obeying the classical laws of general relativity: How could this be so?, asked the physicists. How could the gravitational mass (energy) of an atomic object be computed when the uncertainty principle forbids the simultaneous measurement of its position and momentum?[10]

Wheeler advised his colleagues, as an exercise, to forget about the measurement problem and quantum uncertainty and,

> Imagine what sort of ideas scientists might come up with if they were 'put under torture' to develop a theory that would fully explain all the elementary particles

---

[6] Misner, C. (1957).

[7] DeWitt, C. M. (1957). 18. Note that some of the transcript is not composed of direct quotes but is a summary of what was said, and that not all of the statements in the transcript were directly approved by the participants, although there is no reason to doubt accuracy; see also Dewitt-Morette, C. (2009).

[8] More specifically, he suggested a way of using path integrals to quantize the gravitational field. By quantization he meant generating algorithms to extract quantum numbers from continuous fields of wavy energy. Quantum numbers describe the energy of specific particles, e.g. the energy of an electron orbiting at a certain distance from an atomic nucleus. Wave functions contain this information.

[9] Misner private communication, 5/6/2008, italics added.

[10] DeWitt, C. M. (1957). 82.

and their interactions solely in terms of gravitation and electromagnetism alone![11]

He suggested that, under those conditions, Feynman's sum over histories method was capable of solving these problems.[12] However, many participants were convinced that it was *not* theoretically possible to quantize gravity in this manner. DeWitt summed up a common concern: if gravity is quantized, then *it will be* subject to superposition and wave function collapse. Consequently, "the gravitational field suddenly changes because of a measurement performed on the system." In other words, the theory of general relativity will not hold up if gravitons are allowed to jump around like electrons. Part of the debate (then as now) revolved around the question of whether the wave function is physically real, or simply epistemic, i.e. a description of knowledge.

Feynman speaks

Feynman sketched an experiment on a blackboard showing a ball influenced by a gravitational field while entangled with a superposed quantum system. Taking up Wheeler's suggestion to ignore the collapse postulate, he concluded, "If you believe in quantum mechanics up to any level then you have to believe in gravitational quantization in order to describe this experiment."[13]

Furthermore, said Feynman, he could conceive of the ball existing in a *reversible* quantum mechanical superposition. To say that such a thing is *not* possible:

> There would [have to be] be a new principle! It would be fundamental!...I haven't thought out how to say it properly....I'm trying to feel my way. We know that in any piece of apparatus that has ever been built it would be a phenomenally difficult thing to arrange the experiment so as to be reversible. But is it impossible? There's nothing in quantum mechanics which says that you can't get interference [superposition] with a mass of 1^{-5} gram—or one gram....At the moment all I can say is that we'd better quantize the gravitational field, or else find a new principle.[14]

In effect, Feynman was saying that the Schrödinger equation should apply to macroscopic objects unless a newly invented set of equations could show otherwise![15] By the end of the conference, Feynman was in agreement with Wheeler and Misner that summing over histories was a key to quantizing gravity, but, he said,

---

[11] Ibid. 83.
[12] He called his conjectures "electromagnetism without electromagnetism," and "mass without mass." Prompting Thomas Gold of Cornell University to crack wise that he was proposing, "Answers without answers." Ibid. 131.
[13] DeWitt, C. M. (1957). 137; see also Zeh, D. (2008), which analyzes this conference.
[14] Ibid., 139–140.
[15] "This is a strong argument in favor of Everett's work," said Misner. Private communication, 4/27/09.

Historically, the rigorous analysis of whether what one says is true or not comes many years later after the discovery of what is true. And, the discovery of what is true is helped by experiments. The attempt at mathematical rigorous solutions without guiding experiments is exactly the reason the subject is difficult, not the equations. The second choice of action is to 'play games' by intuition and drive on.... You have nothing to lose: there are no experiments.... In this field since we are not pushed by experiments we must be pulled by imagination.[16]

He drew the line, however, at following after Everett's imaginative leap.

During the last few minutes of the conference, Wheeler returned to the measurement paradox. Knowing that Everett's unpublished long thesis had been privately circulating among physicists during the past year, he suggested a way to simplify quantizing gravity without resorting to wave function collapse:

However, there exists the proposal that there is one 'universal wave function.' This function has already been discussed by Everett, and it might be easier to look for this 'universal wave function' than to look for all the [Feynman] propagators [the particular amplitudes that would be summed over in a theory of quantum gravity].[17]

Feynman was not pleased. He retorted,

The concept of a 'universal wave function' has serious difficulties. This is so since the function must contain amplitudes for all possible worlds depending upon all quantum mechanical possibilities in the past and thus one is forced to believe in the equal reality of an infinity of possible worlds.[18]

And on that credulous note, the conference ended.

Misner speaks

Fifty years after the Chapel Hill meeting, Misner recalled, "I was the first one to use the words 'quantum cosmology.' "[19] Now that's fairly popular. However, that meant you were suddenly tempted to talk about the wave function of the universe. Well, in Bohr's viewpoint this was nonsensical. It was impossible because for Bohr the wave function was something about the information available to an experimenter after he defined an experiment. Well, there were no experimenters or experiments when the Big Bang was going off. Therefore, people had to seriously worry, how could you have quantum mechanics rule the universe at the early times.[20]

Misner explained that, years after Everett's theory was published, discoveries in astrophysics, including cosmic microwave background information about

[16] DeWitt, C. M. (1957). 150.
[17] Ibid., 149.
[18] Ibid., 149.
[19] See: Misner, C. (1969).
[20] Misner interview, 2007.

the early state of the universe, caused some cosmologists to become many worlders.

He observes,

> To interpret that kind of thing and make sense of it, they really had to say what do you mean by the wave function of the universe? That started a bunch of things, not all of which are the same as Hugh's. But, they're all within the viewpoint of believing, as Hugh did, that the standard equations always work and then you just have to understand within that framework how our human, everyday experiences arise.[21]

And that means understanding why we should believe in only one universe, even though Everett could explain why we experience only *one* universe.[22]

---

[21] Ibid.
[22] Misner private communication, 4/28/09.

BOOK 5

POSSIBLE WORLD FUTURES

20 Preparing for World War III

The armament race ... assumes hysterical character. ... The ghostlike character of this development lies in its apparently compulsory trend. Every step appears as the unavoidable consequence of the preceding one. In the end, there beckons more and more clearly general annihilation. ... Within the country: concentration of tremendous financial power in the hands of the military; militarization of the youth; close supervision of the loyalty of citizens, in particular, of the civil servants, by a police force growing more conspicuous every day. Intimidation of people of independent political thinking. Subtle indoctrination of the public by radio, press, and schools. Growing restriction of range of public information under the pressure of military secrecy.

Albert Einstein, 1950[1]

Leaving Princeton

Upon graduation, Everett was scheduled to begin his new job as a Scientific Warfare Analyst with the Weapons Systems Evaluation Group (WSEG) at a top-notch salary of $8,520. But having a PhD in hand was a requirement of the job. He had hoped to receive his degree in the spring of 1956. But after Bohr frowned on his thesis, Wheeler postponed its acceptance, contingent upon it being rewritten. In fact, Wheeler was urging him to forego a career in operations research and "to start working towards a first class academic position." He wrote, "You have something original and important to contribute and I feel you ought not to let yourself be distracted from it."[2] Princeton offered Everett an instructorship, but he declined to accept it as working in academia did not appeal to him.[3]

He was attracted to operations research, and had first interviewed with WSEG in the summer of 1955, mid-way through writing his dissertation. The job required a doctorate because "While such a degree is no absolute measure of the type of man we are seeking, it is a selection device on brain power." Brilliance in applied mathematics and "patriotism ... or other constructive motives are desirable." WSEG's ongoing projects were,

[1] Einstein, A., (1954). "The Pursuit of Peace." 160.
[2] Wheeler to Everett-1, 5/22/56.
[3] Everett, Katharine Kennedy to Hugh Everett III, 5/11/56.

Weapons certificate, 1956.

Various war-gaming computations. Economic, biological, military and engineering effects of installing atomic power in submarines. The application of nuclear power to aircraft. Physical and physiological effects of atomic weapons. These are only a few of the areas involved in our studies. Actually WSEG problems encompass practically all disciplines.[4]

Eager to keep Everett in tow, WSEG allowed him to start with the proviso that he win his degree within the year. By early summer, Everett had received his security clearances and was working at the Pentagon, having been granted an occupational draft deferment after his student deferment expired.[5] In August he took a four day course in "special weapons and guided missile orientation" at Fort Bliss, Texas; and in October he received a certificate engraved with the image of an atomic mushroom cloud for attending the "special weapons orientation advanced class" at Sandia Base, New Mexico.

He was in his element.

Reeling in her man

Meanwhile, Nancy was sunning herself at summer camp in Vermont. She sailed, rode horses, made notes for writing a novel, and conducted a

---

[4] WSEG job description, basement.
[5] Anderson, S. E. Lt. Gen. USAF to Local Board Number 42, 7/18/56.

long-distance relationship with her handsome lover. They fought about their future as a couple over the telephone, and reconciled a few days later by letter. In July they enjoyed a cozy weekend in Nancy's Princeton apartment while her roommates were away. But Everett was thriving in his new job, whereas Nancy had no job prospects beyond the clerical work she hated. The final entry in the diary of her Princeton years reads:

> Hugh – fish or cut bait – but we must define situ[ation] clearly in order to act on it. Everyone else seems to take for granted we are getting married. I try – but can't lose doubts.

She was pregnant.

Everett asked her to get an abortion, but she told him that she was going to have the baby whether they got married or not, *and* that she didn't want to marry him just because she was pregnant.[6] In November, they married in a small ceremony at the Chevy Chase Assembly of God in Bethesda. Everett's father and step-mother attended; he told Katharine about it a few weeks later.

Nancy returned to Princeton to wind up her affairs. In December, they were scheduled to move into Arlington Towers apartments, not far from the Pentagon. While she was away, Everett wrote her a charming letter:

> Dear No. 1 wife,…I wish you were here with me now. I miss you very much.…I think we had best wait till after Christmas for our major purchases (e.g. TV) when the prices should be lower and we have collected more green stamps. In the mean-time we can live on love, and maybe an occasional hamburger.…Still haven't com-pleted work for Wheeler, but hope to have something whomped up to talk to him about. The trouble is I have been working (at work) on a very interesting problem, and have been unable to resist working on mathematical parts at home.

Commenting that a dinner party was coming up, he queried, "What dress will you wear? Lets really put on the dog." And he signed the letter, "I miss you. I miss you. I miss you. I love you. I love you. I love you."

The WSEG problem that he was working on after hours—at the expense of rewriting his thesis—was calculating kill ratios from radioactive fallout. Death rates were calculated per megaton of hydrogen bomb dropped on American and Russian cities. This soon-to-be famous study in operations research was to significantly affect the structure of the nuclear war fighting machine. Plus, he was working on a project to ferret out how many ballistic missiles and atom bombs the Soviet Union had on hand.

Operations research comes of age

During the Second World War, Anglo-American scientists questioned the Navy's rule that German submarines dove to depths of 100 feet after spotting airborne threats. Using probability theory, researchers, especially Bell Lab's

---

[6] NGE diaries, late 1980s.

resident genius, William Shockley, determined that this figure was an over-estimate, and that depth charges should be set for 30 feet.[7] The Allied kill ratio quickly fattened, and military brass was sold on science. At MIT, physicists on war-footing used quantum mechanics and information theory to develop radar detection systems. Wiener's cybernetics gave birth to gun and bomb sight servomechanisms that could pinpoint targets from moving vehicles and airplanes. The tactic of using conventional bombs to ignite annihilatory fire-storms over enemy cities was fine-tuned by operations researchers; the Manhattan Project was a massive operations research project; and the Cold War itself was a product of "OR."

As the nuclear arms race ramped up, researchers at MIT's Lincoln Labs conceived the Whirlwind analogue computer system. It evolved into a digital command and control system labeled the Semi-Automatic Ground Environment (SAGE). SAGE was designed to track and intercept enemy airplanes, and by the late 1950s the computerized surveillance network was operated by the newly created North American Air Defense Command located deep inside Cheyenne Mountain, near Colorado Springs.

Each SAGE computer weighed 300 tons, contained 58,000 vacuum tubes, and sprawled over 20,000 square feet inside a massive building powered by diesel generators and cooled by huge water towers. Each building contained *two* of the digital monsters as mutual back-ups. Two dozen of these SAGE control centers graced the North American landmass. They continually processed radar data, weather reports, missile and airbase status reports, flight plans of civilian aircraft, and reports from the Ground Observer Corps. Feeding information into Cheyenne Mountain, the centers tracked all aircraft traversing North American airspace and were ready to provide interception coordinates to fighter pilots when needed, or so the theory went. In reality, SAGE was massively inefficient and prone to error; nor was it designed to survive a surprise nuclear strike.[8]

Although SAGE was rapidly made obsolete by the invention of intercontinental ballistic missiles, transistors, and integrated circuits, it remained the business template for Cold War command, control, and communication systems. IBM Corporation earned hundreds of millions of dollars building computers for SAGE during the 1950s. RAND programmed SAGE, (eventually spinning off a for-profit company, Systems Development Corporation, to keep it up to date). Working on SAGE, contractors invented (and profited from inventing) many of the core concepts of modern computing, including magnetic memory, graphic display techniques, simulation techniques, parallel processing, digital data transmission, and machine networking.[9]

In a few short years, the science of "OR" had come a long way from out-smarting German U-boat skippers.

[7]   Morse, P. (1977). 182.
[8]   Edwards, P. N. (1996). 75–111.
[9]   Edwards, P. N. (1996). 99–109.

Origin of WSEG

In the afterglow of Hiroshima and Nagasaki, the Department of War was rechristened a "defense" department, although it still operated primarily in offensive mode, engaging in military adventures from Korea to the Middle East. The Pentagon set out to recruit the brightest scientists and computer programmers, but quickly found itself stymied by the ability of private industry to pay higher salaries than federal regulations allowed. And there were scads of industrial corporations vying for contracts to design and build advanced weaponry for the armed services.

Each military service had their own war plan and proprietary weapons systems. And each service jealously guarded access to "their" operations researchers: RAND was tied to the Air Force; the Operations Research Office served the Navy; and Research Analysis Corporation was contracted to the Army. Research reports issued by each "think tank" almost invariably supported the parochial interests of its military sponsor. Science was for sale as each service plotted to influence strategic policy in favor of funding weapons systems that would give it the largest slices of the atomic procurement pie.

Not surprisingly, the Joint Chiefs of Staff were having a difficult time overseeing the technology of war planning because they were getting conflicting scientific advice from the rival services. In 1947, the chiefs had created their own private think-tank, the Weapons Systems Evaluation Group, called by its acronym, pronounced "wessegg." Staffed by military officers from all of the services, and a roster of civilian scientists, WSEG had been expected to deliver research free of the machinations of inter-service rivalry. Questions of the day concerned the relative feasibilities of radiological, biological and chemical warfare, how to fight a "limited" nuclear war, how to decide which of the 35 separate ballistic missile programs currently in production was worth keeping.[10] Unfortunately, the ranking military officers running WSEG also favored the agendas of their own services, and they dominated the group.

Disgusted, President Eisenhower asked Nelson Rockefeller to head a commission composed of military and business men to fix the dangerous situation. The politically savvy scion of the Rockefeller oil dynasty was charged with rationalizing the Pentagon's internal power struggle over control of nuclear policy and budgets. As part of Rockefeller's remake, WSEG was transformed into a division of a newly created non-profit company operated by a consortium of universities, the Institute for Defense Analyses, known as "IDA."[11] University affiliation gave WSEG intellectual cachet, and its non-governmental status enabled it to pay corporate-level salaries to civilian scientists while,

[10] WSEG/Ponturo, J. (1979). 93.
[11] IDA was governed by the presidents of MIT, Case Institute of Technology, California Institute of Technology, Stanford University, and Tulane University. Seven more universities, including Princeton, eventually joined IDA. The non-profit's first president was not a scientist—it was retired Air Force general, James McCormack, Jr. The general and his successors (usually retired generals) owed their primary allegiance to a military service.

presumably, taking it out of the loop of inter-service rivalry.[12] IDA was a secret organization of the national security state, not beholden to civil society.

To smooth the transition from purely military- to somewhat civilian-run, IDA hired Shockley of Bell Laboratories as WSEG director of research. Shockley was to be a co-winner of the Nobel Prize in Physics the following year for his work on the Bell Labs team that invented transistors, but that achievement, and his OR work, are not the only things for which he is remembered. A fervent eugenicist, he loudly promoted his "statistical" theory that white people are more intelligent than non-white people. He also saw a bright side to nuclear war, which he thought had a "fifty-fifty" chance of occurring. He once told a reporter from the *Minneapolis Tribune*:

> But if there is a nuclear war man would at least have to begin to control his own genetics. I think the present situation in the civilized world is anti-evolutionary. The people who reproduce in the largest numbers may be far from the most competent. The more competent people practice birth control and have smaller families. If there were a nuclear war, there would be so much genetic damage that man would be forced to plan populations—yes, control breeding, that's what it would amount to. If we began sensible population measure now, it would make nuclear war less likely.[13]

Working with James R. Killian, Jr., Eisenhower's science advisor, Shockley set about recruiting young scientists of the intellectual caliber of Everett and Misner (who spent a summer working at WSEG). In 1956, thanks to a half million dollar pump-priming grant from the Ford Foundation, WSEG expanded from 25 to more than 100 full-time civilian scientists, supplemented by outside consultants. The group's physicists, mathematicians, economists, historians, psychologists, and philosophers were required to have very high level security clearances. The new blood revitalized WSEG's scientific work, but many of the "best and brightest" were prima donnas—teamwork did not come easily to them.[14]

WSEG was arranged in six inter-locking groups: Tactical Warfare, Strategic Warfare, Air Defense, Command and Control, Costs, and Mathematics (Everett's group).[15] Its top secret-strewn office at the Pentagon was cramped, known as "The Cage." Researchers' telephone and computer lines were sealed off from the rest of the Pentagon.[16] The staff also included 100 military officers, whose role was mostly advisory. Not surprisingly, they spent most of their time promoting the partisan needs of their own services, (so that part of the reinvention was a failure, and resulted in the demise of WSEG in the 1970s).

When Everett joined WSEG, it was analyzing the cost-benefits of global and limited nuclear warfare. Ongoing studies examined nuclear blast and fallout kill ratios; the impact of jamming the electronics of guided missiles and airplanes; and the disturbing problem of the "nuclear blackout," i.e., massive

[12] WSEG/Ponturo, J. (1979). 103, 181.
[13] Shurkin, J. N. (2006). 191.
[14] WSEG/Ponturo, J. (1979). 151.
[15] IDA memo, 3/5/63, basement.
[16] WSEG/Ponturo, J. (1979). 204, 280.

electrical disturbances unleashed by nuclear bombs exploded high in the atmosphere.[17] Other areas of research included undersea warfare, homeland security, and "social studies" that investigated the "effect of civilian morale on military capabilities in a nuclear war environment."[18]

WSEG was also charged with assessing the relative strength of the Soviet Union's nuclear weapons system, and this issue became a hot coal, politically.

Missile gap?

In 1957, Eisenhower empowered a commission packed with military contractors, Wall Street financiers, and media magnates to investigate the nuclear threat posed by the Sino-Soviet bloc. The Gaither Commission was headed by H. Rowen Gaither, Jr., a Ford Foundation official who was also a founder of RAND. Other members included banker Laurance Rockefeller; Frank Stanton, the president of CBS; pollster Elmer Roper; John Cowles, owner of the Cowles Newspapers chain; and Vice President Richard M. Nixon.[19]

WSEG scientists acted as staff to the commission, which made a top secret report that the U.S. was in danger of falling behind the Soviets on the deployment of ballistic missiles. After this startling and completely inaccurate conclusion was leaked to the press, the political hysteria generated by the media's hyping of the so-called missile gap helped propel John F. Kennedy into the presidency in 1960. In reality, however, the gap was reversed: the Soviets possessed the capability of fielding about ten intercontinental ballistic missiles, whereas the U.S. had triple that number and was geared up to manufacture thousands of "ICBMs." Plus, the Strategic Air Command's long-range bomber fleet vastly out-numbered the Soviet's fleet.[20]

Nonetheless, based on a combination of faulty intelligence and pro-business bias, Gaither's report recommended adding $44 billion dollars to the military budget.[21] In doing so, it echoed Wall Street financier Paul Nitze's paranoid National Security Memorandum of 1950, which had also used misleading and inaccurate estimates of Soviet intentions and capabilities to scare Congress into pumping up military appropriations slated for corporate contracts.

And the following year a similar report appeared: the Rockefeller Brother's Report overseen by Nelson Rockefeller's protégé, Henry Kissinger. Repeating the misnomers of the Gaither Commission, Kissinger claimed that the Soviets were beating Americans at manufacturing and deploying nuclear arms.

---

[17] In 1958, the secret ARGUS bomb tests over the South Atlantic revealed that what later became known as the "electro-magnetic pulse" could fry the wiring of unprotected weapons and communications systems. This was to be a major problem for designing a missile defense system based on exploding incoming missiles with small fission bombs, as the resultant pulses could destroy the defense system itself.

[18] WSEG/Ponturo, J. (1979). 167.

[19] Kaplan, F. (1983). 127–152.

[20] Ball, D. (1980). 6.

[21] Kaplan, F. (1983). 145.

Years later, Everett told his friend, Donald Reisler,

that he seemed to have been the one responsible for the missile gap that helped get Kennedy elected. He had done the analysis that shows the missile gap. And it turned out the data was wrong, and there really wasn't a missile gap, but what the hell, he elected him. It was based on the intelligence data, and the data just wasn't right.[22]

Sputnik

Eisenhower was infuriated when the Gaither Commission's alarmist report was leaked, as he had correctly concluded that it overstated Soviet capabilities. The leak occurred shortly after the Russians landed a one-two political punch: they test-launched an intercontinental ballistic missile, followed by blasting Sputnik, the first satellite, into orbit. Politicians, the media, and ordinary folks reacted as if Sputnik's robotic "beep-beep" from outer space—broadcast on corporate radio—signaled the end of days. But nobody in the know was surprised; in fact, RAND had predicted the Soviet achievement almost to the day. And a WSEG official had correctly warned Congress that the Soviets would soon launch a satellite. He even wrote a memo about "risks of psychological damage if the Russians were the first to launch."[23]

Although these Russian achievements appeared to support the erroneous missile gap theory, Eisenhower was secretly gratified by Sputnik's debut because the U.S. was preparing to rocket spy satellites into orbit. He allowed Sputnik to fly over North America without making a serious protest in order to set an international precedent for satellites to circle the planet, snapping photographs at will. And the visceral fear it generated in the minds of the populace eased the passing of ever-larger military research and development budgets through Congress.

In his memoirs, George F. Kennan said that Sputnik, "caused Western alarmists...to demand the immediate subornation of all other national interests to the launching of immensely expensive crash programs to outdo the Russians."[24] One result of Sputnik and the series of paranoid commission reports was that the U.S. detonated 77 nuclear weapons above ground in 1958. One blast, 700 miles from Hawaii, knocked out the local telephone system and observers on the Big Island saw the mushroom cloud, such was the power of the explosion.[25]

Soviet leaders viewed the U.S. escalation with absolute horror. They knew all too well that their nuclear arsenal and delivery mechanism was dwarfed by the power and reach of America's war machine, and that the Air Force was just itching to blast the U.S.S.R. and China into radioactive dust.

It was Everett's job to calculate just how radioactive that dust would be.

---

[22] Reisler interview, 2006.
[23] Dickson, P. (2001). 99.
[24] Ibid. 224.
[25] Ibid. 206.

21 From Wargasm to Looking Glass

Reality resists imitation through a model.

Erwin Schrödinger, 1935[1]

First strike

During the 1950s, the operating nuclear war plan of the United States was all or nothing. General Curtis LeMay, head of the Strategic Air Command, told a Gaither commissioner that a surprise attack by Soviet bombers would destroy the bulk of his B-52 bombers on the ground. He said that the official doctrine of deterrence by threatening a "second-strike," or "massive retaliation," was an improbable dream. He announced that SAC airplanes flew over the Soviet Union 24 hours a day picking up radio transmissions, and, "If I see that the Russians are amassing their planes for an attack, I'm going to knock the shit out of them before they take off the ground."[2] And he intended to do this under his own recognizance, regardless of the opinions of civilian leaders, such as the president. Deterrence, for LeMay meant striking first and without warning.

Neither American nor Soviet war planners suffered from the illusion that a "counter-force" strategy of targeting only enemy military installations would not escalate into blasting cities. Until the development of the Single Integrated Operating Plan (SIOP) in 1961, the only "plan" was for every officer with a nuclear weapon under his belt to fire it at the nearest perceived enemy upon command, or his own recognizance should headquarters go poof. And even if Soviet armies attacked Europe with *conventional* weapons, SAC's intention was to "deter" them with a massive nuclear "retaliation" that would lay waste to Russia and China, as well as most of Eastern and Western Europe.

In 1962, Richard Fryklund reported in his popular book *100 Million Lives, Maximum Survival in Nuclear War*,

> SAC commanders do not relish the prospect of suicide, but no one has shown them a convincing alternative.... The only future they see is to build more and more weapons to destroy more and more of the enemy.... They resent people who tell them they are devoting their careers to arrangements for their own country's funeral. They say the alternative to Devastation is surrender, and they would rather go down killing.[3]

---

[1] Schrödinger, E. (1935A). 137.
[2] Kaplan, F. (1983). 133–134.
[3] Fryklund, R. (1962). 71.

But as the bombs got bigger and more expensive, RAND's Herman Kahn, a pioneer of operations research, and a firm advocate of *winning* a nuclear war, began disparaging the doctrine of massive retaliation as a "Wargasm."

Meet Everett's friend: Dr. Strangelove

Kahn, a physicist, was Everett's friend and colleague for many years. The corpulent, fast-talking RAND consultant was the real-life model for the character of Dr. Strangelove in Stanley Kubrick's 1964 film about "learning to love The Bomb." (*Dr. Strangelove* was Everett's favorite movie, he watched it repeatedly on his home VHS machine.)[4] In 1960, Kahn became famous with the publication of his 650 page tome, *On Thermonuclear War*. In this rambling, disjointed paean to the idea of winning a nuclear war, Kahn did not question the necessity of murdering millions of communists as worth the sacrifice of millions of American lives.

His only question was *how many* American lives?

Kahn's shocking book laid out scenario after scenario of how to prepare, launch, and survive a thermonuclear war. He averred that after the passing of thousands of years, American-style capitalism would re-emerge from the radioactive ash heap, reinvented by genetically altered, but still patriotic, shopping citizens. The gory details underpinning these nightmare scenarios flowed from years of "thinking about the unthinkable" by RAND game theorists, including Kahn, Albert Wohlstetter, Thomas Schelling, and Bernard Brodie. (Daniel Ellsberg, of Pentagon Papers fame, was also part of the dystopian, yet politically influential RAND group; unlike his colleagues, Ellsberg was to disavow the principle that it is rational to consider launching a thermonuclear war.)

Reviews of Kahn's book were mixed, ranging from plaudits in *The New York Times* to James Newman's memorable review in *Scientific American*:

> The style of the book...is by turns waggish, pompous, chummy, coy, brutal, arch, man-to-man, Air Force crisp, energetic, tongue-tied, pretentious, ingenuous, spastic, ironical, savage, malapropos, square-bashing and moralistic. Solecisms, pleonasms and jargon abound; the clichés and fused participles are spectacular...This evil and tenebrous book, with its loose-lipped pieties and its hayfoot-strawfoot logic, is permeated with a bloodthirsty irrationality such as I have not seen in my years of reading.[5]

Kahn's perverse, neoconservative streak was revealed in his suggestion that a benefit of fighting a nuclear war would be "that people would get along with less government services than they did before the war. That is, large welfare programs would be cut back."[6] Kahn was taken very seriously by operations

---

[4] Mark Everett, private communication, 2009.
[5] Ghamari-Tabrizi, S. (2005). 285–286.
[6] Kahn, H. (1960). 93.

researchers charged with designing thermonuclear events; Everett admired him tremendously.[7]

In 1957, Kahn wrote two handbooks for newcomers to operations research, "Military Planning in an Uncertain World" and "Ten Common Pitfalls." In these guides to planning nuclear war, Kahn cautioned researchers against "modelism," i.e. becoming enamored of their computer models at the expense of forgetting about the "real" world.[8]

Researchers, he said, should avoid tackling overly complex problems. Rather than predicting how many airplanes the Russians could produce in a given time, it was more important to predict what percentage of their gross national product the Russians could spend on armaments without bankrupting the state. Describing a core strategy of the Cold War that, not incidentally, benefited American arms manufacturers, Kahn remarked,

> Anything that subjects the enemy to large costs may be worth doing. It effectively reduces his strength by causing him to divert and waste resources.

> For example, people sometimes make the statement that, 'We shall not strike the first blow.' They then take this statement so seriously that they advocate giving up completely all the elements of our offensive strength that are useful only if we initiate hostilities. They forget that the enemy has a tendency to look at your capabilities and not your intentions.[9]

Deterrence, said Kahn, can be measured by the amount of desperation and fear you can cause in the enemy's collective mind; but do not overestimate the enemy's capacity for risk tolerance: "There may be circumstances in which the enemy would not take a 65% chance of suicide but is desperate enough to take a 30% chance."[10] Having a first strike capability, Kahn said, is an invitation for the other side to strike first. But, to survive a first strike and retaliate, a second strike capability has to be at least as strong as a first strike force, so you need a first strike force no matter what (including keeping first strike as a strategic, perhaps preferred option). Consequently, the doctrine of deterrence is structurally "unstable," he admitted.

But all was not lost. The dangerous instability embedded in the arms race would be offset if the enemy went broke trying to keep up with the international Joneses.[11] Provided, of course (the Big If), that war did not break out. But that was a risk worth taking in Kahn's abstract model of a world that he did not truly comprehend, nor, apparently, desire to preserve in all of its marvelous diversity.

---

[7] Interviews with Gary Lucas, 2008, Ken Willis, 2007.
[8] Kahn, H. and Mann, I. (1957A). 38.
[9] Ibid. 28.
[10] Kahn, H. and Mann, I. (1957). 66.
[11] Ibid. 123.

The researcher's job, according to Kahn, was to reduce uncertainty. Referring to an important new work by the Bayesian statistician, Leonard J. Savage, *The Foundations of Statistics*, Kahn argued that,

> uncertainty arises from the fact that people *believe* in different assumptions, have different tastes (and therefore objectives), and are (more often than not) ignorant....It is possible for individuals to assign subjectively evaluated numbers to such things as the probability of war or the probability of success of a research program, but there is typically no way of getting a useful consensus on these numbers. Usually, the best that can be done is to set limits between which most reasonable people agree the probabilities lie.[12]

That preparing for war was the only way to prevent war was the only moral certainty that Kahn recognized. Cooperation between competing societies held no value for Kahn except as a matter of diplomatic expediency masking an intent to launch a first strike at the first opportune moment, deterrence be damned.

However, because the number of possible scenarios for nuclear war are uncountable, Kahn urged mathematicians to derive ways of reducing the number of "possible world futures" to a manageable number of outcomes. Inventing a many worlds-type model, Kahn strove to predict the shape of "The postwar world spun into a desultory tangle of multiple paths issuing from a single stem of the many-branched potentials of civil defense."[13] In other worlds, our future depends upon what we do now, and with the proper planning, a branch of Armageddon could be a passable place to live.

As a proponent of his own many worlds theory, Everett was intrigued by the intersection of probability and belief in a universe where nuclear holocaust was distinctly possible. Like Kahn, he did not view disarmament as a rational option.

In contradistinction to Kahn and Everett, Dr. Herbert F. York, who oversaw WSEG in the late 1950s, was repelled by Kahn's branching Hells—the "balance of terror," he called it. As a major architect of the Cold War during his early career, York turned his back on operations research after deciding that the technical approach was futile.[14] In 1963, he determined that the arms race was uncontrollable, and that only forces external to the defense establishment could prevent war. He stated publicly that there were *no technical solutions* to what was a political problem. "The result [of deterrence] will be a steady and inexorable worsening of this situation.... I believe that there is absolutely no solution to be found within the areas of science and technology."[15] He viewed arms control and disarmament as the only viable road forward, and he urged negotiating with the Russians before it was too late.

---

[12]   Ibid. 157–161. Italics added.
[13]   Ghamari-Tabrizi, S. (2005). 184.
[14]   In 1958, York was appointed Director of Defense Research and Engineering in the defense department, overseeing advanced weapons research. He later wrote a book critical of the nuclear arms race, *Race to Oblivion*. York, H. (1970).
[15]   Ghamari-Tabrizi, S. (2005). 193.

The other side of Looking Glass[16]

In his *Memoirs* (1990), the Soviet physicist and political dissident, Andrei Sakharov, reveals how nuclear scientists lived the Cold War in the bowels of the Soviet military-industrial complex. Sakharov was a primary designer of the U.S.S.R.'s first hydrogen bomb, which was exploded in a test on August 12, 1953. He was also awarded the Nobel Peace Prize in 1975 for his work on human rights. In his later years, he often spoke about the "appalling waste of the arms race."

But after the Second World War had devastated Russia (with its 27 million dead), Sakharov undertook to work on The Bomb for several reasons. Deliberately echoing Enrico Fermi's rationale for enlisting for the Manhattan Project, Sakharov said he welcomed the chance to do "superb physics." But "what was most important for me at the time, and also [for] the other members of our group, was the conviction that our work was *essential*."[17]

Later in life, Sakharov came to admire Oppenheimer for arguing that if the U.S. *did* not make the H-bomb, the U.S.S.R. *would* not.

> If the Americans had not initiated the whole chain of events, the U.S.S.R. would have pursued the development of a thermonuclear bomb only at a much later date, if at all. A similar scenario has been repeated with other weapons systems, including nuclear-powered submarines and MIRVS.

> But in the 1940s and 1950s my position was much closer to Teller's[18] practically a mirror image (one had only to substitute 'U.S.S.R.' for 'U.S.A.,' 'peace and national security' for 'defense against the communist menace,' etc.)—so that in defending his actions, I am also defending what I and my colleagues) did at the time.[19]

Sakharov's attitude could have been expressed by any number of American scientists working in the business of thermonuclear warfare. He said,

> We saw ourselves at the center of a great enterprise on which colossal resources were being expended. We shared a general determination that the sacrifices made by our country and people should not be in vain; I certainly felt that way myself. We never questioned the vital importance of our work. And there were no distractions: the rest of the world was far, far away, somewhere beyond the...barbed wire fences. High salaries, government awards, and other privileges and marks of distinction contributed to the psychological atmosphere in which we lived.[20]

Sakharov and Everett mirrored each other in their appreciation of "cybernetics" and the work of Norbert Wiener, Claude Shannon, and John Von Neumann."[21]

---

[16] "Looking Glass" was the code name for the aircraft from which U.S. presidents were slated to run a nuclear war. Lewis Carroll's sequel to *Alice in Wonderland* is called *Through the Looking Glass*.

[17] Sakharov, A. (1990). 96–97.

[18] Edward Teller and Wheeler "fathered" the American hydrogen bomb.

[19] Sakharov, A. (1990). 100.

[20] Ibid. 116.

[21] Ibid. 131.

One day, Sakharov asked the much-feared Politburo member, Lavrenti Beria, "Why are our new projects moving so slowly? Why do we always lag behind the U.S.A. and other countries, why are we losing the technology race?" This question was asked of Beria at the same time that American experts were baselessly touting Soviet superiority in nuclear technology. "Beria gave a pragmatic answer: 'Because we lack R and D and a manufacturing base. Everything relies on a single supplier, Elektrosyla. The Americans have hundreds of companies with large manufacturing facilities.'"[22]

After the initial Soviet hydrogen bomb test, Sakharov told his celebrating colleagues that he hoped such a weapon would never be exploded over a city. His comment was not well received. A general "squelched" Sakharov's naïve "pacifist sentiment." It was then that Sakharov realized the insidious danger of the "idea of mutual deterrence based in military parity which is only one step away from preventive war."[23]

The halting problem

Everett and his colleagues were well-paid to embrace the claim that nuclear war can be prevented by preparing to wage it. But Pentagon computers were incapable of reducing the complex world in which the cold warriors meddled to a manageable model of reality—and the claim remained unproveable, yet motive.

In his cyber-history of the Cold War, *The Closed World*, Paul Edwards notes that, "the historical trajectory of computer development cannot be separated from the elaboration of American grand strategy during the Cold War."[24] Edwards emphasizes the impossibility of fully modeling a system from within that system. In computerese, this is called the halting problem: no program can decide how long it will take to find a solution (to halt) in a time shorter than it takes to run the whole program. In other words, no computable system can contain a complete model of itself that is separable from itself.

Edwards depicts the melding of mind and machine inside the closed, paranoid, self-mirroring realm of operations research, Everett's world:

> A 'closed world' is a radically bounded scene of conflict, an inescapably self-referential space where every thought, word, and action is ultimately directed back toward a central struggle.... Inside the closed horizon of nuclear politics, simulations became entirely more real that the reality itself, as the nuclear stand-off evolved into an entirely abstract war of position.... Simulations—computer models, war games statistical analyses, discourse of nuclear strategy—had, in an important sense, more political significance and more cultural impact than the weapons that could not be used....

[22] Ibid. 46.
[23] Ibid. 194, 204.
[24] Edwards, P. N. (1996). 2.

The object for each nuclear power was a winning *scenario*—a theatrical or simulated win, a psychological and political effect—rather than to actually fight such a war. *Actual* outcomes no longer mattered, since the consequences had become too enormous to be comprehended and too dangerous to be tested.

The world of nuclear arms became by its very grossness and scale a closed world, a lens through which every other political struggle must be seen. For those who contemplated its strategy, nuclear war could only be understood as a many-leveled game.[25]

And, it must be said, as long as it *remained* a game, all was not lost. The problem with games, however, is that they are *part* of the real world, and that world is not controllable. No model can tame the chaos of the real world, nor predict the fate of what it models inside a closed system. But, depending on the normative and statistical assumptions built into it, a model *can* point to possible futures, even desirable futures.

But inside the closed world of Everett's WSEG the possible futures were explored by inputting data about particular weapons systems into programs set up to predict "pay-offs" in terms of numbers of people killed, and cities buried, in limited or general nuclear "exchanges." Disarmament was not an option—neither input, nor pay-off. And the language of operations research was loaded with dualism. "Defensive" systems, such as anti-ballistic missile batteries, were really offensive systems; as researchers were well aware that strengthening defense caused the other side to strengthen offense. Deterrence was predicated upon maintaining hair-trigger offensive capabilities, and, when tested in war simulations, its halting point was incomputable.

---

[25] Ibid. 12–14.

22 Fallout

> *Theoretical physics forces atomic energy on us; the successful production of the fission bomb forces upon us the manufacture of the hydrogen bomb. We do not choose our problems, we do not choose our products; we are pushed, we are forced—by what? By a system which has no purpose and goal transcending it, and which makes man its appendix.*
>
> Erich Fromm, 1955[1]

Making a mark

After earning his PhD in nuclear physics at MIT in the mid 1950s, George Edgin Pugh was lured to WSEG by the high salary and exciting work. On his first day, he was assigned to work with Everett. The two physicists hit it off immediately over lunch in an executive dining room at the Pentagon. They remained friends and colleagues for the next 15 years.[2]

The newcomers were assigned to an ongoing WSEG project analyzing the effects of radioactive fallout from a U.S. attack on the Soviet Union. The project employed two dozen specialists in nuclear physics, physical chemistry, biochemistry, and meteorology. Using mechanical calculators, the scientists laboriously plotted the lethal consequences of fallout patterns based on prevailing weather patterns and total megatonnage unleashed over "Oblast" areas. Kill ratios were functions of the probability that bombers would reach their targets within a certain margin of error. "Overkill" was built into the attack plans as a percentage of the American bombers would inevitably be disabled by bad weather, or electronic counter-measures, or shot down before reaching ground zero.

Pugh recalls that the Air Force had originally estimated that the most likely scenario for a nuclear attack by the United States on the Soviet Union would

---

[1] Fromm, E. (1955). 87.
[2] Now in his eighties and retired, Pugh had a long career in operations research in both the private and public sectors. In 1977, he published an important book in evolutionary biology, *The Biological Origin of Human Values*. In 2007, the International Astronomical Association named an asteroid for Pugh in honor of his work on the Gravity B Probe, a satellite-based mission designed to measure space-time curvature affected by the Earth and, thereby, confirm Einstein's theory of general relativity. Pugh proposed the mission in 1959, while at WSEG, and the probe was finally launched in 2004. The following year, Pugh wrote a memoir recounting his WSEG days, which is available at the Niels Bohr Archive at the American Institute of Physics.

result in a few hundred thousand fatalities from blast effects. The WSEG fallout team determined that the initial attack would kill at least 4 million people. But when the effect of radioactive fallout was included, Soviet fatalities shot upwards of 100 million!

Everett and Pugh designed a study of fallout kill ratios applicable to any large nuclear campaign, including one against the United States. Everett came up with the notion of maximizing radiation fatalities as a function of the total megatonnage unleashed in the overall attack, but subject to multiple constraints. WSEG kept population data on 60,000 target locations. So, optimizing the allocation of airplanes and atom bombs to different kinds of targets for maximum lethality was an extraordinarily hard calculation.

In those days, large number-crunching OR projects were usually performed by hand, on mechanical calculators by mathematically skilled women, known as "computers." Everett, who headed WSEG's mathematics division, was lobbying his bosses to replace the women with high-speed digital machines. He started thinking about how to construct an algorithm capable of optimizing solutions for the large sets of constraints that he was dealing with in the fallout project, but, in the meantime, the women continued to sweat over their calculating machines.

Questions addressed by the fallout study were formulated in chilling operational terms. "For a given population, distributed geographically in some known manner, how should one distribute a fixed number of weapons in order to maximize the expected casualties."[3] Not too surprisingly, the mathematicians determined that casualties from fallout rose in tandem with the energy of the nuclear blast at ground level—sucking radioactive dirt into the mushroom cloud for dispersal by winds in the upper atmosphere is the most effective way to broadly distribute fallout.

After determining an ideal curve for maximizing fallout fatalities, Everett and Pugh addressed the problem of optimizing the distribution (and cost) of a fixed number of weapon systems to achieve the maximum death curve. In the final section of their report, they showed how fallout fatalities would fluctuate under a variety of targeting doctrines, ranging from the random distribution of bombs inside large, circular areas of countryside to deliberately hitting the most densely populated areas to targeting only military airbases. Lastly, they showed how fatalities from radiation could be limited by civil defense preparedness, primarily a national network of fallout shelters.

Central to the calculation of lethality was setting the fatal dose of gamma rays per person. Not surprisingly, targeting military and industrial hubs in or near densely populated cities was the most cost-effective way to deliver the bad medicine—prevailing winds would waft rural people their share, so it was a waste of "fission yield" to bomb farmers. But for a more expensive, "extremely large-yield" campaign, the kill ratio per weapon was maximized by uniformly

---

[3] Everett, H. III and Pugh, G. E. (1958). 17.

dropping bombs over an entire country, thereby avoiding "excessive over-killing" in any one spot.[4]

The authors noted that their formulas only calculated the death rate for 60 days out from a nuclear holocaust and

> may not be indicative of the ultimate casualties. The delayed effects such as the disorganization of society, disruption of communications, extinction of livestock, genetic damage, and the slow development of radiation poisoning from the ingestions of radioactive materials may significantly increase the ultimate toll.[5]

The sanitized version of "Simple Formulas for Calculating the Distribution and Effects of Fallout in Large Nuclear Weapon Campaigns (With Applications)" concluded that *any* large scale use of nuclear weapons would result in a huge proportion of the population being disabled or killed by fallout. It effectively discredited the prevailing assumption among operations researchers that causalities would be "manageable" in a nuclear exchange. So shocking was this finding to the military high command that Pugh was detailed to brief Eisenhower at the White House in July 1957:

> After I had finished, Sherman Adams [Eisenhower's senior advisor] asked the president if he thought he had understood the presentation. The president responded, 'Yes, it seemed quite clear. In some ways, the effects of radioactivity are like an artillery bombardment. It doesn't matter much where you aim, the important thing is the total fire power that's delivered.' I left with the feeling that we had successfully delivered our message to the president and his staff.[6]

Despite the glory of briefing the president, Pugh was peeved at Everett. He believed that he had done the most work on the fallout project and should receive top billing as co-author. But Everett insisted that their names be listed in alphabetical order: "It was the first of many events in which Everett's brazen grab for recognition and power succeeded in distorting subsequent outcomes," Pugh complained.[7]

Nobel Peace Prize!

So valuable was their fallout report that Everett and Pugh were authorized to sanitize it, stripping out references to specific targets and other top secrets. It was made public during hearings before the Special Sub-Committee on

---

[4] Ibid. 29; In 1956, the B-52 bomber force was about 1,600, each airplane capable of carrying about 40 megatons for a total delivery of 64,000 megatons. This figure did not count bombing by ballistic missile, a capability which was just getting off the drawing boards. Everett and Pugh calculated that the kill ratio of 2000 megatons on an unsheltered population was 22 percent, rising to 40 percent at 2,500 megatons, 80 percent at 8,000 megatons, 95 percent at 10,000 megatons; and mass extinction at 20,000 megatons. But a well-sheltered population was killed at about half that rate, reaching 95 percent dead at 50,000 megatons.

[5] Ibid. 31.

[6] Pugh, G. E. (2005). 16.

[7] Ibid. 17.

Radiation of the Joint Congressional Committee on Atomic Energy. And the March 1959 issue of *Operations Research* published the study with the authors' optimistic foreword:

> It is the hope of the authors…that the results here indicated will illustrate the catastrophic effects of a large nuclear campaign, regardless of specific targeting doctrine. Perhaps the public release of this information will serve to reduce the probability that such conflicts will ever occur.[8]

Linus Pauling credited Everett and Pugh by name in his Nobel Lecture upon receiving the 1962 Nobel Peace Prize for his work on nuclear disarmament, (in 1954, he had been awarded the Noble Prize for chemistry). Referring to their study, Pauling estimated,

> that 60 days after the day on which the war [between the U.S. and the Soviet-Chinese blocs] was waged 720 million of the 800 million people in these countries would be dead, 60 million would be alive but severely injured.

Pauling pointed out,

> There is no defense against nuclear weapons that could not be overcome by increasing the scale of the attack.… The only sane policy for the world is that of abolishing war.[9]

The fallout study was one of the first studies in a growing body of research showing that even a small nuclear war would be lethal beyond all imagining. In 1983, a distinguished panel of scientists determined that the smoke and fires from burning cities caused by exploding the 1,000 bombs SAC planned to drop on the Soviet Union as early as 1953 would have triggered a "nuclear winter" that "enshrouded the earth in darkness and eventually extinguishing all life."[10]

Different worlds

While working with Pugh on the fallout project, Everett explained the premises of his as yet unedited thesis. Pugh was impressed with its brilliance. He asked him if he believed in the reality of the branching universes. Everett replied,

> It is really hard to say what I really believe. In reality, all that we can ever know about any theory is the extent to which it seems to correspond to the real world observations we can make, and to the experiments we can do. Beyond that we never know the extent to which any of our theories capture the *real* reality of the universe, to the actual content of what *really* is out there. We hardly have a clue about what may *really* be out there. So we have no way of guessing how close any

[8] Everett, H. III and Pugh, G. (1959).

[9] Pauling, L. (1963). After the fallout study was published, NATO countries asked WSEG to study the effect of fallout on Europe from U.S. bombs dropped on the U.S.S.R. According to Pugh, the prevailing winds would usually, but not always, carry the fallout away from the NATO countries. Pugh, G. E. (2005). 44.

[10] Herken, G. (1987). 38.

of our theories are to what may really be out there. All we can do is postulate our theoretical idea and then ask how well they correspond to experiments.[11]

Pugh accused him of dodging the question.

Everett replied that he would give it a "70 percent probability" that the multiple universes are physically real.[12] He said that his concept might be ahead of its time. For his part, Pugh thought the theory made sense, but that it was irrelevant because it did not make any real difference to decision making. He suggested that Everett pursue the idea of using his universal wave function to reconcile quantum mechanics and general relativity. Everett agreed that eliminating the idea of quantum jumps (which his theory did) would make that reconciliation easier. But he said that he currently had his hands full with his interpretation of quantum mechanics, and was not anxious to undertake another "theoretical monster." He was more interested in commercializing his mathematical concepts than tackling another fundamental problem in physics.[13]

In fact, Everett and Pugh made a written agreement in November 1956 that,

> the Patent and copyright agreements with the Institute for Defense Analyses, signed by us this day, pertains only to inventions which are both conceived during our tenure of employment, and connected with the work on which we are employed.[14]

A few years later, Everett started up a consulting firm based on an uncopyrighted optimizing algorithm, the one he had been thinking about for the fallout project, and later invented. Known in operations research circles as the "Everett Algorithm," it made both men a lot of money.

Although they were devotees of game theory in their professional lives, Everett and Pugh did not agree about its relevance to real life. Everett thought that game theory was a reliable guide to making real world decisions, and that Pugh was "soft-headed" for believing that human beings are driven by emotions—fear, anger, hatred, love—and not purely by decisional logic. These arguments so negatively affected Pugh that, years later, he wrote and published a study on the biological origins of human values that elaborated his belief that emotions and moral values are genetically encoded in humans. He posited that decisions are based on a mixture of computation, self-preservation, and social altruism.[15]

From the point of view of social engineering, Pugh thought that unless science could offer a better alternative, it was best not to tinker with religious faith, which served to comfort and anchor people in their daily lives. Everett disagreed, calling organized religion "a fraud on the people perpetrated to garner resources for the religious professionals."[16] He loved to argue.

---

[11] Pugh, G. E. (2005). 21–22.
[12] At other times and places, as duly noted, Everett expressed even firmer support for the physical reality of his worlds.
[13] Pugh, G. E. (2005). 22.
[14] Basement papers.
[15] Pugh, G. E. (2005). 23; see: Pugh, G. E. (1977).
[16] Pugh, G. E. (2005). 23.

BOOK 6
CROSSROADS

23 A Bell Jar World

We are not in danger of becoming slaves any more, but of becoming robots.

Adlai Stevenson, 1954[1]

July 1957 was a red letter month for the Everetts. They settled into a small house in Alexandria, recently purchased. The relative states theory was published in *Reviews of Modern Physics*. The paper on recursive games was published in *Annals of Mathematics Studies*. And daughter Elizabeth Ann was born. Everett was charmed by baby Liz, but, consumed by his career and the pursuit of leisure, he left the nurturing to Nancy. If she wanted to raise children, that was between her and Dr. Spock.

Like millions of middle-class white women in Cold War America, Nancy's psyche was pummeled by the culture of conspicuous consumption. Relieved by machinery from many household drudgeries, the housewife-as-consumer was an advertising target for thousands of frivolous household products and handy-dandy appliances. Radio and television shows, films, books, popular songs, and romance novels celebrated the removal of women from intellectual life, glorifying some as sexual objects worthy of capture, while denigrating those who were not. On television, Jack Benny, Jack Parr, Jackie Gleason, Ricky Ricardo, Henny Youngman, and Johnny Carson regularly insulted an entire gender: endlessly cracking bad jokes about blithering, chocolate-eating, pink housecoat-wearing house wives who burned the pot roast after crashing the car and bouncing a check at the lingerie shop. It was popularly believed that women were not physically and mentally capable of driving a car or operating heavy machinery. As baby-making machines, they were basically barred from entering professions that Nature had supposedly reserved for men: science, medicine, business, law, academia and politics—except, of course, as secretaries.

Desiring independence and a meaningful career, Katharine Kennedy Everett had rebelled against the convention of marriage, but social institutions generally frowned upon woman who dared to profess *ambition*. Empowered by credit cards to shop for brightly packaged frozen food and (carcinogenic) beauty products and expensive bedroom sets, many college-educated women

[1] Speech at Columbia University, quoted in Fromm, E. (1955). 102.

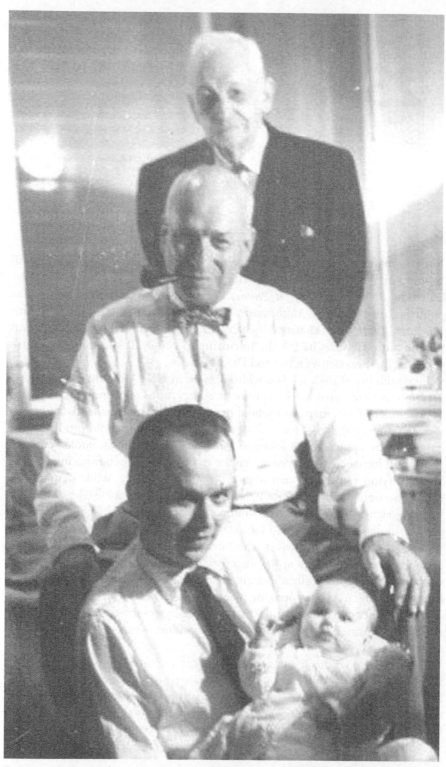

Four generations of Everetts: Hugh Sr., Hugh Jr., Hugh III, Elizabeth, 1957.

resigned themselves to serving the needs of her male "breadwinner." The standardized American housewife of the mid 20th century was designed to remain a perpetual child married to a grown-up with a career.

Nancy fretted in the trap. Everett worked long hours and traveled frequently, leaving her home alone to vacuum the carpets and change the diapers. As his work was top secret and highly technical, she was excluded from gossiping with him about goings on at the office. The WSEGers socialized at each other's homes, but spouses were appendages, not principals. Bored, she volunteered with the local chapter of the League of Women Voters, editing the monthly bulletin.

She penned a semi-serious letter (which she probably never mailed) to television talk show host, Jack Parr:

Dear Mr. Parr: HELP! Please do something to keep my marriage from coming to a premature end. Its only our fifth anniversary, and I find out my husband can't go to sleep at night without you on TV. Its not so bad for him; he eventually has the good sense to drift off to sleep. I however am distracted by your guests....I became addicted at the same time my husband did (what could I do, sleep in the guest room?) but my life has been miserable since....I hope you will see what you can do about changing your time slot say to noon...I know it is too late to change my husband's habits, so unless you want an unhappy marriage on your conscience, PLEASE come up with something!

It might make me feel a little better if you invited a League of Women Voters woman on your program. Well, why not? You've had everything else but the kitchen sink.[2]

Tucked into her Princeton diary was a thought she had typed up, written perhaps after meeting Everett, (it does express his philosophy of life):

Someone prophesied that 90 percent of people will be ruled by 10 percent of people. These 10 percent will be the most highly intelligent. They will guide people into doing most constructive and healthy activities. But they will have power over peoples' minds. They will hammer into their heads what is 'right,' what they must do for their own good and for the good of all others. They will accomplish this in [the] same way [as] the science of advertising a lot of ignominious, trifling necessities of everyday life puts all out of proportion their importance.

Advertising will be used for mass control. People won't want to listen, and they will hate it, but they will listen and grow used to it. Just as the child revolts against losing his freedom. But when he becomes accustomed to receiving and carrying out orders from a few elite, wise counselors, he will not know what to do if they should cease.[3]

She saw her marriage as unequal, writing (after Everett's death),

[2] NGE files, basement.
[3] Ibid.

I must accept responsibility for being unassertive, childish, (dumb), slow, dense ?!? .. With Hugh I never asserted myself on purpose – 'I will never be the cause, I will never give him reason for our break-up' (if such should threaten). If there is to be one it will be because your spouse wants it – 'don't stay with me on my account, I want you to feel free to leave' – [I was] masochistic – too giving.

I've never developed <u>close</u> support systems with women – never nec. I looked down on nitty gritty home making things women were relegated to and talked about, menus, cleaning, etc. – not for me. If I had a man to share life with – the rest falls into place ok.... I think it was O.K. marriage – really, it grew – but on some levels – I could never be outgoing, gracious, loving.... Why must I always be the understanding one of other's needs?[4]

Hot shot at the Pentagon

In late 1957, Everett was promoted to head WSEG's Strategic Analysis Group, the team responsible for mathematical analysis and computerization. He quickly became a virtuoso at programming in FORTRAN.

In 1958, the Institute for Defense Analyses contracted with the Department of Defense "to perform basic research in communications theory, mainly in mathematical areas, for the National Security Agency."[5] At that time, the very existence of the NSA was a state secret.[6] IDA offered Wheeler the directorship of the NSA's "cryptographic laboratory" at Princeton, but he declined.[7] Among Everett's papers is a blank NSA security clearance application form. It lists several hundred organizations that "have interests in conflict with those of the United States." Job applicants must not have been associated with any of these groups, which included American Women for Peace, Committee for the Protection of the Bill of Rights, and the National Negro Labor Council.[8] It is almost certain that, through IDA, Everett was involved in the early years of software development at the NSA's Princeton lab, which used the most advanced computers to design and run decryption and encryption programs. Then as now, the black-budget-funded "puzzle palace" scooped up domestic and global electronic communications for intelligence data mining, often illegally. After Everett left WSEG, his Lambda Corporation held an NSA contract.[9]

At the Pentagon, Everett soon convinced his bosses to purchase WSEG's first computer: a $47,000 LGP-30 manufactured by the Librascope company of Glendale, California. One of the first commercial "desk" computers, the machine encased 133 vacuum tubes inside its 740 pound bulk. Its brain knew

[4] Ibid.
[5] Ponturo, J. (1979). 190. IDA also administered the newly formed Advanced Research Projects Agency, which was charged with designing futuristic weapons, high-tech surveillance systems, and counter-insurgency tactics for fighting guerilla armies waging national liberation struggles.
[6] U.S. Congress, Office of Technology Assessment. (1995). 2.
[7] Aaserud, F. (1995). 213, 222.
[8] Everett, H III files, basement.
[9] Lambda spreadsheet, basement.

only 16 commands, and its output was a paper tape: it was more of a super-smart typewriter than a machine capable of running war simulations, which was Everett's ultimate programming goal. Pugh and Everett flew to California to vet the Librascope. While sharing a hotel room, Pugh was amazed when his friend pulled a large bottle of bourbon whiskey out of his suitcase. He said this was the first time he had noticed Everett's dependence on alcohol, an addiction which was to progressively worsen.[10]

Everett's dream machine, the IBM 650, priced at $115,000, finally arrived at WSEG in early 1958. Taller than a person, it needed its own room with a specially built floor to accommodate the mainframe, four tape drives, printer, and punch card sorter. With the assistance of WSEG chemist, Larry Dean, Everett invented software platforms capable of running his optimizing algorithms. WSEG used the new computer to project the fighting capabilities of the U.S.S.R. nuclear forces, using information on the Soviet economy and its military-industrial infrastructure gathered by the Central Intelligence Agency and the Defense Intelligence Agency.[11] The burning issue of the day remained accurately evaluating ballistic missile parity.

Wheeler, the lobbyist

In early 1958, Wheeler lobbied hard on Capitol Hill to escalate the development of nuclear armaments by creating a civilian-run National Advanced Research Projects Laboratory; the purported missile gap was central to his arguments. He kept Everett apprised of his progress, sending him a copy of a letter he had written to Eisenhower's science advisor, Dr. James Killian, with a cover note:

> This is a copy of what I sent off after listening to a classified briefing on the terrifying lead the Russians have on us. I can't see any escape from out having to set up some such central advanced projects research lab.[12]

Wheeler worried that excessive security precautions were keeping scientists from getting all of the information they needed to invent advanced weaponry, including anti-ballistic missiles (ABMs). He told Senator Lyndon Johnson's Preparedness Subcommittee in 1958,

> Anyone who visits the laboratories in which some of our national defense effort is going on is deeply impressed and disturbed by the fragmentation of effort... Able and devoted men in these groups are unable to have access to each other's efforts... to gain important information about the nuclear characteristics of the device that they are supposed to be shooting down. Men in this work are supposed to do the job with one hand tied behind their backs and the other hand all taped up except for one finger.[13]

---

[10] Pugh, G. E. (2005). 16.
[11] Pugh, G. E. (2005). 32.
[12] Wheeler to Everett, circa 1958.
[13] Wheeler to Everett with mimeographed enclosure, circa 1958, basement.

Wheeler was particularly concerned that research on solid propellant fuel for intercontinental ballistic missiles was not centralized. He wanted the new lab to be closely connected to the Institute for Defense Analyses' ARPA and WSEG divisions. And in his testimony he put in a substantial plug for the weapons research division of the Du Pont corporation, although he did not mention that he regularly consulted with Du Pont. He also had long time consulting ties to Convair Corp., which manufactured Atlas missiles, and had mounted a public relations campaign to promote the claim that the Soviets possessed vastly more ballistic missiles than the United States.

In the summer of 1958, the nationally syndicated columnist, Joseph Alsop, pushed the phony missile gap story; he went so far as to fabricate a table of figures showing that the Soviets would keep ahead of the U.S. by more than a factor of ten for the foreseeable future; when, in fact, the U.S. was ahead of the Soviets— and responsible elements in the intelligence community knew that to be true.[14] As the politics around the missile gap got hot, heavy, and contentious, Wheeler abruptly stopped lobbying for the laboratory. He retreated to academia and his theoretical work. After Kennedy was inaugurated president, top officials were compelled to drop the missile gap claim because, blessed with irrefutable intelligence data from U-2 spy plane flights over of the Soviet Union, the truth was reported to them by WSEG and Everett in a blacked-out study called Report 50.

Wheeler, the coach

Shortly after Everett's thesis was published, Wheeler dropped him a note:

> Dear Hugh, I am very eager to see you and talk with you and learn the latest information. I saw General James McCormack [IDA president] at the Bohr Atoms for Peace ceremony in Washington Oct. 24 and asked him how you were getting along. He said you were worth your weight in PU 239 and that you were one of the top people in the whole organization in his view.[15]

Wheeler believed that Everett's talent for quantum theory was being wasted at WSEG. He set Everett up with the offer of a job teaching physics at the University of Wisconsin.

> I hope that Lou Sachs will succeed in luring you into quieter and more reflective areas at Wisconsin because I think you really have a lot of original things to give to the world which you can't do through the present set-up. If you are hell bent on staying in Washington at any price why don't you let me see if George Washington University couldn't make a really attractive position for you? And for the love of Mike please wire or phone Bohr long distance and make a series of dates to chew on what you now have formally in print – and what I hope you will soon further augment in print.[16]

---

[14] Ball, D. (1980). 6–8.
[15] Wheeler to Everett, 10/30/57.
[16] Ibid.

Wheeler convinced the physics department at Boston University to solicit Everett for a teaching position.[17] But the long distance struggle with Bohr had seriously soured Everett on the perils of academic discourse and, regardless, he was captivated by the world of operations research; he fully intended to use WSEG as a catapult into the lucrative arena of private defense consulting where a young man with a taste for the good life could make some real money and change the world.

One of the most important assignments ever undertaken by Everett was his leading role in making a global assessment of the U.S.'s offensive options in nuclear war. Known as Report 50, this historic report made operative the concept of "assured destruction" that became the strategic posture of both the Soviet and American blocs. Everett made a mathematical breakthrough that was a lynchpin of the report, which, among other achievements, debunked the notion of a missile gap. But prior to hunkering down to invent a new way of computerizing calculations for the report, Everett finally complied with Wheeler's request: He traveled to Copenhagen and met with Bohr in the spring of 1959—with unintended consequences.

[17] Siegel to Everett, 4/16/59.

24 A Vacation in Copenhagen

If Bohr had discussed his view on language before quantum physics was developed, many might have thought he had been bitten by a mad mathematician. After all, how could there possibly be any connection between the real world and the abstract mathematical schemes that interested Bohr so much? On the contrary, such schemes were merely logical fictions created by the human imagination.

But an abstract mathematical scheme is the basis of quantum physics.... This makes the radical character of Bohr's point of view stand out more clearly. If one were to ask Bohr how we can live with this abstract quantum world, he would answer: 'There is no abstract quantum world. There is only an abstract quantum physical description! The task of physics is not to find out how nature is, but to find out what we can say about nature.'

Aage Petersen, 1968.[1]

Fock in Denmark

In Everett's files there is an article by Soviet physicist Vladimir A. Fock chronicling his visit to Bohr at his Institute for Theoretical Physics in Copenhagen in 1957. It sets the scene into which Everett walked two years later to defend his theory. Fock's account may have given him cause to believe that Bohr could be open-minded about the relative states theory.

Fock observed,

Everywhere in Copenhagen one feels the closeness of the sea. Outside my windows in the hotel there is a canal (the entrance to the inner harbor) over which there is a bridge which can be opened for boats. The central part of the city is on the northern side limited by some long narrow lakes which give the impression of a river or canal, but which unexpectedly end. On these lakes there are gulls, duck and swans, and in the whole of Copenhagen, and especially in the older part of town, the pointed towers are particularly characteristic. In the town there are many parks and open places, but they make an unpleasant impression by the circumstance that they are filled with military huts with concrete walls rising

---

[1] Petersen, A. (1968). 188.

from the ground with portholes. These huts were built partly during the German occupation and partly after the end of the war. Considering the fact that in military respect they are useless, it is difficult to avoid the thought that the reason for their construction is to keep up the atmosphere of the 'cold war'.[2]

An extended visit to the Institute was structured to encourage the informal exchange of ideas between the 30 or so physicists on the premises at any one time. Fock explained,

All the physicists come to the Institute every day. Every one of them has his own—if not very big—office. At the Institute there are also experimental laboratories.... The lunch room serves as Common Room, where everybody comes together twice a day for tea or coffee (one has to buy sandwiches in a nearby shop and bring them back). It is characteristic that besides paper napkins there are sheets of paper on the tables upon which one can write formulas.

During my stay, i.e. about a month, everyone gathered together twice for an evening's entertainment, where one showed movies and where the participants themselves entertained (accompanied by general laughter, the theoreticians had to perform very simple experiments), and a modest evening meal was had. My work at the Institute consisted of attending other scientist's lectures and speeches and above all conversations with Niels Bohr about the fundamental problems of quantum mechanics.

Fock admired elements of Bohr's philosophy and was thrilled to talk with him about the foundational questions:

I have become aware that even if he is more than seventy he is spiritually young, that he can get excited and talk passionately, but that he always speaks honestly. He doesn't try to impress you by his authority, but he is convinced that he is right and he considers a patient exposition of his point of view as a weapon in a discussion.... Above all, there was the difficulty that Bohr became so engaged in the formulation of his thoughts that it was difficult for me to enter into the conversation.

Fock's professed admiration was counter to the official stance of Soviet physics, which considered Bohr's artificial separation of the quantum and classical worlds to be positivism.[3] Marxist-Leninists disparaged positivism because it denied the existence of a knowable, objective reality. But according to Fock, "Bohr declared from the beginning that he is not a positivist and that he simply endeavors to consider nature as it is."[4] This is an interesting statement considering Bohr's dictum, as later reported by Petersen, "The task of physics is not to find out how nature is, but to find out what we can say about nature."[5]

Fock continued:

[2] Fock, V. A. (1957). 1.
[3] The positivists of the "Vienna Circle" asserted that we can only understand the world through verification of experimental result. It is meaningless to ask questions about an underlying reality, they said.
[4] Fock, V. A. (1957). 2–3.
[5] Petersen, A. (1963).

I pointed out that many of his formulations suggested a positivistic interpretation, which obviously he did not at all wish they should. I stressed the necessity of giving all quantum mechanical concepts 'a rational foundation' as reasonable abstractions on the basis of his own interpretation of the experiment. He answered that in no way did he reject their lawfulness. Our points of view gradually became closer; in particular, it became clear that Bohr fully recognizes the objectivity of atoms and their properties. He realizes that one has only to neglect determinism in the sense of Laplace, but not causality in general, that the expression 'uncontrollable interaction' is inadequate and that all physical processes are controllable. Perhaps it should be said that the similarity of our points of view only became clear through our conversations, but that it had existed before and independent of the conversations.[6]

Clearly Fock considered causality to be consistent with Bohr's interpretation;[7] which might have given Everett hope that the great man would extend his generalization of causality to include the uninterrupted flow of the Schrödinger equation. In any event, he went out of his way to meet with Bohr and try to hash it out.

Dining with Bohr

Everett, Nancy, and baby Liz arrived in Denmark on March 17, 1959. They had planned a six-week European vacation, spending two weeks in Copenhagen, so that Everett could confer with Bohr. After that meeting, they were scheduled to meet up with Katharine, who was to accompany them on a tour of France. Meanwhile, they ensconced themselves at the luxurious Hotel D'Angleterre.

Misner was also in Copenhagen, spending spring and summer at the Institute, where he had already met his Danish fiancée, Susanne Kemp. Her father, a prominent lawyer, was a financial supporter of the Institute, and Bohr's long time chum.[8]

Misner, Susanne, and the Everetts attended a small dinner at Bohr's palatial home. The mansion had been built by the founder of Carlsberg brewery company, which donated its use to Bohr, who was considered to be a national treasure. The downside was that the house adjoined the brewery and stank of fermenting hops. After a salmon first course and the main dish of pork roast, Everett pulled out a cigarette. Susanne, horrified, quickly informed him that in Denmark polite people do not smoke at the table. He refrained, but in the years

---

[6] Fock, V. A. (1957). 3.

[7] Beller references correspondence between Bohr and Fock during 1957 showing that their struggle was more intense than Fock indicates in "The Journey to Copenhagen." In particular, "Both Fock and Born were amazed by Bohr's rejection of quantum concepts and frustrated with his resignation from the attempt to comprehend the underlying quantum reality." Beller, M. (1999). 183. See also Camilleri, K. (2009) on how Fock tried to reconcile Bohr's philosophy with materialism.

[8] What follows in this account is from an interview with Charles and Susanne Misner in 2004.

Nancy, Everett, Liz in Copenhagen 1959.

Visiting Charles and Susanne Misner in Copenhagen 1959.

to come, whenever he lit a cigarette in her presence, he exclaimed, "Oh, Hugh, but *we* do not *do* that *here!*"

Susanne recalls, "Hugh nearly got kicked out of the hotel for putting cocktails and food on the outside window sill. He was sloppy and had a cigarette *all the time*. He ate very rich food and was fat and never had any exercise."

Misner adds, "And he drank a lot, too."

Everett did not invest much time hobnobbing with the other visiting physicists. Because he hated public speaking, he did not make a formal presentation on his theory, but he pitched it in private to Bohr and four other physicists, including Misner and Rosenfeld during the course of a couple of afternoons.

Bohr and Everett listened to each other, but Bohr was very difficult to understand. Misner recalls:

> He would look down and think deeply and begin a sentence and halfway through the sentence his pipe would go out and he would go to the blackboard where there was a little gadget you would expect to find in a tobacco store and he would relight his pipe and it would last for three or four sentences, and then he'd do the whole thing over again. He was hard to hear. You had to lean close.

And that was it. There was no great debate, no wielding of equations, no logical traps or conundrums about relativity and uncertainty to negotiate (as there had been between Bohr and Einstein). There was simply a polite hearing and a lot of mumbling. Misner says, "Bohr's view of quantum mechanics was essentially totally accepted throughout the world by thousands of physicists doing it every day. And to expect that on the basis of a one hour talk by a kid he was going to totally change his viewpoint would be unrealistic."

Nor was Everett open to abandoning his theory on the basis of Bohr's opposition. Disgruntled, he spent most of his time sight-seeing, or drinking beer in the bar of the Hotel Østerport. And in that bar, he conceived another bright idea. It had nothing to do with quantum mechanics, but it revolutionized operations research and affected the course of the Cold War.

Looking for the magic

In the late 1700s, an Italian mathematician, Joseph Louis Lagrange invented the Lagrange Multiplier method. The Lagrange multipliers are variables signified by the Greek letter λ (Lambda). The multipliers are used to optimize our ability to predict the consequences of change. For example: a farmer wants to fence a circular pasture. He knows the number of square feet in the area of the circle. He knows that 100 feet of fence will enclose the circle. Using Lagrange's multiplier method he can determine how much land can be enclosed by adding one foot to the fence.

Military operations researchers deal with much more complex problems than fencing in a circle—problems that have thousands of continuously changing variables representing resources and constraints that feedback on each other, such as weather patterns affecting maximum engine speeds, or the explosive yield of atomic weapons versus the hardness of enemy missile silos. Even a small change in the value of a single variable alters possible outcomes, e.g. adding one megaton of explosive yield to a weapon may be the puff that breaks the missile silo's back. When Everett and Pugh constructed their fallout model, they used Lagrange multipliers to calculate percentages of casualties as a function of the number and yield of nuclear bombs dropped under different wind conditions.

The computations involved in designing the fallout model were relatively easy compared to the computational power demanded by the task force working on Report 50. Everett had been thinking about how to solve this formidable problem ever since working on the fallout project, but with the necessity of constructing Report 50, WSEG researchers were desperate to find mathematical shortcuts through vast thickets of interlocking variables. With the acquisition of the IBM 650, WSEG could run huge numbers of multiplier-type calculations, but without an optimizing algorithm the computer would be trapped in feedback loops and overwhelmed by the exponential numbers of possible relationships between the variables.

The power of beer

This brings us back to Everett's vacation in Copenhagen: One afternoon, he wrote a letter on Hotel Østerport stationery to his WSEG co-worker, Bob Galiano. The historic letter began, "While drinking beer yesterday several ideas came about our maximization problem." He went on to describe and christen the *Generalized* Lagrange Multiplier Method. And he called his λs "magic multipliers."

The magic multiplier method was a computational shortcut. It used a simple but powerful algorithm to break up extremely complex problems into sets of smaller problems. As the values of λ are changed by single units, the computation gradually converges on a optimal solution (if there is a solution). When programmed into Everett's favorite language, FORTRAN, the multipliers find solutions that are guaranteed to be true—balancing resources and constraints.

Operations researcher Gary Lucas, who worked closely with Everett at Lambda Corporation, explains that before the Generalized Lagrange Multiplier Method was invented,

> Classical Lagrange multiplier theory was great for theoretical development in classical physics, but it was not very useful for large real world problems. There was not enough paper. Everett's multipliers moved optimization into the computer world. It is entirely a product of that world. It would never have been developed without the computer. You could therefore say that Hugh made the

transition from the classical optimization world of Newton to the modern number crunching world of the modern computer. This is quite a leap![9]

Although the method itself left operations researchers gasping at its simplicity and mathematical beauty, its practice was a high art. It turned out that only Everett and those who worked with him consistently enjoyed success using the new method, which, in part, depended upon making educated guesses about what λ variables to insert into the equations. One had to have a feel for the method to make it work.

Leon Lasden at the University of Texas at Austin notes that the repercussions of Everett's breakthrough are still being felt. His revolutionary contribution, says Lasden, was extending the use of the Lagrange multipliers from problems with continuously linked variables to problems with discreet variables that "jump" around. Before Everett, these "non-linear" variables resisted algorithmic solution.[10]

In August 1962 Everett wrote up his invention in WSEG Research Memorandum 25. A year later, it was published in *Operations Research* as "Generalized Lagrange Multiplier Method for Solving Problems of Optimum Allocation of Resources." With a bow toward his former teachers, Kuhn and Tucker, Everett noted in his introduction, "The basic theorems upon which the techniques to be presented depend are quite simple and elementary, and it seems likely that some of them may have been employed previously. However, their generality and applicability do not seem to have been well understood at present (to operations analysts at least)."

When Everett returned to the Pentagon from Copenhagen, his new tool in hand, the success of Report 50 was assured.

---

[9] Lucas interview, 2008.
[10] Lasden interview, 2008.

BOOK 7
**ASSURED
DESTRUCTION**

25   Everett and Report 50

A thermonuclear war is not impossible. As long as the weapons exist in plenty, such a war could explode at any time—its trigger pulled by a deliberate attack on the United States, by a massive invasion of Europe, a cold-war gambit pushed too far, a ballooning little war, a nuclear accident, or the act of a reckless officer. No one of these events seems likely to occur, but when all of the possibilities are added together and then multiplied by the number of days of cold war ahead, the chances of nuclear war in our lifetime become fearfully great.

Richard Fryklund, 1962.[1]

The unthinkable

In the months leading up to the presidential election of 1960, Democrats accused Eisenhower of being unprepared to wage and win a nuclear war. Pressured by public opinion, and by the fact that striking second was not a viable strategy in the age of nuclear submarines and ballistic missiles, the Joint Chiefs instructed WSEG to study offensive capabilities. In September 1959, 30 staffers, including Everett, were assigned to spend the next year researching and writing *WSEG Report 50, Evaluation of Strategic Offensive Weapons.* Long kept under classified lock and key, redacted portions of Report 50 are now publicly available.[2] These are extraordinary source documents, revealing the origins of the doctrine of assured destruction, warts and all.

The historic report illuminates dark shadows in the history of the Cold War, particularly concerning the strategic focus upon striking first, which formed the back bone of the U.S. nuclear posture for many decades.

Locked inside The Cage, WSEG researchers were given access to top secret National Intelligence Estimates, and other closely held data on the weapons capabilities of the U.S., its allies, and its enemies. According to Pugh,

---

[1] Fryklund, R. (1962). ix.
[2] Report 50 was composed of 10 major "enclosures," each with multiple volumes and appendixes. Of the five publicly available sections of the enclosures, two are concerned with how global economics and politics interact with the U.S.'s evolving nuclear strategy. Three are of a technical nature, concerning the development of specific weapons systems, and they are severely redacted.

Everett's contributions were absolutely central to the effort—for without Everett's generalized Lagrange optimization methods the whole Report 50 project could never have been attempted. So although Everett was not the 'project Leader,' and was not officially responsible for the project results, his presence and his contributions were in reality <u>the</u> critical element that made the project possible.[3]

WSEG's official history notes that Report 50, "became a basic source document, used for orienting incoming officials and initiating fundamental reappraisals of ongoing defense programs." The command and control section became "something of a 'best-seller' contributing to an upsurge of interest and concern in command and control."[4] No wonder: it exposed the second strike doctrine as an iron glove lacking a fist.

Decapitation

WSEG determined that the United State's nuclear forces were vulnerable not only to a Soviet first strike, but also to a retaliatory strike, despite the fact that the U.S.S.R. seriously lagged behind the U.S. in manufacturing nuclear arms. Squelching the idea of a missile gap, military intelligence assessed that the Soviets were *not* interested in launching a first strike, as they were aware of their own strategic weakness, and they knew that the U.S. was poised to blow them away if provoked. Therefore, WSEG warned against predicting Soviet capabilities by a "simple mirror-imaging of our own offensive development program."[5] Unfortunately, the implementation of the preventive-preemptive orientation[6] that emerged from Report 50 compelled the Soviets to mirror the first strike capability of the U.S., making the world that much more dangerous.

WSEG shocked officials by reporting there was no possible defense against submarine-launched missiles or intercontinental ballistic missiles (ICBMs) that arrived in "clusters" surrounded by clouds of decoys thwarting defensive radars. Furthermore, high-altitude explosions of atomic weapons would fry the electrical circuits of retaliatory missiles with electromagnetic pulses.[7] With a 15-minute warning window, the bulk of SAC's bombers would be destroyed

---

[3] Pugh, G. E. (2005). 35. Incidentally, the report must have held a certain nightmare fascination for Everett as, according to his universal wave function model, global nuclear war was bound to occur in some universes. It is interesting to speculate in an "anthropic" sense that if nuclear holocaust was highly probable, as so many experts believed during the Cold War, then the fact that we now live in a universe where it has not (yet) occurred could be construed as a validation of Everett's theory because it would be improbable that a nuclear war did not occur if there was *only one* universe!

[4] Ponturo, J. (1979). 176–177.

[5] WSEG Report 45. (1960). 148. This report, issued on September 23, 1959 was incorporated into Report 50.

[6] *Preemptive* warfare is defined by the U.S. military as "An attack initiated on the basis of incontrovertible evidence that an enemy attack is imminent." *Preventive* warfare is "A war initiated in the belief that military conflict, while not imminent, is inevitable, and that to delay would involve greater risk." The distinction between these two aggressive doctrines becomes meaningless if intelligence findings are cooked and manipulated.

[7] Building expensive ABM [anti-ballistic missile] systems would be militarily ineffective, said WSEG, but could be justified "principally on political or psychological grounds," i.e. psyching the American people into believing that they were safer than they were. WSEG Report 45. (1959). 8.

in their hangers by a surprise missile attack. Report 50 made it clear that a handful of Soviet ICBMs could decapitate the U.S. high command: "[The national political and joint military command structure]...is highly vulnerable and could not be counted upon to complete its minimum essential retaliatory functions if attacked."[8]

WSEG urged national security officials to construct a command and control system capable of surviving a sudden attack and mounting at least a limited nuclear response. Certainly, the public, Congress, and most government officials, had no idea that the announced U.S. strategy of massive retaliation was built on operational quick sand: The chain of command and most of the population would be dead or dying before a second strike could be launched.

Using the Everett-Pugh fallout study, the report showed that after an attack by either superpower on the other, the majority of the attacked population that survived the initial blasts would be sterilized and gradually succumb to leukemia. Livestock would die quickly and survivors would be forced to rely on eating grains, potatoes and vegetables. Unfortunately, the produce would be seething with radioactive Strontium 90, which seeps into human bone marrow and causes cancer.[9]

Global politics

Henry Kissinger, one of the first advocates of fighting "limited" nuclear wars, was consulting at WSEG during the writing of Report 50, which echoes his Metternichian brand of *real politick* by treating nations as expendable pawns in the game of global dominance. Aware of this approach, Europeans were terrified of becoming collateral damage in the superpower struggle. Only Great Britain, Italy, and Turkey allowed the U.S. to deploy air bases and nuclear weapons in their territories.

With military projection of nuclear force restricted by lack of compliant allies, WSEG was afraid that, "the United States would find itself isolated in a sea of Communist continents," as the Sino Soviet bloc moved without opposition into developing markets.[10] National determination for colonized peoples was *not* synonymous with America's national interest: "Had it not been for the tide of colonial emancipation, most of Africa as well as India, Burma, Ceylon, Indonesia, Vietnam, Laos and Cambodia would today be parts of allied territory."[11] Underdeveloped countries absorbed by their own economic struggles had a "strong incentive to avoid involvement in the cold war."[12]

---

[8] WSEG Report 50. (1960). "Enclosure C, National Command and Control Vulnerability" as summed up in Wainstein *et al.* (1975). 241. Partial text, original classification: top secret. Available at National Security Archive.
[9] WSEG Report 50. (1960). "Enclosure A, Evaluation of Programmed Offensive Systems 1964–1967," Third Volume, 58–62.
[10] WSEG Report 50. (1960). "Enclosure I, Changes in the Free World." 14.
[11] Ibid. 57.
[12] Ibid. 14–21.

WSEG also foresaw a danger that,

> authoritarian governments [e.g. South Korea, Turkey] now militarily allied to
> the U.S. will be replaced ... by regimes more responsive to public opinion. ... It is
> doubtful if the United States will be able to count on receiving as much support
> in these countries as it does today. ... All the forces that make for antagonism
> against the West, against the present or former colonial powers, against 'eco-
> nomic imperialism,' against the white man–particularly if he is known to dis-
> criminate against colored people–militate against alignment with the West.[13]

Given the inability to wage a credible second strike, the logical strategic
policy was to maintain credible first strike capabilities. Report 50 swung
the technological foundation of the basic war plan toward a preemptive
focus.

> Nothing short of a first-strike or counterforce capability could serve as a reliable
> deterrent to such overseas aggression as would leave the U.S. strategic force
> intact and with the option to intervene. A strategic force capable only of a strike
> at urban targets in the Soviet Union–possibly the only strategy that a second-
> strike force could employ–would fail to provide a credible deterrent against
> Sino-Soviet aggression overseas, because its very employment would invite the
> destruction of substantial areas of the United States.[14]

In the final days of the Eisenhower administration, the secretary of defense
authorized the new Single Integrated Operating Plan (SIOP) "to replace the
independent and often competing war plans of the military services."[15] Creating
the SIOP was an invitation to the Soviets to mirror the core U.S. strategy of
maintaining a massive first strike threat capable of surviving an enemy attack
and retaliating. This was the doctrine of deterring war though assured destruction, and if its logic seemed circular: it was.

Feedback loops

Assured destruction was Prisoner's Dilemma played with a nuclear button
that, if pushed by one player, was tantamount to the commission of national
suicide—*if* the other player's retaliation option worked. After WSEG determined that the second strike option was a mirage, it proposed to beef up a
first strike option that could also function as an effective second strike. Fear
of the second strike might cause each player to refrain from "defecting" and
pushing the first strike button, or so the logic of assured destruction went. But
fear of a first strike could also cause a player to defect first, hoping to be able
to defang the other player's second strike capability, which was really a first
strike capability. And around and around the atomic mulberry bush went the
circular reasoning. The phrase "assured destruction" was quickly morphed

---

[13] Ibid. 21.
[14] Ibid. 30.
[15] Herken, G. (1987). 126.

into "mutual assured destruction" by its critics, as the "MAD" acronym was irresistible.

Unfortunately, the major sociological lesson to be learned from Prisoner's Dilemma—that altruism and cooperation pay off better in the long run than defecting—was completely lost on WSEG, which was not tasked to promote cooperation. In fact, Report 50 identified the populace's yearning for peace as an *obstacle* to strategic planning:

> One such factor of mounting importance is a growing public awareness of the devastation likely to be inflicted on any participant in a general nuclear war. This has increased popular fears of military conflict in any form and created particular aversion to the use of nuclear weapons.... There are indications that sectors of the population might be willing to pay an extremely high price to avoid involvement in war.... The existence of nuclear weapons with their hitherto inconceivable destructive power has introduced an emotional element into the debate which has sometimes clouded the more rational considerations, particularly in nonofficial circles.[16]

WSEG quoted the finding by a 1958 Gallup poll that, by a two to one margin, the population of Great Britain preferred coming to terms with the Soviets *at any price* over fighting a nuclear war.[17] WSEG worried that disarmament proposals supported by prominent Britons, including the mathematician, Lord Bertrand Russell, were undermining the moral foundation of operations research as expressed by the slogan, "We shall prevent war by preparing for war."[18]

On the other hand, preparing for war could hasten it.

> The threat to the U.S. should not be measured solely by the strength available to actual or potential enemies. The seriousness of this threat is also affected by the intention and resolution of enemy nations to employ their strength against us. It is therefore appropriate to take into account the factor of the willingness of the enemy to accept the risks of modern war.[19]

Report 50 assumed that if the enemy was willing to accept risks, then we must. But in the case of deterrence through assured destruction, a risk-bearing enemy would be mirroring our own risk-taking agenda, so the real question was are *we* willing to accept the risks of modern warfare? And, if not, would it not be logical to disarm? But WSEG did not question its core assumption that the socialist bloc should be destroyed one way or another. That a death-defying high-wire arms race should exist was a forgone conclusion: WSEG's job was to run it cost-effectively.

Report 50 noted that the Soviet Union's annual economic growth rate of 6 percent outstripped the United State's growth rate of 3.5–4.5 percent and that,

[16] WSEG Report 50. (1960). "Enclosure I, Changes in the Free World." 43, 102.
[17] Ibid. 18. Italics added.
[18] Ibid. 127.
[19] WSEG Report 50. (1960). "Enclosure J, Strategic Implications of Possible Changes in the Nature of the Threat." 2.

"economic growth will enable the U.S.S.R. to carry the burden of competitive armaments...and increase Soviet leverage in world affairs."[20] The report identified Iran, Iraq, and the Pakistan-Afghan tribal areas as probable sources of superpower competition and conflict, along with Latin America, Northern Africa and Southeast Asia. Contrary to the rhetoric streaming out of the White House and the Pentagon, the researchers did not find that the Soviets were willing to risk general war to protect these interests, quite the opposite. However, "Soviets may consider U.S. strategic posture indicative of offensive intent."[21]

> There is apparent consensus among the Soviet leadership that strongly favors policies that stop short of general war, and that discourage lesser wars also, partly at least, from fear that they might get out of hand. Russian leadership appears to have nearly come full circle, and almost to have resumed the previously condemned views of Malenkov concerning the disastrous probable consequences of thermonuclear warfare. There is also a doctrinal legacy that deplores adventurism.... We do not know, of course, what views and plans Soviet officials may have for the use of their strategic offensive weapons.... What may be inferred from their actions, and from the repeatedly expressed views on the destructiveness of nuclear warfare suggest a rather amorphous view that the most profitable role of Soviet strategic power is to serve as a counter-deterrent.[22]

WSEG's analysis of Soviet intentions implicitly recognized that there was an opportunity to back-down from the arms race and that the Soviets might very well have responded to *genuine* disarmament proposals. But neither the Joint Chiefs nor WSEG were interested in ending a global nuclear arms race, rather, they hoped it would *spread*. Contemplating a war with China, the authors stated that a major objective for the next decade should be:

> to foster conservative attitudes on the part of China and Russia toward a general nuclear war with the United States, but also to foster the divisive factors in the Sino-Russian alliance.... It could produce a situation in which a war between the U.S. and China, with the U.S.S.R. remaining neutral, is imaginable.... It is conceivable that, just as the Chinese Communists might upon occasion *feel* it *desirable* to involve the U.S. and the U.S.S.R. in a war, *sane* Russian leadership might come to *feel* that a war between the U.S. and Communist China, if not *desirable*, might be turned into an opportunity to get rid of the unwelcome elements of Chinese communism and weaken the U.S. as well.[23]

That statement is psychologically instructive: Reflecting the zenophobic mentality of the typical Cold Warrior, WSEG suggested that the Soviets or the Chinese were inclined to launch a world war on the basis of "feeling" and "desire" or dubious "sanity," unlike the Americans who were prepared to ignite the planet on the basis of pure reason.

[20] Ibid. 5.
[21] Ibid. 52.
[22] Ibid. 9–10.
[23] Ibid. 13–14. Italics added.

The mandate of Report 50 was to calculate the benefits and costs of implementing the policy of deterrence through assured destruction, but the authors recognized that,

> There are limits to what may be achieved by policies of deterrence, and when these limits are exceeded, deterrence is likely to fail. It is likely to fail because it becomes incredible, or because it appears to the enemy intolerably oppressive or threatening. It may appear to be incredible because it does not appear that the potential gains to ourselves are equivalent to the risks involved in invoking the deterrent force.... It may appear threatening... because the technical or strategic characteristics of our deterrent suggest that general nuclear war is inevitable or highly probable. This could serve to justify assumption of the risks of preventive or pre-emptive attack upon us as the lesser of two evils.... Improvements in strategic offensive posture cannot forcibly prevent the Soviets from destroying from half to nine-tenths of our people and wealth in a general war. *This suggests that the problem cannot be solved solely by improvement of the military posture.*[24]

One of the most important conclusions of Report 50 was that achieving strategic stalemate through deterrence (in which neither side *feels* that it can attack the other first), "will curtail drastically, and perhaps eliminate, our ability to project U.S. strategic power... into foreign areas."[25] In other words, the success of assured destruction will (and did) encourage the use of limited, conventional, proxy warfare by superpowers fighting over market turf *because* the consequence of waging a general war was too destructive. Yet, limited wars are fuzes for nuclear conflagration, said WSEG.

Recognizing that the U.S. military was becoming increasingly embroiled in fighting wars in third world countries, WSEG analyzed that tactical nuclear weapons were useless against mobile guerilla armies of rice farmers (as in Vietnam).[26] There was another danger with using tactical nukes, whether in third world countries, or in Europe:

> It would be difficult, if not impossible, for contestants to know at once whether nuclear strikes were occasioned by tactical bombs or strategic bombs, whether missiles were tactical, intermediate range, or even intercontinental; or whether to expect the next salvo to be the strongest blow of all—an all-out intercontinental strike. In a situation so grave, the stakes would be so high that either side might, with plausible reason, launch its intercontinental attack in desperation.[27]

Trapped inside the feedback loop of Cold War Reason, the authors were clear that the *threat* of general nuclear war encouraged both sides to engage in limited wars, including limited nuclear exchanges. However, WSEG noted, limited war increases the "probability of general war by accident or miscalculation and thus erode[s] the deterrent effect of the strategic posture." In other words: *Deterrence erodes deterrence.* And, due to lack of adequate defense against a

[24] Ibid. 20. Italics added.
[25] Ibid. 21.
[26] Ibid. 48.
[27] Ibid. 52–53.

first strike, "a favorable outcome of a general war does not appear attainable" for either side. But this did not matter to WSEG's calculus because even if one power has no intention of launching a first strike, it might be attacked, therefore it *must* build a hugely redundant second strike capability, "which is bound to include a fearful first strike capability."[28] But a first strike by either superpower would not bring victory, even if it destroyed 90 percent of the enemy's nuclear force, as there was no way to destroy *enough* enemy missiles and bombers to prevent a debilitating retaliatory strike by the remnants of a force designed to strike first, said WSEG.

After having identified the considerable political and technological weaknesses of assured destruction, the Report 50 authors recommended what appears to have been a forgone conclusion: despite the risk of inviting a first strike, the U.S. should continue to manufacture an expensive first-strike capability so horrifying that it will continue to "deter" the enemy into "deterring" us. They also recommended developing a new generation of conventional weapons with which to fight limited wars, supplemented by tactical nuclear bombs, despite having exposed the danger of limited warfare.

The cost-benefit of Armageddon

Parts of Report 50 are classic Everett (at Lambda he authored several similar reports). Using his multiplier method, the report measured "single shot kill probabilities" for *integer* numbers of bomb payloads delivered to specified types of sites (military bases, factories, population centers) as a function of how many bombs could be spared from inventory to achieve a maximum kill ratio.[29] Thousands of constraints had to be considered in setting up the kill formulae, including the probabilities of missiles and B-52s reaching the target, the ability to retarget in mid-flight, the probability of dud bombs, the variable damage caused by shock waves, overpressure, fire, and radiation. Under Everett's guidance, WSEG's new computer sifted through vast quantities of data to construct alternative world futures containing ranges of cost-benefit solutions.

For example, the future was affected by WSEG's determination that nuclear bombs were relatively cheap to make. It was the cost of the delivery system that mattered.

> The unit cost of bombs and warheads, after deducting the salvage value of nuclear materials [oralloy, plutonium, tritium], is relatively low as compared with the unit cost of the weapon system. In most cases the net cost of the warhead and/or bombs is less than 10 percent of the cost of its carrier.[30]

---

[28] Ibid. 3.
[29] WSEG Report 50. (1960). "Enclosure A, Second Volume." 22. The generalized Lagrange multiplier method allowed computer simulations to reflect real world scenarios in which bombs and missiles were deployed in integer amounts, not fractions.
[30] WSEG Report 50. (1960). "Enclosure F, Estimated Costs of Strategic Offensive Weapons Systems." 2.

The authors suggested scrapping development programs for the Atlas and Titan missiles, which were cumbersome and extremely expensive. They recommended relying on Minuteman batteries and the Polaris submarine fleet, in conjunction with air-borne B-52 bombers on "flying alert," as the preferred strategic weapons system for waging deterrence. In other words, they called for maintaining an around-the-clock first strike capability. Using the Everett-Pugh fallout model, WSEG suggested basing the Minuteman forces in "lightly populated areas in the northern mountain states" to minimize fallout on more heavily populated areas when the Minuteman launch sites were bombed.[31]

Members of the joint chiefs of staff were not happy with Report 50. It exposed fatal weaknesses in military planning; it debunked the politically useful missile gap; it spelled the end of certain boondoggles. They succeeded in keeping the report out of the hands of the Eisenhower White House, but when Robert McNamara, incoming President Kennedy's new secretary of defense arrived on the job, he demanded a briefing.

Everett briefs a top dog

On January 26, 1961, Everett, Larry Dean, and the Report 50 coordinator, George Contos, scheduled a 90-minute briefing for Secretary McNamara. He was so fascinated by the nitty-gritty that the session lasted all day. Everett was consulted repeatedly by high-ranking administration officials over the next two months as Kennedy and McNamara sought to find a way out of the doctrine of Wargasm.[32]

One result of the briefing was that previous estimates of the force levels necessary to prevail in a hot war were scrapped. McNamara agreed with WSEG's arguments for relying on Polaris and Minuteman (bombers were essentially obsolete, but kept for show and to keep SAC quiet). In fact, missile-laden submarines were the most effective weapon wielded in the Cold War. They hid beneath oceans and functioned as both a first strike and second strike weapon—assuring destruction.

McNamara was momentarily drawn toward the concept of "no cities/counterforce," which targeted only military installations in initial salvos. But WSEG favored deterring the enemy from provoking or attacking America by threatening the destruction of enemy cities and vital infrastructure. In January 1963, influenced by WSEG, McNamara officially endorsed assured destruction. RAND's Albert Wohlstetter, a proponent of limited nuclear warfare, criticized the doctrine as a "balance of terror."[33]

It certainly did not balance budgets. With the costs of the U.S. war on Vietnam increasing, McNamara told Congress, "Every hour of every day the Secretary is confronted by a conflict between the national interest and the

---

[31] Pugh, G. E. (2005). 45.
[32] Ball, D. (1980). 37; Kaplan, F. (1983). 258–262.
[33] Ball, D. (1980). 39.

parochial interests of particular industries [and] individual services."[34] By 1964, the corporate supply-siders were winning. McNamara started investing in hugely expensive civil defense and anti-ballistic missile defense projects,[35] even though he and his experts, including Everett, knew that neither system could work as advertised.

Historian Desmond Ball assessed that despite the sober view of Soviet intentions and capabilities presented in Report 50, the Kennedy administration grossly inflated the danger of the Soviet nuclear threat for domestic political reasons.[36] The by-word of the administration became "flexible response," to replace the Wargasm, but Ball points out:

> Beneath the 'shifting sands' which characterized the McNamara strategy throughout his years as secretary of defense was a constant, although not always explicit, acceptance of a particular version of deterrence as a national strategic policy. This was the necessity for the United States to have the capability, at all times and under all circumstances, 'of destroying the aggressor to the point that his society is simply no longer viable in any meaningful 20th century sense.'[37]

Ball also notes that McNamara seriously considered the option of launching a preemptive first strike on the Soviet Union and China.[38]

In the aftermath of Report 50, Everett's brilliance was legendary in black budget operation research circles.[39] In support of his path-creating work, his bosses authorized the purchase of a "super-computer," the new Control Data Corporation 1604, which was adept at running complex war simulations, including anti-ballistic missile defense models. WSEG was regularly asked to evaluate contractor-proposed ABM system designs, but they all had to be rejected, "because the cost of any effective defense was . . . 1000 to 1 higher than what a determined opponent would have to spend for a larger or more capable offensive force that could simply overwhelm the defense."[40] Nonetheless, the

---

[34] Ball, D. (1980). 250.

[35] Kaplan, F. (1983). 320–321.

[36] In 1980 Desmond Ball published *Politics and Force Levels, the Strategic Missile Program of the Kennedy Administration.* He interviewed Everett and Pugh and many other players in the operations research that underpinned the adoption of assured destruction. Among Everett's papers is an early version of the book, a typescript.

[37] Ball, D. (1980). 171; interior quotation from McNamara, R. S. (1967).

[38] Ball, D. (1980). 185.

[39] At an international conference in 1968, national security expert, Jan M. Lodal, summarized Everett's contribution to MAD: "Mathematically, an Assured Destruction problem can be framed as a two-sided, zero-sum game. The initial attacker should allocate both his offensive and defensive weapons to minimize the maximum amount of damage the other side could do in return, assuming that the other side would allocate his weapons to maximize the same measure of damage. However, the set of possible strategies is so large that the usual methods of solving two-sided games cannot be used. Consequently, we have reduced the problem to two separate resource allocation problems by making use of an intermediate set of values for U.S. weapons, and have used a modified Lagrange multiplier method [Everett's] to solve the allocation problems. We have thus used [Everett's generalized] Lagrange multipliers to approximate minimax strategies for a two-sided game with discrete strategy sets." Lodal, J. M. (1969).

[40] Pugh, G. E. (2005). 35.

wraith of ABM continued to haunt defense budgets,[41] materializing primarily as private profits for contractors, which soon included Everett.

[41] Report 50 asserted that building a successful ABM system, combined with the sheltering of hundreds of millions of people from blast and fallout effects, could save 40–60 million lives. Depending on the mix of urban and rural sheltering, the cost of sheltering a person reached as high as $1,400; the per capita costs of an ABM system were similarly prohibitive. WSEG Report 50. (1960). "Enclosure A, Third Volume." 82.

26   Everett and the SIOP

> *In the councils of government, we must guard against the acquisition of unwarranted influence, whether sought or unsought, by the military industrial complex. The potential for the disastrous rise of misplaced power exists and will persist.... Yet, in holding scientific research and discovery in respect, as we should, we must also be alert to the equal and opposite danger that public policy could itself become the captive of a scientific technological elite.*
>
> Farewell address by President Dwight D. Eisenhower,
> January 17, 1961.

Calming the Wargasm?

In November 1958, President Eisenhower was briefed on the Air Force plan to launch all of its bombers at the major population centers of the U.S.S.R., China, and their allies the minute that the president (or the Strategic Air Command) believed that the Soviets were preparing to attack U.S. interests, even with conventional forces. Appalled, Eisenhower commented that the number of targets was beyond those needed to destroy the Soviet's will to fight. The U.S. does not require, "a 100 per cent pulverization of the Soviet Union," he said. He asked the Pentagon to formulate a way of selectively taking out military targets, while sparing cities. He wanted targeting options because, he observed, there is a "limit to the devastation which human beings can endure."[1]

The result of his order was the Single Integrated Operating Plan—a project so secret it was assigned its own security classification, "extremely sensitive information." It was intended to integrate the duplicative nuclear bomb targeting plans of the Army, Navy, and Air Force into a single plan capable of being reprogrammed—capable of flexibility, of surgical strikes (with hydrogen bombs).

The Joint Chiefs entrusted the development of the SIOP to SAC because it already had a command and control system in place, as weak and undependable as it was, and some understanding of the problems involved in launching and coordinating nuclear attacks. But the chiefs were not naïve. They did not

---

[1] National Security Council. (1958). 5.

fully trust SAC—nor the other armed services, nor the weapons contractors who benefited from manufacturing instruments of war—to put the interests of the country before their own. So, they ordered WSEG to work with SAC.[2]

Although many details of this history remained sealed, there is no doubt that Everett, as WSEG's chief mathematician and resident computer wizard was a leader of the group detailed to design software for the SIOP, as well as the command and control systems of the National Military Command System, which was the Pentagon's 24-hour-a-day war room. WSEG staff worked with the

> actual plans, data, and procedures with a high degree of both national security and political/administrative sensitivity ... [The work] was conducted under special access and reporting arrangements.... Results were not published in official WSEG studies but were reported either informally, without a distinctive written product, or in the form of memoranda [transmitted through] a 'quiet' reporting channel.[3]

The very concept of a SIOP was kept secret from the public until 1976, when it was first referred to in news reports, glancingly. It was not until near the end of the Cold War that scholars were granted access to government documents outlining the war plan.[4] It is clear that civilian officials had hoped that the first SIOP would provide them with political flexibility by allowing for surgical operations, but, by the time that the military got through beefing it up, it was more sledgehammer than scalpel.[5]

The center of gravity of SIOP-62 was its first strike capability. It was geared to instantaneously shoot off the entire U.S. nuclear arsenal of 3,423 bombs, obliterating more than a thousand targets and 285 million people in the U.S.S.R., China and Eastern Europe.[6] And that damage was to come from blast effects alone; the effects of firestorm and radioactive fallout were not included in the damage calculations. When these were taken into consideration, the potential dead numbered over a billion.[7] The SIOP's retaliatory option alone promised to throw 1,706 bombs against 725 targets. The Wargasm mentality had not been shed. The plan made little or no distinction between communist countries that were at war with the United States and those that were not; it was poised to incinerate them all in a massive nuclear belch regardless of any threat they posed at the time of launch.

Where SIOP-62 principally differed from Wargasm was in attempting to synchronize the uncoordinated strike plans of the Army, Navy, and Air Force. This process was monstrously complex and cumbersome, generating hundreds of new

---

[2] Ponturo, J. (1979). 242–244.
[3] Ibid. 250–251.
[4] The National Security Archive at George Washington University has unearthed a cache of previously classified documents, including the Pentagon's own internal history of the creation of SIOP-62, that sheds considerable light on its birth pangs.
[5] National Security Archive Electronic Briefing Book No. 130. (2004). 3.
[6] Kaplan, F. (1983). 269.
[7] Rhodes, R. (2007). 88.

acronyms as it stitched together thousands of tactical instructions into a single strategic plan. Eisenhower was reportedly aghast at the unconstrained violence of the newly integrated attack plan, but he was unable to change it.[8] Kennedy was also appalled at the rigidity of the new plan, and he requested more flexibility in the next version, SIOP-63, which did introduce the ability for a commander-in-chief to "withhold" some major target areas. Nonetheless, the five "options" in the first decade of the SIOP were three first strike and two second strike spasms.[9]

The SIOP was supposed to enhance flexibility, but as it was designed to operate cybernetically, automatically, it provided minimal opportunity for human intervention. Basically, it updated Wargasm to include more advanced weapons systems and greater firepower. Under the minimum option available, at least 80 million people would die.[10] And the system was set on a hair-trigger, with tremendous potential for premature launch. According to the National Security Archive,

> Policymakers understood the dangers associated with preemptive attacks—the warning of the enemy attack being preempted might be inaccurate and preemptive attack on another nuclear power could not prevent tremendous destruction to the United States."[11]

City blasting

Within military circles, thinking differed greatly on "how many and what kind of targets should be destroyed."[12] The chiefs were well aware that Soviet capabilities were overestimated by generals and politicians who sometimes massaged intelligence data to reinforce their own agendas. But the chief's mandate was to prepare for winning a war, so they strove to maintain the clear superiority of the U.S. in war-fighting forces, even at the risk of nuclear overkill.[13]

A core question was how best to allocate weapons between cities and military installations, (although striking military targets would certainly impact cities). Those who favored striking directly at population centers argued for that strategy as a necessity because it was impossible to find and destroy *enough* Soviet missiles and bombers to prevent their retaliatory launching, and the situation would only worsen when the Soviets mirrored the U.S. and armed their submarine fleet with nuclear warheads. So, it was best to threaten the jugular of society: the cities. Adding weight to this argument was that if the Soviets attacked first, it would be impossible to know which of their forces had been left in reserve—especially in the years before powerful reconnaissance satellites were available—so cities were the only stable targets.

[8]   National Security Archive Electronic Briefing Book No. 130. (2004).
[9]   Burke, A. "Special Edition Flag Officers Dope," 12/4/60; National Security Archive Electronic Briefing Book 173. (2005).
[10] National Security Archive Electronic Briefing Book 173. (2005). 1.
[11] National Security Archive Electronic Briefing Book No. 130. (2004). 6.
[12] Twining, N. F. "J. C. S. 2056/131," 8/20/59. 1149.
[13] Burke, A. "Memorandum to Chairman of JCS from Office of Chief of Naval Operations," 9/30/59.

"Counterforce" proponents, on the other hand, believed that selectively destroying a few hundred military targets would decapitate the Soviet response capabilities:

> If we were to strike certain urban and control centers, the Soviets would be incapable of prosecuting the war, and the United States would emerge on top. Therefore,... it is a waste of money to build a strategic delivery system capable of attacking more than a few hundred targets.[14]

Counterforce was a surprise attack aimed at preventing a Soviet strike. Of course, the United State's offensive warheads could equally be destroyed by a Soviet counterforce strike. So, in a counterforce war, the winner would be the nation that shot first—which undermined the logic of deterrence by assured destruction, as there would be no mutual destruction if one side's forces were decapitated before they could retaliate in a meaningful manner.

Critiquing counterforce, the Navy noted,

> In a surprise attack it is not inevitable that enemy missiles will land before our weapons are launched, but the chances are great that they will. Under these conditions, it would be injudicious to launch the remainder of our greatly depleted forces against a primary target system of empty bases and missile sites, even though we should know their locations, which we probably won't.[15]

Counterforce, no-cities was out.

On paper, if not in cruel reality, the SIOP was designed to operate after a surprise attack due to the sheer number of warheads deployed, especially on submarines. But it was meant to function most effectively as a first strike mechanism, whether preemptive or preventive. And that was about the limit of SIOP-62's flexibility.[16]

Using Everett's algorithm

WSEG's main task for the SIOP project was to write programs that minimized the expenditure of weapon resources while maximizing damage to the enemy. Everett's λ multipliers were undoubtedly used to calculate the trade-offs of deploying specific weapons systems as a function of "kill ratios," or destruction probabilities. Optimizing destruction required predicting the amount of damage caused by various explosive yields.

The "damage function" was also constrained by a "circular area of probability," i.e. the radius around a target in which a bomb was likely to fall if its "delivery vehicle" managed to penetrate enemy air defenses. Expected destruction was a function of the height at which the bomb exploded. Surface bursts delivered more fallout and localized destruction than air bursts, but required more

---

[14] Twining, N. F. "J. C. S. 2056/131," 8/20/59. 1149.
[15] Burke, A. "Memorandum to Chairman of JCS from Office of Chief of Naval Operations," 9/30/59. 1301.
[16] Burke, A. "Special Edition Flag Officers Dope," 12/4/60.

bombs to ensure a high probability of kill over a wide area. The amount of toleration allowed for "collateral damage" to friendly forces or untargeted populations provided another constraint.[17]

Differing values for these variables could be plugged into Everett's multiplier formulae and the "price" for maximizing desired results would pop out as a probability measure, i.e. according to λ, X amount of dollars will buy Y amount of damage on target T with probability P.

Air Force v. Navy

The SIOP was based on a list of targets drawn from the ultra secret National Strategic Target List. Intelligence data for each Desired Ground Zero was fed into a sealed folder to be opened by its keeper only during time of war. The effectiveness of the SIOP was premised on the data for the ground zeros being automatically updated as new intelligence became available, or when political objectives changed, or when budget constraints required tweaking the optimal mix of weapons and targets. But the Joint Chiefs could not easily agree on the who, what, where, when and why of targeting.

The Air Force favored configuring the SIOP to surprise and obliterate the Russians. But as Navy forces were more likely to survive a surprise attack on the United States than were the SAC bombers, the Navy favored configuring the SIOP to employ its aircraft carriers and submarines in a more graduated response.

The heads of these services disagreed, almost violently, on how to set standards for destruction:

> It can be stated that the Army and Navy favor a lower level of destruction while the Air Force favors a higher level of destruction... because of [its] experience that it is almost always cheaper to destroy a target in the initial attack, even if it requires more force, than to have to reattack the same target.[18]

Adjudicating that particular question, said Air Force Major General Nathan F. Twining, the chairman of the joint chiefs, "can be better handled by analytical and mathematical techniques than can other aspects of the problem."[19] Twining was eager to reduce friction caused by inter-service rivalry by tasking WSEG to demonstrate choices using war games.[20] Everett invented a war game for the SIOP, as well as a RSIOP, the Pentagon's mirror of the Russian war plan; and his simulations were used by the joint chiefs to monitor SAC.

Much of what we know about the SIOP comes from memos written by Navy admirals who were incensed that control of the targeting and, consequently, the coming world war, had been handed over to SAC. The Navy argued that SIOP-62 was inflexible because it was a "pre-conceived plan (seldom is such a plan executed)." Consequently, argued the Navy,

---

[17] Twining, N. F. "J. C. S. 2056/131," 8/20/59. 1149–1153.
[18] Ibid. 1153.
[19] Ibid.
[20] Ibid. 1153–1156.

We would forfeit the flexibility that is inherent in the decentralized execution of the strike plans by several unified commanders. The military logic of retaining this flexibility is overwhelming. In preparation for World War II France had a single pre-conceived plan that she thought was foolproof, but it was virtually worthless.[21]

As planning was held hostage to a struggle for control of billions of dollars in weapon procurement funds, WSEG's analytical capabilities were sought by SAC's opponents. Trying to curtail SAC's influence on targeting strategy, the Navy suggested that,

> We should subject the target lists and damage criteria to analysis by machine and mathematical techniques. A major objective of this analysis would be to arrive at an estimate of 'how much is enough.'[22]

Recognizing that excessive fallout could harm its own forces, the Navy suggested that operations researchers determine whether,

> Carefully planned utilization of radioactive fall-out [can] be exploited where practicable to contribute it the destruction of enemy government and military controls and to the general deterioration of the Sino-Soviet bloc.[23]

In other words, if fallout from bombing Russia wafted over to China, which was not at war with the U.S., so be it!, as long as easterly winds didn't shower the U.S. Pacific fleet with radioactive dust.

Mirror, mirror

In August 1960, Admiral Arleigh Burke had a long talk about the SIOP with Secretary of the Navy William Franke. There is a transcript of the super-charged conversation, in which Burke threatened to quit the Navy if SAC was kept in charge of the war plan. He was afraid that his carriers would be assigned "insignificant targets."[24]

Burke noted that the group putting together the SIOP under SAC control had some Navy officers in it, "But it is just like putting a little bug in a piece of plastic. The bug does not control the plastic. The plastic encases the bug. In other words, these people could be absorbed."[25] Burke's concerns provides us with a very important insight into how the military works:

[21] Burke, A. "Memorandum to Chairman of JCS from Office of Chief of Naval Operations," 9/30/59.1304.
[22] Ibid.
[23] Blouin, F. J. "JCS memo SM-679-60," 7/15/60. 7 (spreadsheet).
[24] Burke, A. "Admiral Burke's conversation with Secretary Franke," 8/12/60.10.
[25] Ibid. 4; Burke compared the Air Force command to communists: "It's the same way as the Communists, it's exactly the same technique. As a matter of fact their textbooks, originally about 10 years ago, were built on the textbooks of the Communists, how to control these things. They put one out by RAND, which is a good book to read. I read it…how to deal with Communists, the operations of the Politburo, but it was written in such a way that—the methods of control, how you control organizations—could be put in any organization."

If SAC gets control of this thing, the number of atomic weapons will be tremendous and they will be the wrong kind of atomic weapons. The number of horses [bomber planes] will be tremendous. There will be thousands and thousands of Minutemen. They will control the budget. They will control everything, and they will wreck...everything.... And the Joint Chiefs of Staff and the Secretary of Defense and nobody else can stop it because they [SAC] are the ones that have the figures [target data]...In a year of this stuff—you can never undig it. Grave harm. And the President won't have the guts any more than the past Presidents have had the guts—because these people will be entrenched. The systems will be laid. The grooves will be dug. And the power will be there because the money will be there. The electronic industry and all of those things. We will wreck this country.[26]

In a remarkably forthright condemnation of SIOP-62, Burke wrote a series of top secret memos accusing SAC of ignoring Eisenhower's instruction to prioritize retaliation over preemption. He said SAC was overkilling—assigning many more hydrogen bombs to Moscow, Stalingrad, and Kalingrad than Eisenhower had requested. Burke suggested that the high percentage of surface bursts included in the SIOP could result in "world-wide contamination" from radioactive fallout. He questioned the policy of casually obliterating cities whose only sin was that they lay under air corridors leading to more high value targets. He commented that, "misses will kill a lot of Russians and Chinese even if the specified objective is missed, therefore misses should not be given a zero in the box score of damage achieved." He asserted that the number of targets was "unnecessarily large," and that the "damage criteria are excessively high," and that war gaming independent of SAC should be used to "assess validity of the plan."[27]

He then damned the doctrine of first strike:

Preemptive, preventative or initiative strikes will not prevent serious damage on United States because first, we do not know where their land-based missile sites are and never will know all of them. And, second, we will not be able to destroy all enemy seagoing missile forces simultaneously before they can get off their missiles. In addition, manned aircraft are of no use in a pre-emptive war. If we were to try to coordinate missiles and bombers so that the strikes arrive on targets at the same time such tremendous universal effort by United States would be known by Russia several hours before bombers could arrive and Russia would launch their strikes against us. If we use missiles first, the landing of the first missile would cause all bombers and missiles in Russia to be launched and our bombers arriving hours later would be bombing empty sites and empty airfields. It is surprising that the Air Force is sponsoring a pre-emptive strike...[28]

---

[26] Ibid. 17.
[27] Burke, A. "NAVAL MESSAGE TO CINCPAC," 11/22/60; Burke, A. "U.S. NAVY EYES ONLY," 11/24/60.
[28] Blackburn, P. P Jr. & Burke, A. "EXCLUSIVE U.S. NAVY EYES ONLY," 2.

Burke was worried that the size of the target list "will determine force levels in the future and also will have a great deal of impact on the types of weapons systems which will be procured." He complained that raw intelligence was being "cranked" into the plan without having been evaluated by experts. As the Navy viewed SAC as not properly using its computers, Burke, echoing Twining, called for war gaming by WSEG to offset SAC's technical incompetence. He signed that memo with a handwritten note, "None of the above has anything to do with 'Merry Christmas' which I hope you all have. Arleigh."[29]

In the last days of Eisenhower's administration, his science advisor reiterated many of Burke's criticisms of SIOP-62. George Kistiakowsky concluded that it "follow[s] rather closely the earlier War Plans of SAC," rather than limiting the potential for nuclear holocaust. He observed that better mathematical procedures were needed to optimize targeting, because the planners were prioritizing targets based more upon human judgment, than computer output and, therefore, erring on the side of overkill (and larger budgets). Kistiakowsky noted,

> The [SIOP] staff is making extensive use of computers, but I believe that their programming could be improved and that the most competent people (such as available in WSEG, for instance,) should become [more] involved.[30]

As the head of WSEG's computational efforts, Everett was in a position to generate optimal weapon and target mixes, but it was up to the high command to prioritize targets. In early February, 1961 McNamara was briefed on the new SIOP. He evinced displeasure that the target lists for first and retaliatory strikes were essentially the same—reproducing Wargasm. Military advisors told McNamara that that a more optimal mix of first strike and second strike targets could be generated if *more* weapons were deployed, which would cost more money, of course, but would also increase the number of possible targets and thereby, flexibility in targeting—or so their circular reasoning went.[31]

The Marine Corps spoke up. Commandant David Shoup argued that the SIOP's single list of targets was illogical, not to mention unfair. It called for obliterating the U.S.S.R., Eastern European, and China in one fell swoop, with almost no room for withholding any individual targets. SAC general Thomas Power had told Shoup that the plan was to be executed as a whole, and that leaving any country out would "screw up the plan. Even tiny, Communist Albania, which had disassociated itself from the U.S.S.R., would be wiped out."[32]

[29] Burke, A. "Special Edition Flag Officers Dope," 12/4/60; For report on SAC's computer use deficiencies, see: Blackburn, P.P Jr. "Office Memorandum U.S. Navy Eyes Only," 10/26/60.
[30] Kistiakowsky, G. "Annex" to "J.C.S.2056/208," 1/27/61.1915.
[31] "Memorandum for the Record, Secretary McNamara's visit to JSTPS," 2/4/61: Parker, Adm. "NAVAL MESSAGE EXCLUSIVE FOR ADMIRAL BURKE," 2/6/61.
[32] Shoup, D. "J.C.S. 2056/220," 2/11/61; Kaplan, F. (1983). 270–271.

At Kennedy's insistence, the original SIOP was revised.[33] No doubt, Everett and his colleagues did their best to cost out the risks and benefits of a variety of attack plans, but flexibility was an illusion. In the end, despite the logical foundations of the SIOP, the plan was more about pumping up military budgets than keeping the world safe for democracy—just as Eisenhower had warned.

[33] The revised SIOP-63 included five model strikes, escalating in degree of violence: a *preemptive* strike against Soviet and Chinese nuclear facilities outside urban areas and military and political control centers (ALPHA); a *preemptive* strike against ALPHA and non-nuclear Soviet and Chinese conventional military capability outside urban areas (BRAVO); a *preemptive* strike against ALPHA, BRAVO and Soviet and Chinese nuclear weapon capabilities in urban areas, plus 70 percent of the urban-industrial sector (CHARLIE); and two retaliatory strikes: ALPHA-BRAVO-CHARLIE and ALPHA-BRAVO. SIOP-63 contained limited options to withhold certain targets, but as thousands of ground zeros were pre-selected, the plan was necessarily automated and inflexible. The basic ALPHA, BRAVO, CHARLIE mixes were the only flexible responses, and they were Wargasmic. National Security Archive Electronic Briefing Book No. 130. (2004). National Security Archive Electronic Briefing Book 173. (2005).

BOOK 8
TRANSITIONS

Behind Closed Doors

I feel very strongly that the stage physics has reached at the present day is not the final stage. It is just one stage in the evolution of our picture of nature, and we should expect this process of evolution to continue in the future, as biological evolution continues into the future.... One can be quite sure that there will be better stages simply because of the difficulties that occur in the physics of today.

P.A.M. Dirac, 1963.[1]

Attacked by Bohr's man

After its publication, Everett's relative states theory appeared to be dead on arrival. The paper was only cited twice in scientific literature before 1962 (it did not take off until 1982).[2] However, lack of publicly expressed interest did not mean that it was being ignored. On the contrary, it was causing some of the best minds in physics to debate the measurement problem behind closed doors.

And because Everett's theory was not dead, he had made a powerful enemy, Leon Rosenfeld, Bohr's amanuensis. For many years, he campaigned against the relative states theory on the basis that it contradicted Bohr.

The son of a Belgian electrical engineer, Rosenfeld made significant contributions to quantum theory, mostly in collaboration with Bohr. He was timid and kind-hearted—except when defending his mentor, when he became caustic and back-biting. He frequently disparaged theories with which Bohr disagreed as "theological," as based upon faith, and not upon experimental science.

Wolfgang Pauli affectionately labeled Rosenfeld, "$\sqrt{\text{Trotsky}} \times \text{Bohr} = \text{Rosenfeld}$."[3] The Belgian, a Marxist, considered Bohr's philosophy to be a form of dialectical materialism (although Bohr was unconscious of this deep connection, Rosenfeld asserted). But try as he might, he could not define complementarity:

> It must be realized that it is impossible in principle to write a text-book about dialectics, since this would be to fix a mode of thought which is essentially flowing. It is exactly the same with complementarity (which is the modern form of

---

[1] Dirac, P. A. M. (1963).
[2] ADS-SPIRES Citations History.
[3] Jacobsen, A. S. (2007). 3.

dialectics): you cannot give a 'definition' of it, but only understand what it is by *re-thinking for yourself* the typical cases in which it occurs.[4]

Rosenfeld rejected the Wigner-von Neumann notion that subjective consciousness is the ultimate reality, holding that, "There is an external world independent of what we think and which is the ultimate origin of all our ideas."[5] But personal loyalty to Bohr appears to have been Rosenfeld's criterion for interpretive validity. According to historian Anja Skaar Jacobsen,

> [Rosenfeld took] up the fight against all disbelievers of complementarity, whether Soviet or Western Marxist physicists or just supporters of the causal program with no Marxist agenda. It was a fight in which he used all possible means, including polemical papers, book reviews, and personal connections. In addition, he served as consultant or referee in matters of epistemology of physics and the like at several well-reputed publishing houses and the influential journal *Nature*. In this capacity he used his influence effectively, and several books and papers, among them some by Frenkel, Bohm, and de Broglie, were rejected on this account.[6]

What makes Rosenfeld's mini-jihad against Everett remarkable is that he viewed himself as a champion of new ideas in physics.

> Rosenfeld suggested that... the introduction of a new idea in science will happen only as a result of a veritable fight between the pioneer who made the discovery... and the conservative scientific tradition of the scientific community which he confronts....If the pioneer is not able to convey his ideas...he remains a forerunner, and it is then necessary to wait....[If the] time is not ripe for the assimilation of a discovery it will have to be rediscovered at a later stage.[7]

In the early 1970s, as Everett's interpretation was gaining credence, Rosenfeld attacked not only the theory, but also the theorist. But long before that, in 1959, he took issue with Everett for daring to question Bohr's assertion that the quantum world must be explained with purely classical notions. In a letter to a colleague, Saul M. Bergmann, who had inquired about his opinion on Everett's theory, Rosenfeld observed,

> This work suffers from the fundamental misunderstanding which affects all attempts at 'axiomatizing' any part of physics. The 'axiomatizers' do not realize that every physical theory must necessarily make use of concepts which cannot in principle be further analyzed since they describe the relationship between the physical system which is the object of study and the means of observation by which we study it....It is clear that in the last resort we must here appeal to common experience as a basis for understanding. To try (as Everett does) to include the experimental arrangement into the theoretical formalism is perfectly hopeless, since this can only shift, but never remove, this essential use of unanalyzed

---

[4] Rosenfeld, L. (1963), quoted in Jacobsen, A. S. (2007). 12.
[5] Jacobsen, A. S. (2007). 14.
[6] Ibid. 23.
[7] Jacobsen, A. S. (2008).

concepts which alone makes the theory intelligible and communicable.... The fact, emphasized by Everett, that it is actually possible to set-up a wave function for the experimental apparatus and a Hamiltonian [the energy of a system] for the interaction between system and apparatus is perfectly trivial, but also terribly treacherous; in fact, it did mislead Everett to the conception that it might be possible to describe apparatus + atomic object as a closed system.... This, however, is an illusion.[8]

Rosenfeld did not explain how or why Everett's "perfectly trivial" reasoning was incorrect, excepting that it did not echo Bohr's reliance on classicality to explain quantum phenomena. But not all physicists went along with Rosenfeld. In October 1962, a select group met privately at Xavier University in Cincinnati, Ohio to discuss the measurement problem among themselves. And shortly after the conference started, they asked Everett to make a presentation.

Everett goes to Xavier

In early 1959, Everett discussed his theory with Xavier University physics professor Boris Podolsky in New York City. Podolsky asked Everett for a copy of his long thesis. Everett said he would send it to him after he returned from "argue[ing] about it with Bohr for a month or so."[9]

Two years later, Podolsky invited Everett to speak at the Conference on the Foundations of Quantum Mechanics at Xavier sponsored by the Office of Naval Research and the newly formed National Aeronautics and Space Administration. The meeting was held behind closed doors because, Podolsky explained in his opening remarks, "We want the participants to feel free to express themselves spontaneously... without things getting out in the newspapers."[10] The conference secretary, F. G. Werner, later reported on some of the discussions (barely mentioning Everett) for *Physics Today*. But it was not until 2002 that a transcript of the proceedings was released to the public. It is fascinating document, showing an intimate, sometimes angry colloquy between scientific heavyweights. So desperate had these thinkers become about the paradox of measurement, that they were willing to entertain the idea of multiple universes if it provided a possible solution.

Everett despised public speaking, but he could hardly have dreamed up a smarter, more informed group of physicists to address. It was the first of only two known occasions in which Everett explained his theory to a public gathering of his peers.

---

[8] Rosenfeld to Bergmann, 12/21/59.
[9] Everett to Podolsky, 3/12/59.
[10] Werner, F. G. (1962). Despite Podolsky's precautions, *The Cincinnati Enquirer* got wind of the affair and ran a story headlined "They Tackle Tangled Mess, World of the Atom." It featured pictures of the empanelled physicists and began, "'The world of the atom is a mess,' P. A. M. Dirac said yesterday." CI. 10/2/62. 34.

The stellar panel was composed of Eugene Wigner,[11] P.A.M. Dirac,[12] Yakir Aharonov,[13] Wendell H. Furry,[14] Podolsky, and Nathan Rosen.[15] Fifteen physicists from various universities and national laboratories were allowed to sit in the room and listen to the discussion, speaking only when specifically asked for their thoughts. It is important to remember that no one except Everett, Wheeler, and a few others had read the original thesis; the group's prior understanding of Everett's argument was limited to the edited version published in 1957. This was his chance to expound upon his idea in depth.

The morning of the first day was spent discussing action-at-a-distance, particularly as it related to the contradiction between quantum non-locality and special relativity as posed by EPR.[16] Searching for an explanation of EPR, the participants kept returning to the measurement problem with its inexplicable corollary of wave function collapse, and the topic of Everett's paper arose.

ROSEN: According to Everett, it is not necessary to worry about the problem of the reduction of the wave packet, because all the different possibilities after measurement are on an equal footing. The various possible results of a measurement correspond to a kind of branching so that if you get one result it means that you are just on one of the branches. But since all of the other branches exist on the same footing, one describes all of the possible measurements as one huge tree. Each time after a given result is found, one simply goes along one of the branches and from this branch one continues into further branching by making another measurement, and so on.

PODOLSKY: Oh yes, I remember now what it is about—it's a picture about parallel times, parallel universes, and each time one gets a given result he chooses which one of the universes he belongs to, but the other universes continue to exist.

ROSEN: I just have some recollection of the paper. It's not a question of mathematics, it seems to me, but rather a question of interpretation. The mathematics involved is very simple, you expand a wave function as a linear combination of the [possible states] of the observed quantity.... The usual belief is that when the

---

[11] Everett's spiritually minded teacher, Wigner, we have met—the Hungarian ex-patriot who wished to treat the universe as quantum mechanical, but as a manifestation of human consciousness.
[12] As a founder of quantum mechanics in the 1920s, Dirac had long been a legend. Shy and self-effacing, he seldom spoke. When he did it was usually to the point at hand and with scary precision.
[13] In 1959, Aharonov co-discovered the Aharonov-Bohm effect, which showed that electromagnetism was not a purely local phenomenon.
[14] Furry, a Harvard professor of physics, had worked in operations research during the war. He was a political maverick in this group, having survived an attempt by Senator Joseph McCarthy to get him fired after he took the Fifth Amendment about his Communist Party membership.
[15] Podolsky and Rosen co-authored with Einstein in 1935 one of the most famous papers in the history of physics, "Can Quantum-Mechanical Description of Physical Reality Be Considered Complete?," known as the Einstein, Podolsky, Rosen thought experiment, or EPR paradox.
[16] The conference was held several years before John Bell published his EPR-related theorem showing that quantum mechanics is self-consistent *and* non-local.

measurement is over, one of these terms is singled out and the others are thrown away. That is what is referred to as reduction of the wave packet. The other point of view, that of Everett, is to keep all of the terms.

AHARONOV: There seems to be a problem here. It raises the question is time reversible? If you look on the process of branching you see that it has a definitely preferred direction of time. You never experience any collection of past branching connected together with one observer in the present.

Aharonov was concerned that as, in physics, action is theoretically reversible, Everett would have to explain the apparent irreversibility of his branching model. A phone call was made to the Pentagon. Everett agreed to fly to Cincinnati for the next day's session.[17]

That evening, Wigner presented a talk, "The Concept of Observation in Quantum Mechanics." He distanced himself from Bohr, saying that it is not possible nor desirable for physics to reduce the study of quantum nature to measuring collisions between particles:

WIGNER: Fundamentally it is not enough because the world is constantly in a collision with us, and there is a constant interaction between matter. Unless we make it the purpose of physics to describe only certain carefully made experiments, but not more than that, we can't get along entirely with just the collision matrix. It is not true that everything is only a collision. The world continues. For instance, a gas constantly exerts a pressure on the wall. There are many similar examples which show that it is not really possible to reduce everything to a collision.[18]

Wigner then distanced himself from Everett, too, declaring that unless physics gives up the superposition principle, "The fact is quantum mechanics does not permit objective reality."[19] Everett was persuaded otherwise: his branching universes emerged from cascades of collisions between superposed objects in an objectively real *quantum* environment.

The next morning, with Everett in the room, Rosen began the discussion, "I would like to think that the world has an objective reality independent of whether there are people present to observe it or not." He acknowledged Everett's presence and characterized his view of the measurement problem.[20]

ROSEN: He does not accept the reduction of the wave packet.... He does not want to distinguish between the actual result as obtained in a given case and the other possible results which might have been

[17]   Werner, F. G. (1962). MON-AM session. 12–14.
[18]   Werner, F. G. (1962). Wigner paper. 12.
[19]   Ibid. 1–3.
[20]   All quotations in this section are from Werner, F. G. (1962). TUES AM.

obtained, so that even after the measurement he has the series of terms, instead of one term. He thinks of the wave function as changing only in accordance with the Schrödinger equation, in a continuous way, without the possibility of this sudden change in the wave function, which we call the reduction of the wave packet. My own feeling is that such a point of view is tenable and consistent, but should be interpreted as referring not to what one observer finds but what many observers carrying out the same sort of measurement on the same sort of system would find.

Rosen was attracted to keeping intact the logical consistency of the Schrödinger equation, but he was not ready to accept branching universes.

EVERETT: I think you said it essentially correctly. My position is simply that I think you can make a tenable theory out of allowing the superpositions to continue forever, *even for a single observer*.

From the floor, Abner Shimony said that Everett's theory suggested two possibilities regarding consciousness. One was that "Ordinary human awareness is associated with one of these branches and not with the others."[21] The other possibility, said Shimony, was that a separate awareness is associated with each branch.

Rosen asked Everett to briefly describe his theory.

EVERETT: Well, the picture that I have is something like this: Imagine an observer making a sequence of results of observations on a number of, let's say, originally identical object systems. At the end of this sequence there is a large superposition of states, each element of which contains the observer as having recorded a particular definite sequence of results of observation. I identify a single element as what we think of as an experience, but still hold that it is tenable to assert that all of the elements simultaneously coexist.

In any single element of the final superposition after all these measurements, you have a state which describes the observer as having observed a quite definite and apparently random sequence of events. Of course, it's a different sequence of events in each element of the superposition. In fact, if one takes a very large series of experiments, in a certain sense one can assert that for almost all of the elements of the final superposition the frequencies of the results of measurements will be in accord with what one predicts from the ordinary picture of quantum mechanics. That is very briefly it.

PODOLSKY: Somehow or other we have here the parallel times or parallel worlds that science fiction likes to talk about so much.

---

[21] This statement is akin to the so-called mindless hulk theory of Everett's many worlds suggested decades later by philosopher David Albert, until he retracted it.

EVERETT: Yes, it's a consequence of the superposition principle that each separate element of the superposition will obey the same laws independent of the presence or absence of one another. Hence, why insist on having a certain selection of one of the elements as being real and all of the others somehow mysteriously vanishing?

FURRY: This means that each of us, you see, exists on a great many sheets or versions and it's only on this one right here that you have any particular remembrance of the past. In some other ones we perhaps didn't come to Cincinnati.

EVERETT: We simply do away with the reduction of the wave packet.

PODOLSKY: It's certainly consistent as far as we have heard it.

EVERETT: All of the consistency of ordinary physics is preserved by the correlation structure of this state.

PODOLSKY: It looks like we would have a non-denumerable infinity of worlds.

EVERETT: Yes.

PODOLSKY: Each proceeding with its own set of choices that have been made.

FURRY: To me, the hard thing about it is that one must picture the world, oneself, and everybody else as consisting not in just a countable number of copies but somehow or another in an undenumerable number of copies, and at this my imagination balks. I can think of various alternative Furrys doing different things, but I cannot think of a non-denumerable number of alternative Furrys. (Podolsky chuckles.)

EVERETT: I'd like to make one final remark here.

What follows uses some technical terminology, but is mostly couched in ordinary language. It is worth quoting in its entirety as Everett never published a word on quantum mechanics after July 1957, and his Xavier comments shed considerable light on how he contrived his probability measure, and his view of the physical reality of the "splitting" worlds.

Imagine a very large series of experiments made by an observer. With each observation, the state of the observer splits into a number of states, one for each possible outcome, and correlated to the outcome. Thus the state of the observer is a constantly branching tree, each element of which describes a particular history of observations.

Now, I would like to assert that, for a 'typical' branch, the frequency of results will be precisely what is predicted by ordinary quantum mechanics [i.e. the Born rule]. Even more strongly, I would like to assert that, as the number of observations goes to infinity, almost all branches will contain frequencies of results in accord with ordinary quantum theory predictions. To be able to make a statement like this requires that there be some sort of a measure on the superposition of states. What I need, therefore, is a measure that I can put on a sum of orthogonal [i.e. separate, non-interfering] states. There is one consistency criteria which would be required for such a thing. Since my states are constantly branching, I must insist that the measure on a state originally [the probability distribution for an electron to be found at position X or Y or Z...] is equal to the sum of the measures on the separate branches after a branching process, [i.e. probability is conserved].[22]

Now this consistency criterion can be shown to lead directly to the squared amplitude of the coefficient [Born probability], as the unique measure which satisfies this. With this unique measure, deduced only from a consistency condition, I then can assert: indeed, for almost all (in the measure theoretic sense) elements of a very large superposition, the predictions of ordinary quantum mechanics hold.

Now I could draw a parallel here to statistical mechanics where the same sort of thing takes place. Here we like to make statements for almost all trajectories. They are ergodic [i.e. the influence of initial conditions is preserved over long runs of time] and things like that. Here also you can only make such a statement if you have some underlying measure that you regard as fundamental, since any such statements would be false if I take a measure that had only non-zero measure on the exceptional trajectories [i.e. my probability measure must apply to all physically possible events on the average].

In both the original and edited versions of his thesis, Everett had made the mathematical argument that it is not necessary to postulate the Born rule in quantum mechanics, because there is a statistical equivalent to it in classical mechanics that emerges naturally as a measure of probability for observers confined to a single branch (our branch). But, for mathematical consistency, he needed to show how this measure conserves the seeds of probability in a universe in which everything that is physically possible happens. Stepping away from the quantum mechanical Born rule, he analogized a classical probability measure placed across sets of separate, non-collapsing branches *after* an interaction—as a measure of the total weight of the multiple results flowing from that interaction (adding up to 1, or 100 percent).

He continued,

In statistical mechanics it turns out there is uniquely one measure of the phase space [coordinate frame in which entire systems are represented by points]... being essentially the only measure giving the conservation of probability [i.e. the probabilities for the various positions of an electron at any given moment in all branches add up to 1, or 100 percent]. It is precisely this analogue that I use on the branching of the state function and I can therefore assert that the probabilistic interpretation of quantum mechanics can be deduced quite as rigorously from pure wave mechanics as the deductions of statistical mechanics.

[22] A property of the Schrödinger equation is the conservation of probability (called unitary evolution).

Although Everett's explanation of his derivation of a measure that was equivalent to the Born Rule—thereby explaining why probability appears in "our" branch without the collapse postulate—was not totally convincing to panel members, it held enough promise that they did not outright scoff at his argument that we subjectively experience probability in a universe of universes in which everything happens—i.e. all elements of a superposition are equally "real," even if they are not equally probable.

When Everett was finished, Shimony said, "You eliminate one of the two alternatives I had in mind. You *do* associate awareness with *each* of these [branches]."

EVERETT: Each individual branch looks like a perfectly respectable world where definite things have happened.

SHIMONY: What, from the standpoint of any of these branches, is the difference within a branch, between your picture of the world and one in which there are stochastic [random, probabilistic] elements?

EVERETT: None whatever. The whole point of this viewpoint is that a deduction from it is that the standard interpretation will hold for all observers. In addition, however, one can, within this viewpoint, get some hold on *approximate* measures and this type of thing.

And with that, Everett sat down.

Rosen reflected that in the light of these interpretive problems, quantum mechanics was incomplete and physicists should either "reformulate" it, or "look for a different theory." Clearly referring to Bohr and his inner circle, he said,

> I am sometimes a little annoyed at the attitude of some quantum mechanicians because of a certain dogmatism that they display in these discussions. There is an old saying that the revolutionary of yesterday is the conservative of today. Some people even refuse to consider the possibility that there can be any other valid point of view than that which corresponds to the orthodox interpretation. Of course, nobody here in this discussion is considered to be guilty.

The rest of the morning was spent discussing (and largely dismissing) Bohm's theory of hidden variables, which does not allow for superpositions and eliminates the measurement problem.

Parallel interpretations

That afternoon, questions from the audience were entertained. One question pinpointed the contradiction at the heart of the Copenhagen interpretation: "Is it justified to make a theory ignoring at the outset questions of the measuring process, and then expect to obtain, by means of that theory, a description of the measurement process?"[23] Restating the problem: because the Copenhagen interpretation posits a moveable boundary between the quantum and classical worlds, and because there is no definite place in the chain of interactions and correlations for a definite measurement (the "cut") to occur, quantum mechanics cannot say exactly what a measurement is, nor why one result rather than another emerges, nor exactly when a wave function collapses. And, yet, it is incumbent upon any coherent interpretation of quantum mechanics to explain why measurements work (they have predictive power).

Wigner opined that the collapse, "takes place only through the act of cognition." Rosen disagreed, saying that a machine could carry out a measurement. Wigner tried, but could not explain why only human consciousness could collapse a wave function and a machine could not. Furry, Rosen, and Podolsky mocked Wigner (politely) for his failure to explain.

Furry said he thought there was something to be said for considering entangled, interfering quantum systems as "coherent," and that they lose that coherence at the macroscopic level of interaction, becoming "incoherent."[24] He was struggling to find a physical process underlying wave reduction that was not populated by hidden variables or multiple universes. Aharonov remarked that in a closed quantum system [such as Everett's] there is no such thing as collapse and that all possibilities would exist at the same time. It was more or less agreed that without the necessity of explaining *how* observers observe, quantum mechanics would be consistent.

A member of the audience said that everyone seemed to agree that quantum mechanical systems are multi-valued [superposed], but "where we disagree is, *if* we can select out one of these values, and *when* the selection is made." Wigner commented that Everett's position was that no selection is made. When asked if he would like to comment, Everett said, "Yes. Well, what he said pretty much covers it." If he had anything more to add, the page that contained his further response is unfortunately missing from the transcript and cannot be recovered.

At one point an audience member asked Podolsky, "What is reality?" He answered, "Something more than just subjective information."

Furry then loudly and furiously attacked Wigner for his subjectivity:

> I'm really too old to believe in the branching that Mr. Everett believes in—in the parallel universes of Mr. Everett and things like that. But for instance, if I were to take cosmic rays that come down right through the air of this room rather frequently—they are leaving trails of ionized molecules. The fact that we haven't set up the right conditions of super-saturated vapor to render them visible doesn't mean they aren't really there. But according to the point of view that puts all the emphasis on cognition, they aren't even in the cloud chamber

---

[23]   Quotes in this section are from Werner, F. G. (1962). TUES PM.
[24]   Shades of decoherence theory!

unless you take a picture! (Furry shouts) And they are not even in the cloud chamber or in the picture then unless you look at it! (Furry shouts until Wigner finally speaks again.)

WIGNER: It is done. It is surely agreed that it is done. We will surely admit that it is done.

FURRY: I can't go that far, somehow.

The conference adjourned for the day.

A big bang

On the third day, Dirac spoke. Famously taciturn, he had refrained from entering the philosophical debates. He made a highly technical presentation on relativistic quantum mechanics, and the tone of the conference cooled.

On the fourth day, Shimony brought the topic back to philosophy. Returning to the measurement problem, he suggested, as had Everett in his thesis, that solipsism would be a solution, albeit an "unhappy" one.[25]

Puzzled, Dirac asked, "What are solipsists?"

Shimony kindly explained, and the panel members joked about how solipsists could kill each other and it would not matter since each was a product of the other's imagination. Shimony concluded that if the quantum formalism was to be kept intact, "there are many, many blind alleys; and I, for one, do not see a way out."

The conversation returned to Everett's theory. Shimony pointed out that an observer in one of Everett's branches would experience one reality—even though "ultimately the universe has one state, and its propagation is governed by the Schrödinger equation."

SHIMONY: [Everett's] claim is that the theory he is proposing is more logical. Well, I don't know what this means. I think that if you have two statistical theories each equally consistent, you can't claim one is more logical than the other. It seems to me that in some sense these are equivalent ways of talking about the same thing.

Aharonov said he didn't see any inconsistencies in Everett's theory; and Shimony could not pinpoint any. Shimony fell back on a stance that is often used by Everett's critics: "I think one should evoke Occam's razor: Occam said that entities ought not to be multiplied beyond necessity. And my feeling is that among the entities which aren't to be multiplied unnecessarily are histories of the universe. One history is quite enough."

The discussion circled back to the central questions of the conference—whether the mathematics of quantum mechanics suffices to describe the real world, is the world observer-independent, or is it a figment of the imagination of a solipsist? Nothing was resolved.

[25] Werner, F. G. (1962). FRI AM.

A few minutes before the conference ended, the recorder noted: "(There is an extremely loud explosion outside; Bang!!! followed by fifteen seconds of silence.)"

AHARONOV: Are we all agreeing that there was something, an explosion here, or (laughter)...Is everybody here on this same branch (referring to Everett's theory)?

The problem of probability

The transcript does not reveal Everett's coordinates at the moment of the explosion, but 11 years later he talked about the Xavier conference in the letter to Max Jammer:

> The unwillingness of most physicists to accept this theory, I believe, is [due] to the psychological distaste which the theory engenders... Thus the theory was not so much criticized, as far as I am aware, but simply dismissed.

> Subsequent to the publication of the paper, I had informal discussions with a number of physicists concerning the subject (including Bohr and Rosenfeld in Copenhagen, in 1959, Podolski [sic] and Wigner and a number of others active in the field at a conference at Xavier University several years later). I was somewhat surprised, and a little amused, that none of these physicists had grasped one of what I considered to be the major accomplishments of the theory–the 'rigorous' deduction of the probability interpretation of quantum mechanics from wave mechanics alone. This deduction is just as 'rigorous' as any of the deductions of classical statistical mechanics, since in both areas the deductions can be shown to depend upon an 'a priori' choice of measure on the space. In classical statistical mechanics this measure is standard Lebesque measure on the phase space whereas in quantum mechanics this measure is the [Born Rule].

> What is unique about the choice of measure and why it is forced upon one is that in both cases it is the only measure that satisfies a law of conservation of probability through the equations of motion. Thus, logically in both classical statistical mechanics and in quantum mechanics, the only possible statistical statements depend upon the existence of a unique measure which obeys this conservation principle.[26]

There are physicists and philosophers, today, who think that Everett was on or very close to the mark with his derivation of probability. Physicist Max Tegmark agrees with Everett's derivation as presented in the long thesis.[27] And philosopher Simon Saunders says

> If you assume it is legitimate to talk about probability, given you have all these branches, and the question is 'what is the right probability measure over branches,' then Everett's derivation of the Born rule is pretty good, in my opinion. Maybe

---

[26] Everett to Jammer, 9/19/73.
[27] Tegmark, private communication, 8/18/08.

he did nail it. Probability is no worse off in Everett than it is in any other physical theory (and its fine just to inductively infer the Born Rule).[28]

But physicist Dieter Zeh, whose work on quantum decoherence has been inspired by Everett, points out that physical facts cannot be derived from mathematics alone:

> Pointing out the existence of a mathematical structure of a 'measure' is certainly far from giving a proof of its physical meaning. Apparently, Everett hoped to find a justification for the probability interpretation (and perhaps the Born rule) in the mathematics. This would be a typical mistake of a young scientist. The formal rules for probabilities were derived or postulated to describe classical ensembles. In particular, the correlations he is discussing in this connection are mostly those characterizing classical ensembles (incomplete information). The main problem of measurement is precisely the transition from entangled superpositions to (apparent) ensembles. So it would not suffice to point out some formal similarities.

Zeh views Everett's argument as *justifying* his probability measure, but not as having *derived* it: "However, one must also expect that he was in some state of confusion, which is normal when you start thinking about a completely novel idea that has strange consequences. Just compare it with the inconsistent claims of the founding fathers of quantum theory!"[29]

---

[28] Saunders, private communication, 8/21/08.
[29] Zeh, private communication, 8/21/08.

28 Death's Other Kingdoms

These are those who have passed through our modern materialistic, ultra-rational dream world, with their eyes directed upon death's other-world kingdom. These are those who have retained faith in the face of our modernism and have the ability to see reality, through our illusory world. To them we, who outwardly appear satisfied of our own self-sufficiency and ostentatiously are in need of no God, appear as hollow men, as stuffed men.

Herbert O. Horn[1]

The cold spot

While Everett was in Cincinnati, his mother lay dying. Katharine chronicled her battle with cancer in a letter to one of her nurses, written days before she died.[2]

Her doctor had discovered a lump in her left breast shortly before she vacationed in Europe with Hugh and Nancy in 1959. But she could not feel the lump herself, so, regrettably, she did nothing about it. Eighteen months later, her left arm swelled up, and the nipple of her left breast ulcerated. Six months *later*, she went for a physical. The doctor said she had a tumor.

She checked into George Washington University hospital for a biopsy, or so she thought.

> However, the following morning I was operated on and woke up after 5 hours in the operating room to find out that they had performed a radical left mastectomy with skin graft.

She was dizzy and in constant pain and had trouble breathing; nonetheless, she returned to work doing research and managing an office for an anthropologist, Edward T. Hall, who had a government grant to study the effects of crowding on the sensory perceptions of humans.[3]

During the next six months, she was

---

[1] Horn, H. O. (1951). In 1951, Everett's WSEG colleague, Herbert O. Horn, wrote a paper on T. S. Eliot's poem "The Hollow Men" for an English Lit class.
[2] Everett, Katharine Kennedy to Nurse Gantley. 11/22/62.
[3] Hall's classic study of the perceptual relationships between private and public space, *The Hidden Dimension*, appeared in 1966.

extremely fatigued all the time and though I could type and use my left hand normally from the elbow down…I could never lift it above shoulder level and when I tried to follow the [therapeutic] exercises suggested, the donor site of the skin graft [on her shoulder] would break open and cause me excruciating pain.

She asked her doctor if the tumor had metastasized. He said, "No, Ma'am."

I became disgusted with the whole process and began to study Christian Science. I secured the services of a practitioner, but she could not seem to help me in alleviating the pain in the shoulder, the fatigue, shortness of breath, and feeling that the skin graft was pressing on my lungs and heart. Also, new lesions appeared that bled.

Abandoning Christian Science, she returned to the hospital and saw a different doctor. She asked him if her mastectomy had been competently performed. He said, "yes," and prescribed radiation.[4] After a long course of X-ray treatments, "the arm began to swell enormously and I began to feel nauseated." She lost her $375 a month job; her savings "dwindled and melted." Social Security paid $100 a month, and she received free house-cleaning and cooking services. But her rent was $97, groceries were $30; she scrounged for cab fares to get to medical treatments.

In late September, she was back in the hospital.

I broke the lease at the apt.…and Nancy came and supervised the moving of my furniture to her home.…She agreed to help out financially and see me through the crisis.

But after Everett returned from the Xavier conference, he, Nancy, and Liz went to Kitty Hawk for a vacation. A friend drove Katharine to the Kensington Gardens Sanitarium in nearby Maryland on October 12, where she collapsed in despair.

I often cried and sometimes became hysterical and could not eat.…I lay for three nights in a torn, blood-stained slip. I had little or no nursing care. Then Barbara Smith, a night nurse came, took off the filthy slip, secured a pink-flowered gown and bathed me, washed my back & cleaned me up. I felt like human being again – not a sick animal.

---

[4] In *Under the Radar, Cancer and the Cold War* (2009), Ellen Leopold documents how during the late 1950s and early 1960s cancer treatment centers experimented on terminally ill women (without their consent) by unnecessarily administering post-mastectomy radiation treatments. Data on loss of motor skills and survival rates from varying degrees of radiation exposure was gathered for military purposes. It was applied to "radiological warfare" planning, including trying to determine if pilots of nuclear-powered airplanes could be shielded from radiation (they could not and the project was dropped). According to public records at the Department of Energy, WSEG, as a recipient of data on radiation experiments done on animals and human "volunteers", was deeply involved in analyzing "the effects of ionizing radiation on population groups." WSEG was a key agency in the R&D on the nuclear power aircraft. And in 1957, one of WSEG's tasks was look at radiation data in the light of the possible "racial deterioration that may follow increases in the load of deleterious mutation" in the wake of a nuclear war. "Thus, while military objectives are being achieved, a biological condition conceivably could be produced which would result in precipitous racial degradation and possible species disaster." Dunham, C. L. to Henshaw, P. S. (1957). Katharine Everett's mistreatment mirrored that of many, many women who were cruelly radiated after involuntary mastectomies, including Rachel Carson. Whether or not Everett's mother's treatment ended up in WSEG's data set is unknown. See: Leopold, E. (1999).

"Things gradually fell into a pattern here. Actually, I like it tremendously and now get excellent care from the staff," she told Nurse Gantley. She asked for some kind of narcotic that would not make her vomit, perhaps Excedrin, or Nytol. She asked for an oxygen tent.

> Sometimes at nights I cannot breathe and am aware of an extremely cold spot, about the size of a silver dollar on the heart (lungs?). Covering my chest with a blanket does not take away the cold spot.

New lumps appeared on her right breast and neck. Her mouth was "frozen" with abscessed teeth. She lost the ability to speak, but she tried to be optimistic, communicating by writing on yellow legal pads.

> I like it here at Kensington Gardens so much and do feel I am getting better in spite of the nausea and inability to take food.

Here the account abruptly ends.

Katharine lingered at the sanitarium for ten weeks. Nancy and Elizabeth visited a few times, bringing jars of baby food and air freshener and stationery. There is no record of Everett, who had an aversion to hospitals, visiting his mother. But his paternal grandmother (Ma Maw), who was also in a nursing home, dropped Katharine a card:

> Nancy tells me that you like where you are.... I think contentment is the greatest blessing we can have.... Hope you see Elizabeth often, she is such a darling, talks to me over the phone and will sing to me a song sometimes when she feels like it.

Katharine was also visited by literary and political friends. One of her visitors was Dame Adelaide Lord Livingstone, a world-renowned peace activist. Of her dying friend, Livingstone told Nancy that, "She knew beauty, but she didn't know reality."

As she declined, Katharine wrote letters to old friends from her college days. She told them about her temporary indisposition (she planned to be cured by Christmas), and how swell her family life was, and how proud she was of her son.

> [He] has been married for some time to a sweet New England girl...and she is dear as a daughter to me. I also have a darling grandchild, Elizabeth (aged 5) and another child expected next Spring!

> You know my son did very well.... He got his PhD from Princeton before he was 30 years old and has a fine job now!... at the Pentagon, where he is head of the mathematical techniques section of WSEG (IDA) and has a 'cosmic' security clearance. He does computer-oriented research and travels all over.... He is one of the new 'space' planners.... My life is full with my son's family and my work!

One friend wrote back:

> Glad you are recuperating rapidly.... It is thrilling to think that that small mite of humanity of 31 years ago (HE III) is now in the rocket-computing business.... Yes, Ken, you should be very proud in having made such a contribution to civilization.

Everett and child Liz, circa 1962.

Days before she died, Katharine received a report she had requested, "The Ways and Power of Love," from the Harvard University Research Center in Creative Altruism. The organization promoted the concept that only love and non-religious altruism will change the world because,

> Tomorrow the whole world could become democratic, and yet wars and bloody strife would not be eliminated because democracies happen to be no less belligerent and strife-infected than autocracies.

Among Katharine's effects was a flyer, "Women Strike for Peace," calling for an end to nuclear testing due to the discovery of Strontium-90 in milk. Also on her bedside table was a flyer calling for the abolition of the House Un-American Activities Committee. And a letter from a nursing aide:

> Mrs. Everett you have been great to me, teaching me how to write a book in which I am still working on.... I only wish I had a mom like you.

In a one-paragraph last will and testament, Katherine left the only possession she valued, her literary estate, to her son. She bequeathed him any royalties from publication, and asked that he and his heirs preserve her papers "for a period of at least one hundred years."

And she wrote a final poem, "The Heart's Landscape":

> Who will chart the road to the heart's secret fortress,
> Through tangled forests, weary lands...
> The Mother waits: the lost Child stumbles home.

By Christmas morning, Katharine, 59, was dead.

Travels with Q

Everett was a constant traveler. One week, he was in Austin, Texas at a computer conference. Then it was off to RAND in Santa Monica to think the unthinkable. He appears to have been a regular visitor to a lab in Princeton that served the National Security Agency.[5] At Sandia National Laboratory in New Mexico, he worked on artificial intelligence, helping to create a path-breaking "associative memory" system that mimicked the operation of human memory.[6]

One morning, "Mr. & Mrs. Q. Everett" flew to Hartford, Connecticut; "Q" was Everett's security clearance level, so traveling under that name was not very secure, but flaunting it must have amused him in the same way that taking a photo of a White House courier would tickle him a decade later.

In June 1963, he attended another closed door science conference, "The Nature of Time," at Cornell University, sponsored by the Air Force Office of Scientific Research. It was attended by high caliber physicists: Wheeler, Mr. X (aka Richard Feynman), Fred Hoyle, Hermann Bondi, Misner, Rosenfeld and others. But it was an odd, unproductive event.

---

[5] Travel vouchers, basement.
[6] Simmons, G. J. (1963); Gary Lucas private communication, 10/2/08.

The main topic of discussion was the irreversibility of time. Rosenfeld, applying Bohr's complementarity model, propounded that time's arrow was imposed on the universe by observers and "the conditions of observation." Others said the arrow existed independently of observers. One reviewer of the conference papers commented,

> Indeed it is disappointing to find that the discussion of philosophical issues by a group of outstanding physicists is in such a low level.[7]

Conference organizer, T. Gold, reported, sadly,

> It is an embarrassment for a scientist who has concerned himself with the basic nature of the physical laws to have to admit that the coordinate system in which the laws are embedded is mysterious.... Introspective understanding of the flow of time is basic to all our physics, and yet it is not clear how this idea of time is derived or what status it ought to have in the description of the physical world.

> Most young persons beginning work in a field believe it to be fairly systematic and well understood, and as they learn more are disappointed at the muddled thinking, the ignorance, and the uncertainty among experts. This disillusionment is an essential part of the learning process. It is this that usually gives the student the courage to enter the fray himself.[8]

Wheeler delivered a paper suggesting that our perception of the direction of time is a function of the fact that human observers are subject to the laws of thermodynamics, i.e. entropy governs our perception of time. And then he made a remark that must have pleased Everett:

> The universe is not a system that we can observe from outside; the observer is a part of what he observes. Observation under these conditions presents new features; Everett's so-called 'relative state formulation' of quantum mechanics does provide one self-consistent way of describing such situations. Although we are very far from having seen our way through these problems, there is a well-defined formalism in which to carry out the analysis.[9]

Mr. X ruminated,

> I have never found myself able to use the concept of probability except in an inconsistent, indefensible way.[10]

And,

> It's a very interesting thing in physics that the laws tell us about permissible universes, whereas we only have one universe to describe.[11]

One can imagine Everett grinning, superciliously, from the safety of the back row.

---

[7] Lacey, H. M. (1969). 88–89.
[8] Gold, T. (1967). vii–ix.
[9] Wheeler, J. A. (1962). 106.
[10] Gold, T. (1967). 109.
[11] Gleick, J. (1992). 126.

Winding down WSEG

Young Everett had the fate of the world in his hands; he was not inclined to waste his time on scholarly frays. He was, however, looking for new opportunities. WSEG was caught up in an internal power struggle at the Pentagon, and Everett's job was becoming increasingly unpleasant and precarious.

After the disastrous Bay of Pigs invasion in 1961, Richard M. Bissell, the CIA planning director in charge of the ill-conceived plot, resigned and was appointed president of IDA. The non-profit corporation was growing into one of the country's largest, most prolific, most secretive military think tanks. Bissell wanted to use scientists like Everett on non-military projects to generate more income. Concerned about the security of IDA's private sector work, Pentagon officials suggested that WSEG's access to Q level information should be restricted. And Congress started asking hard questions about whether or not WSEG scientists were worth the premium salaries they earned.[12] Concerned, the Joint Chiefs slashed WSEG's budget.

In the spring of 1964, Everett and six scientists[13] resigned en masse from WSEG to form Lambda Corp. Everett was a principal owner and the chairman of the board of directors. The company started up with a portfolio of lucrative contracts to do much of the same work that they had been doing at WSEG. Pugh, who was by then working at the Arms Control and Disarmament Agency, came on board bearing an ACDA contract. Besides performing SIOP testing, ABM modeling, and civil defense planning, Everett was determined to turn other projects he had worked on at WSEG—including a prototype of a word processing program—to commercial advantage. He was soon earning a salary of $35,000—$213,000 in 2008 dollars.

Having fun

Like many suburban couples during the Cold War, the Everetts were immersed in a culture of consumerism through which flowed a current of justifiable paranoia. If the world was going to end in a radioactive bang, better have some "fun," reasoned many middle-class Americans. Or so observed Erich Fromm, who explained the psychology of consumerism in *The Sane Society*:

> Leisure time consumption is determined by industry, as are the commodities [the consumer] buys; his taste is manipulated, he wants to see and to hear what he is conditioned to want to see and to hear; entertainment is an industry like any other, the customer is made to buy fun as he is made to buy dresses and shoes. The value of the fun is determined by its success on the market, not by anything which could be measured in human terms.[14] ... So people do worry, feel

---

[12] U.S. Congress, Office of Technology Assessment. (1995). 2.
[13] Larry Dean, Bob Galiano, Paul Fitzpatrick, Joanna Frawley, Neil Killalea, Betty Jo Ellis; Lambda information from basement files.
[14] Fromm, E. (1955). 136, 166.

inferior, inadequate, guilty. They sense that they live without living, that life runs through their hands like sand.[15]

During their first years of marriage, the Everetts enjoyed a vibrant sex life. Their bedside library bristled with pornography and sex manuals. Birth control pills took the stress out of fornicating at will. But the luster wore thin when Everett started what Nancy called his "side things." She decided to cast aside the pills and get pregnant again.

> I was determined to have a 2nd child to validate our marriage and love etc etc – I was very happy with 2nd pregnancy – and felt great physically…Hugh was not particularly interested in having a 'family.' But maybe it grew on him, like the pets.[16]

And, so, Mark Oliver Everett was born on April 10, 1963, at the tail end of the Baby Boom.

Baby Mark Everett, 1963.

[15] Ibid. 166.
[16] NGE dairies. Nancy had a miscarriage between Liz and Mark, so this was, technically, her third pregnancy.

The WSEG-Lambda group was tightly knit professionally and socially—they worked and played and flirted and consumed together. But not long after Mark's birth, the Everetts stunned their fun-loving circle of friends by declaring an open-marriage—they were free to have fun affairs—with their friends' spouses.

One gets the impression from Nancy's dairies and letters that she participated in the swinging lifestyle to keep intact a modicum of self-respect in the wake of her husband's compulsive philandering.

> I was playing [a] traditional, helpless role. You put up with whatever nonsense and go on to greater goal. I truly believed we were growing and getting on with mutual understanding....I had not taken seriously husband's side things. They were always side things. No real threat to marriage to me and they weren't.

Communication was not the hallmark of their marriage, yet they were loyal to each other. Nancy observed,

> My life is too defined by parameters of those I touch. Too much, I lose my identity in others....My husband used to say he couldn't tell how I felt. I didn't speak up. What did he want from me? For me to be less childish. My reactions were childish he said. I ran away blah. [But] Hugh was there to take care of me when other [situations] fell apart.

There was anger and regret in her motivations, however:

> Like Ann Dean used to say—if Hugh wasn't my husband, I wouldn't have 'played around' with others spouses as I did in retaliation to H. Just because we were having an 'open marriage' didn't mean the rest of the social set appreciated this. But they all excused my behavior because they knew I had been hurt by H.

But the first time she cast aside her natural shyness, kissing another man in front of her husband, she was surprised that he got upset.

The kiss

It was the summer of 1964. The Cuban missile crisis had come and gone, terrifying Everett and his colleagues with the near miss of nuclear war. Then, Kennedy had been assassinated. Goldwater (whom Everett disliked) was running for president on a hard-line platform against Lyndon Johnson (who was himself no slouch as a militarist). The Greeks and the Turks were fighting a war over the island of Cyprus that could spread to the superpowers. The constantly escalating war on Vietnam was more of a disaster every day. The impoverished inner cities were about to explode in riotous protest.

So, the operations researchers had a party.

Nancy had a few drinks and started "smooching" with "L," one of Everett's business partners. On the drive home, Nancy confessed that she had fallen in love with "L," although they had not yet had sex. Everett exploded.

The next day, Nancy flew with Mark and Liz to Amherst to visit her parents. She and her angry husband wrote letters back and forth—only hers have survived. Her letters provide an insight into their marriage, and probably the marriages of millions of suburban couples searching for sexual identity in an increasingly rootless society. As the Fifties turned into the Sixties, affluent, literate consumers who read Henry Miller, Anais Nin, Lawrence Durrell, and Norman Mailer, people like the Everetts, openly questioned the value of monogamy.

Nancy wrote,

> There are so many ways of looking at the demise, or temporary failure of our joint prospect. I still feel a shred of justification on my part, in spite of also feeling like a damn fool...for 'having a crush'...I simply said I had a meaningful experience, tho if there is some one specific point you can say was the experience, I don't know what it was....I was in love with myself, I suppose. I had been granted a gift.

Nancy continued her liberation manifesto,

> This whole experience was not oriented to any one personal relationship. I knew that all the things that generally hold one back DON'T MATTER, that one just goes on being oneself totally (wholly or whorely as the case may be). It doesn't matter if your 34 and look like dishwater—you just go ahead and do what you feel....I felt grateful to L. for having such a grand effect on me – party-drink-smooching-wise, but it stopped there.

Nancy *is* considering going all the way with "L," but only if his wife, one of her best friends, consents.

> One culmination of such a situation could be to give oneself personally. On one level it seems harmless and also beneficial to all involved. On another level it seems entirely outside confines of convention and full of danger! (possibility of changing status quo). The fact that everyone just seemed to have been expecting this to occur months ago, when we finally did lay down our hands, makes me now wonder whether I have been untruthful to myself about my personal feelings. Are they just altruistic?

She ended the letter reminding Everett that the diaper service came on Monday and that he should clean the cat litter box. The next day, Nancy wrote to her husband,

> When I awoke this am it all seemed quite clear to me that we both had indeed gone off the track somewhere. We seemed to become almost obsessed with the subject of sex, etc. rather than following your recent advice of moderation in <u>all</u> things....

> I'm a thwarted out-door-type girl especially noticeable when I visit my father's camp every summer without you. They think: what have we in common? And I think we have all the important things in common—philosophy of life, completely satisfactory sex life, sense of balance, humor, philosophicalness about life, easy understanding of each other.

> But I always regret that it is a little lopsided in that I always take my problems to you, but somehow you never admit to having any.

The next day, she asked,

> Do you suppose everyone can forgive and forget? I'll try if you all will (no more affairs with friends—just other types!)

> I really can't conceive of someone being unable to entertain the possibility of one's spouse straying for just a mere fling.... So, everyone involved seems to have goofed a little. So a major goof has been averted—for the present! I'm not going to feel like the guiltiest or the goofiest party to the 'crime' (if it is a crime to love life, then I'm guilty) and I'll try not to blame you or anyone else either. You would probably say my only goof was in talking to [L's wife] – but I never could do otherwise in this case. (Unless it was dreadfully casual and open and aboveboard and everyone else would be doing same thing (orgy?!) Oh my.

She asked Everett to drop a card to Liz at the Gore's summer camp. "She said last year she wrote you, but you didn't even get it out of the mail."

Liz was having a hard time. She did not want to participate in camper activities; (although she liked to kiss boys). Nancy wrote that Liz was afraid of "showing how bad she is at whatever, (damn PopPop!)".

It appears that Everett's father had teased Liz, as he had teased Pudge.

Liz wrote to her father:

> I miss you very much. I want you to right to me soon.... How are you and the cats. Lots of times I forget about you and the cats. But you are stil my boy frend. and the cats are too.

On August 5, Nancy wrote that she had watched CBS's special report on Vietnam. The special featured Oregon Senator Wayne Morse's fierce criticism of the war as illegal and unwarranted. "What did you think of maverick Morse's criticism—sounded similar to what you had said—but I guess he exaggerated a little."

Regarding the *almost* sexual love affair with "L," she added,

> I think it was most remiss of me to get so involved with an employee of yours! ... I imagine that he had as weird psychological reasons for involvement with me as I did.

She enclosed receipts for a dress and a purse that she had purchased on sale. And added that she agreed with

> what [Secretary of State] Dean Rusk said about how impossible it is to even convey your meaning to a Red Chinese (because their conception of reality is not the same as ours).... They simply want war—need it for good of their country—it's the only way they know—so I guess we can oblige if they insist. (But shouldn't U.N. be settling it, not us?) (Rah, rah maverick Morse). More anon LOVE (more) Nancy!

In her last letter from Amherst before returning home, she worried that birth control pills were not good for her health, but, "what if I talk to the o.b. and he says to stop taking them? How will I stand it?"

Everett family on patio, circa 1969.

She apologized for running off with Everett's paperback *Catch-22*, which he had not yet finished reading (and he *was* mad). But, "the only love that is <u>real</u> to me, is the love I feel for you!" As an afterthought: "Do get 'your' computer working on giving the Greeks and the Turks all the rights they want."

The family was soon reunited, and the marriage became increasingly "open," and centered around Everett's romantic needs more than Nancy's (she really desired only him). Years after Everett's death, she explained to herself in her diary,

> In the past people had called me because they knew what Hugh was up to (fine reason)....and that he wouldn't object or couldn't (blackmail fear – but he wouldn't care) and that I would be amenable. So I chose people I wouldn't feel that I must have – could take or leave – so from that I felt I knew what was going on, that I could handle relationships. Those I could, but not others later [after Everett's death] with no one there to p.u. pieces.

> No calls going to come now.

BOOK 9
BELTWAY BANDIT

29 Weaponeering

In modern war, one individual can cause the destruction of hundreds of thousands of men, women and children. He could do so by pushing a button; he may not feel the emotional impact of what he is doing, since he does not see, does not know the people whom he kills; it is almost as if his act of pushing the button and their death had no real connection. The same man would probably be incapable of even slapping, not to speak of killing, a helpless person.

Erich Fromm, 1955.[1]

Son of WSEG

Lambda Corporation was founded on Everett's brainstorm, the magic multipliers, and it was tailored to suit his desire for independence. WSEG's mental cream, including its secretary, who was cleared to do top secret work, hastened to follow Everett when he moved into a glass office building in downtown Arlington not far from the Pentagon. Aside from support staff, Lambda hired only people with doctorates in physics or mathematics. It offered great benefits and profit sharing to young PhDs coming out of Ivy League schools, as well as to ex-CIA officers and RAND veterans.

Start-up capital of $60,000 was provided by a California-based defense contractor, Defense Research Corporation, in return for a 50 percent ownership stake. As chairman of the board, Everett held the second largest block of stock; Larry Dean, Robert Galiano, and other WSEG alumni held significant chunks. To interface with the Pentagon's war gaming computers, Lambda bought a used IBM 1604: it was the new company's pride and joy.

Thanks to Everett's connections to McNamara's inner circle, the firm sprang into being blessed with a million dollars in sole sourced contracts to continue projects that Everett had been doing at WSEG. Lambda also contracted with IDA, the Advanced Research Projects Agency, Sandia Corporation, and the National Security Agency. The tie to the NSA stemmed from Everett's work with IDA's Communications Research Division at Princeton University. His later work for the NSA was probably on associative memory and pattern

[1] Fromm, E. (1955). 119–120.

recognition programs that would have interested code-breakers and electronic surveillance experts.[2]

One of Lambda's prime contracts was with the Office of Civil Defense to operate the BRISK/FRISK II Damage Assessment System. This computer program projected the lethal effects of fallout under various war scenarios. But Lambda's *pièce de résistance* was a contract from the Assistant Secretary of Defense for Systems Analysis to operate a top secret program to shadow the SIOP. Called QUICK, it allowed Pentagon war-gamers to keep tabs on the Strategic Air Command by running their own simulations of nuclear wars. In essence, QUICK mirrored the SIOP, and, when set in motion, it automatically generated nuclear attack plans down to the smallest targeting details.

Mid-day cocktails

Physicist Ken Willis, was one of Lambda's first hires. He had studied physics at the University of North Carolina with Bryce DeWitt, who had already played a significant role in Everett's life, (and was soon to play an even bigger role). At Lambda, Willis ran QUICK simulations of the SIOP. Despite the flexibility that advanced software supposedly brought to nuclear war planning, he recalled, "When they ripped open the envelope there would only be one of four basic plans to execute."[3]

When Lambda set up shop in Arlington in 1964, local restaurants did not serve cocktails. Everett regularly drove across the Memorial Bridge to Washington D.C. for his liquor-soaked lunch. Pugh and others were not big drinkers, but Willis was happy to join him, especially after Virginia's Blue Laws were rescinded and they could eat across the street. Everett usually drank three Perfect Jack Daniels Manhattans with lunch; they packed a wallop and opened him up.

"We talked a lot about physics and multiple universes," Willis says.

> Hugh was disenchanted with the academic community and its protection of the status quo. I tended to stick to the Einstein school that something was missing from quantum mechanics in terms of causality. We just bullshitted about it, but he viewed it as a joke that the mathematics takes you to multiple universes if you take quantum mechanics at face value.

> We talked a lot about girls. Hugh was definitely a swinger. But, you know, people often thought that he was impersonal and insensitive and did not care much for people. But I was in San Francisco with him on business and we were walking out of the hotel and this woman came up to him and *kissed* him. He told me the story later: She was a prostitute and he had hired her one night. The next day he walked out of the hotel and she ran up and gave him a hug and a policeman arrested her for soliciting. Hugh said 'Hey, wait a damn minute, she is not

---

[2] The NSA says it has no records on Everett.
[3] Willis interview, 2007.

Liz and Mark, circa 1969.

soliciting, she is a friend of mine!' And the cop arrested her anyway. So Hugh hired a lawyer and flew back to testify on her behalf. So, *that* was rather caring!

Everett was a right-of-center Cold War liberal with a libertarian streak. And he loved to argue for the sheer fun of taking a contrary position. Says Willis,

> He looked at life as a game, and his object was to optimize *fun*. He thought physics was fun. He thought nuclear war was fun. Well, not so much the idea of *experiencing* it in any universe. But he was amused by the fact that the world had come to the point where we were boxed into such an irrational confrontation. We both viewed it as crazy that this business we were in even existed, and that it should *not* be possible to find some mutual agreement where you could reduce the number of nuclear weapons.

Disarmament, of course, was not on Everett's business agenda. Willis continues,

> We visited Herman Kahn at the Hudson Institute several times. Herman was a big, fat guy and he and Hugh liked each other a lot. We had good fun at the institute's catered lunches, lots of wine and beer.

> I remember Kahn complaining about environmentalists who were beefing about drilling in Alaska for oil. He said, 'For God's sake, if there is anyplace you are going to have to despoil, that should be Alaska!'

> I never met Hugh's kids, and he never talked about them.

Trouble in paradise

In 1965, the Everetts sold their Alexandria town house and built a home in the newly developed suburb of McLean, Virginia for $54,500. The two-storey house was not a mansion, but it included all of the modern conveniences: air conditioning, dishwasher, garbage disposal, and to Everett's delight, the developer allowed him the choice of a small ballroom or indoor swimming pool. He ordered the pool. It was tiny and divided from the living room by sliding glass doors. For a few years, he swam laps in it almost every night.

Downstairs in the basement he made a small home office and stored doomsday supplies on shelves in another room, (it was not a real fallout shelter and later served as a wine cellar for Everett's homemade brew). In third grade, Mark settled into the basement bedroom, which he later painted black. When Liz was a teenager, a grateful boy friend built a wooden deck off her upstairs bedroom, down the hall from her parent's second floor bedroom. The spacious homes and kid-friendly neighborhood were typical features of the upper-middle-class, white suburbs ringing the nation's capital city and prospering in the Cold War economy. But the children born into this culture of privilege were not guaranteed happiness.

Signs that Liz, known to her friends as "Lizard," was in trouble appeared in early adolescence. The principal of her elementary school wrote a note to Nancy complaining that Liz, "was running around the building and playing among the cars," when she was supposed to be watching a "Drama Performance." The Everetts were not disciplinarians: they believed, along with Dr. Benjamin Spock, that children should be allowed to express themselves freely. They basically put no boundaries on the behavior of their children. Testing social limits in her own way, Liz enjoyed grossing out her friends. The parents of one of her girlfriends became incensed when they discovered a couple of mildly scatological drawings Liz mailed to their daughter. One was of a boy tooting a horn while emitting a musical fart; another was of a wickedly grinning lad shooting a marble between a girl's spread legs. But there is an edge to these drawings— almost like she was saying, "I dare you to stop me!" Before long, Liz was smoking pot, drinking, and going through boyfriends like Kleenex as she ran with rougher and rougher crowds.

QUICK and the dead

In 1966, Everett attended a NATO operations research conference in The Hague, Netherlands. He talked about the latest in operations research by day, and sampled the rich cuisine and other delicacies of Dutch life by night.[4]

At the meeting, one of McNamara's "Whiz Kids," Alain Enthoven, claimed that increasing the American defense budget by $27 billion would reduce American lives lost from a Soviet first strike to 80 million from 115 million, (which was, not to be *too*-editorial, bad news for the 80 million, but good news for American weapon makers). He explained that the job of operations research was to "illuminate the alternatives" for decision-makers according to "the economist's theory of consumer choice."[5]

It turned out that traditional game theory did not effectively model nuclear war, because, for one thing, there was almost no information about the practice of nuclear war to include in a game set-up. And in a real life conflict, the players do not possess the error-free information required by game theory, nor are they guaranteed to act rationally. Political life is woven from complexities and uncertainties that defy capture by standard game theory techniques. But—and here is where Everett's magic multipliers were so vital to war planning—game playing computers could reduce hideously complex problems to sets of manageable options. QUICK could not tell you whether or not you *should* push the button, but it could tell you what was likely to occur if you did. Everett's multipliers were vital to determining the shape of wars to come, and he was much appreciated at the conference.

---

[4] After the conference, Everett complained to a former WSEG colleague, Joseph M. Clifford, that the Dutch prostitutes he had employed did not use condoms. Clifford interview, 2009.
[5] Enthoven, A. (1966).

Shopping for nukes

For its non-expert customers, Lambda described the magic multiplier method by analogizing it to how a housewife shopping for groceries maximizes the overall utility of the items she picks from the shelves as limited by her household budget. While shopping she assigns a utility value—a generalized Lagrange multiplier, or λ—to each item in the store, a kind of utility *price*. The sum of all the utility values in a shopping cart full of goods should not exceed the housewife's budget. This

> technique allows the shopper to make decisions on an item by item basis, rather than considering the whole store at once. The question of whether to buy eggs is separate from whether to buy peas, and this separation makes the procedure easier. However, not all items are independent, for certainly the 17th can of peas has lost considerable value to us if we have 16 cans in our cart.[6]

The canny shopper learns through experience how to assign the λ values to maximize the synergy between utility benefit and monetary cost each time she fills her cart. Otherwise, she will return again and again to the check stand with a cart full of items she either cannot afford, or that are not very useful. Substitute "hydrogen bombs" for "cans of peas" and you get the idea.

Like the grocery shopping housewife, QUICK estimated the cost-benefit of destroying a target with aircraft or ballistic missiles or cruise missiles by inserting λ variables that reflected bomb yield, delivery vehicle range, velocity, weapon reliability, circular error probability, penetration probability, and probability of destruction before launch. And there were other variables reflecting weather factors, defender systems to circumnavigate, the effects of bad intelligence, etc.

QUICK used 500 λ multipliers to calculate the utilities and costs of achieving quantifiable amounts of destruction using a range of weapons systems. For example, to destroy 40 percent of Soviet industrial floor space in Kiev on a first or second strike, QUICK looked for the most effective, least expensive mix of gravity bombs, submarine-launched warheads, and land-based missiles.[7] It accomplished this by using Everett's *non-linear* multiplier method, which broke down complex problems into manageable pieces and made appropriate trade-offs between cost and utility.

Rather than try to keep track of every possible weapon-target permutation, an incomputable task, QUICK randomly *sampled* the vast range of probable outcomes to select the most probable results. And without this automatic sampling method, there would have been no solutions, as trying to solve these problems one step at a time—linearly–was technologically impractical. It took tremendous skill to operate the QUICK simulation. Every time that the computer made another pass over the data the program learned from the experience,

---

[6] Lambda Report 3. (1966). 22–23.
[7] Flanagan interview, 2007; Decision Science Applications Report 39/1038. (1995).

gradually optimizing the deployment of weaponry. Part of the high art involved from the human end was originally assigning values to the multipliers in such a way as to avoid becoming stuck in recursive feedback loops.[8]

Lambda's Paul Flanagan was the project director for QUICK in 1969. He programmed it through several generations of missile systems, including Polaris and MIRVed Minuteman missiles (MIRV stands for "multiple independently targetable reentry vehicles"). He says that QUICK mirrored the SIOP, so it had to access the same databank of information about targets, damage functions, kill ratios, etc. It was constantly updated with hard intelligence on the capabilities and locations of American and Russian nuclear forces. But it was not *the* SIOP, i.e. Lambda staff did not know the precise content of the targeting plan at any given moment.

QUICK also ran its twin: the RISOP program, which was the mirror of the Soviet version of a SIOP. Flanagan says that Everett once came up with a RISOP configuration that completely wiped out America. "If the Russians get this, we are toast," Everett told his colleagues.[9]

Kill probabilities

The Advanced Research Projects Agency wanted to know how many warheads should be dropped on an urban area when the exact location of a target, such as a steel factory, or a radar installation, a government building, or a food supply warehouse, was uncertain. Demonstrating how to do this, Everett and Galiano authored, "Some Mathematical Relations for Probability of Kill." Using the multipliers, the idea was to figure out the probability that a random point can be destroyed by aiming a variable number of weapons at the city center, with civilians to be treated as "collateral damage".

After parameterizing constraints such as aiming errors, dud possibilities, air or ground burst choices,

> The remaining step in the calculation of expected damage to the city is then a summation or integration of the damage at all points in the city weighting each point by its value. Thus, the focus enlarges from individual buildings to all the buildings in the city, finally encompassing all structures or people in the city.... We shall accordingly adopt this general function as our destruction function, *subject to subsequent empirical verification.*[10]

Galiano wrote a related paper on how to defend cities against ballistic missile attacks that include decoys. He addressed the overriding problem with ABM planning (then and now): it is difficult to justify deploying an expensive ABM defense system because it costs far less to fool it with decoys or to overcome it with a shower of missiles than it costs to build and maintain it. The upshot of

---

[8] Pugh. G. E. (2005). 63.
[9] Flanagan interview, 2007.
[10] Lambda Paper 6. (1967). Italics added.

Galiano's paper was that by using a "preferential defense," where selected targets within a region are preferentially defended, and the enemy does not know *which* ones will be defended, an ABM defense *could* be cost-effective.[11]

The next year, Everett wrote a game theoretic paper showing the ABM concept to be ineffectual when there is a limited supply of missile interceptors—the enemy can just keep shooting at the target until the defender's supply is exhausted. The problem in defense is, of course, that all it takes is for one missile in a salvo to penetrate the defense and the ABM radar and interceptors are history. Everett concluded that "the defense is in a quandary," because if it shoots at all incoming missiles it will exhaust interceptor supplies and fall prey to the next wave of attack. But if it holds back, the chances of an incoming missile penetrating the defense vastly increase.[12] This conclusion is common sense, but the Pentagon often paid experts to back up common sense with expensive calculations, and billions of dollars were at stake.

Modeling possible deployments of blast and fallout shelters as a cost function of lives to be saved was another bread and butter job. For the cheapest shelter, the cost of a life came in at $194, rising to $381 for providing a cozier, harder habitat; the premium shelter cost $38 billion for 100 million lives (five percent of the gross domestic product). Lambda then compared the strategic benefits of protecting industry from blast damage versus *not* building shelters for the population. In the end, massive sheltering was shown to be worthless if it resulted in the enemy deploying bunker-busting hydrogen bombs.[13] Shelters were never built in any meaningful way in the United States, although Everett kept his ad hoc basement shelter stocked with canned food and guns, just in case.

The game of sex

Flanagan remembers Everett, Dean, Pugh, and Galiano obsessively playing Kriegsspiel chess in a room set aside for that purpose. Kriegsspiel was game theory brought to life and nobody played regular chess at Lambda. In Kriegsspiel, each player has a separate chess board. Neither player can see his opponent's board. A referee announces whether proposed moves are legal or illegal. After a while, the game had to be banned from the office, as it was cutting into billable man-hours.

"Everett was the smartest man I ever met," says Flanagan. "He was an incredible out-of-the-box thinker. But he had a terrible attitude towards women. He treated them as objects. He would have affairs in the office, and there would be fallout and the women would quit." But that did not bother the boss.

> Hugh viewed life as a mini-max game. Unfortunately, his objective function didn't include emotional values. I don't think he meant to be bad to people, he just did not think about the emotional impact of what he did on others. He was

---

[11] Lambda Report 5. (1967).
[12] Lambda Paper 17. (1968).
[13] Lambda Report 3. (1966).

not warm or friendly. He hurt many people by how he treated them. And he drank too much.

We talked about his multiple universe theory once. There seemed nothing wrong with it except it was a bit goofy.

Flanagan quit Lambda in 1971 because it was starting to bother him that "targets" were, in fact, people. He no longer wanted to spend his days thinking about whether it was more cost-effective to kill 160 million or 180 million people. There were too many fingers on the trigger, and an accidental holocaust could be caused like the firing of the first shot at Concord, Flanagan worried. So, looking for a less explosive solution to the world's problems, he took a job at the Christian Broadcasting Network for, "the *second* smartest man I ever met: Pat Robertson."[14]

Although Everett relished his role as the CEO of an important company, he left the day-to-day operations to Dean and Pugh and Killalea. But his sexism trickled down. Killalea often hired "Miss America-type" secretaries, and then hit on them for sexual favors.[15]

Everett's co-workers do not remember him slurring his words, or being falling down drunk, but it was obvious to many of them that his heavy drinking was debilitating. Returning to the office after lunch, he would take a nap, arising refreshed and productive. But he gradually became pasty-faced and obese; his breath reeked of alcohol, his clothes of tobacco smoke; he became increasingly intolerant of people he considered to be unintelligent. And even as the feminist movement gathered steam in American society, he gleefully treated women as sexual objects, which repulsed some of his colleagues.

"He attacked women on a mental, not a physical level," recalled one of his closest friends, Don Reisler. "Think about the Selfish Gene and the fact that men and women operate differently and somehow he convinced them that having sex was the logical thing to do. Plus, he spent a lot of money and seemed to know a great deal about the fine points of life.

"Plus, he was really quite charming."[16]

Gender wars

Born in 1928, Joanna Frawley was a first-generation operations researcher and computer programmer. With her Q-clearance at WSEG and, then, Lambda, she worked with Everett on weapons systems for many years.[17]

The office atmosphere at the Pentagon in the early 1950s was treacherous for a woman: "I initially did research papers for a man who simply erased my name and put his own name on them." But Frawley's brains could not be sto-

---

[14] Flanagan interview, 2007.
[15] Lucas interview, 2008.
[16] Reisler private communication, 3/2/2009.
[17] Frawley interview, 2007.

len, and by the time Everett arrived, she was appreciated for her skills. "Those were the days when computers and programmers were rare, so everything you did was extraordinary. Everything had to be written from scratch." She quickly rose to be WSEG's Chief Applications Programmer. She modeled aircraft designs, programmed nuclear war gaming systems, and built nuclear blast and fallout damage models. One of her first big projects was writing a program to graph the lethal relationship between megatonage, fallout patterns, and radiation dosages. Recognizing her knack, Everett included her in his Strategic Analysis Group, which spearheaded the statistical research for Report 50, and she became proficient at wielding his multipliers.

Frawley remembers her boss as brilliant, but sad. And he did like his cocktails. "Everybody smoked and went to poker and dinner parties in each other's homes where we drank martinis and old fashioneds." And there was wife-swapping. "This was the LSD era when people were becoming more emancipated and wilder and an awful lot of things went on." But Frawley had her limits: "I would never have had an affair with Hugh because I found him to be a physically unattractive, smoking drunk. Not that he didn't try, but he never pursued it after the 'no.' Obviously there were no hard feelings, because I worked for him many years."

30   The Bayesian Machine

Science, business, politics, have lost all foundations and
proportions which make sense humanly. We live in figures
and abstractions; since nothing is concrete, nothing is real.
Everything is possible, factually and morally. Science fiction is
not different from science fact, nightmares and dreams from
the events of next year. Man has been thrown out from any
definite place whence he can overlook and manage his life
and the life of society. He is driven faster and faster by the
forces which originally were created by him. In this wild whirl
he thinks, figures, busy with abstractions, more and more
remote from concrete life.

Erich Fromm, 1955.[1]

Nuclear uncertainty

In early September, 1969, the United States struck the Soviet Union with 3,700 nuclear warheads exploded on 1,300 targets delivered by a combination of five types of aircraft and three types of missiles. Some of the weapons experienced unexpectedly high failure rates causing the "catastrophic failure of the war plan."

Fortunately for the planet, the attack was a QUICK simulation aimed at teasing out hidden variables in destruction scenarios. From the computer game, Lambda determined that "overkill" might be wasteful in some circumstances, but, overall, it provided a "hedge against uncertainty."[2] Overkill capacity gave the military confidence in the war plan.

The QUICK study was based on "frequentist" probability theory. Frequentism is tied to how often an event has occurred in a series of experiments (i.e. relative frequencies of past events). But frequentism limits the scope of war games, because it is not possible to determine relative frequencies by experimenting with nuclear wars. As the sixties came to an end, Everett turned to another type of probability theory, called Bayesian inference, to glimpse the future.

Thomas Bayes was an 18th century mathematician who formulated an evidential theory of probability. The advent of "Bayes Rule" set off two centuries

---

[1] Fromm, E. (1955). 120.
[2] Lambda Paper 34. (1969). 2.

of debate on the fundamental nature of probability, which, despite its reliability as a calculational tool, remains ephemeral, abstract, almost indescribable.

Frequentists assert that probability has meaning *only* as a measure of the relative frequency with which a particular event has occurred over time: probability *is* frequency. Frequentists focus on events, such as the flip of a coin, that can be repeated many times. They assert that the probability of a particular outcome occurring—heads, say—is the number of times that heads has reportedly occurred in the past: one half of the flips.

Bayesians, by contrast, assert that probability is more than relative frequency. It is a function of a *belief* that an event will reoccur with a certain frequency. They measure probability by using Bayes' rule, a formula that updates a prior belief about the probability of a recurring event with new information, new evidence in addition to relative frequency records. The new information might be the frequency with which similar events have occurred, or new data about changed conditions in the environment. For Bayesians, rational people make decisions based upon evidence-based beliefs about the likelihood of an event occurring. A frequentist might not find this method to be rational—but we will leave that debate to the philosophers.

J. P. Morgan gets mad

The construction of what Everett called his "Bayesian Machine" began after John Barry, a former member of WSEG, went to work for J.P. Morgan & Co. in New York. In 1971, Everett had a remarkable idea for building a computerized stock market timer; Barry convinced Morgan to front Lambda $10,000 to develop it. The prototype program, called Predictor, strove to optimize portfolio returns by quickly identifying new trends in financial markets according to Bayesian probability theory.

The program calculated the probable paths of stock prices as a function of current prices and historical trends. Like QUICK, Predictor reduced the computational scope of the problem at hand by selecting statistical samples, in this case from an enormous, branching universe of probable price paths. But Predictor added a new twist: it used Bayes rule to strip away the most improbable paths and uncover the presence of pricing trends, rising or falling. It aimed at unveiling turning points in market trends in time for investors to react.

There is, of course, no such thing as a foolproof stock market timer; the market is incredibly complex and random variables such as weather, psychology, and politics are driving factors in performance. Barry recalls that Everett's timer was efficient, as far as timers go, but, "He refused to disclose the method to Morgan. He refused to give us the computer code and insisted that Lambda be paid for market forecasts." Barry was very angry and considered suing Lambda after Everett started using the innovative program to do military research. He held his grudge for decades, saying, in 2002, that Everett was, "a brilliant, innovative, slippery, untrustworthy, probably alcoholic, individual."[3]

Barry needn't have worried; Everett was unable to make Predictor generate better than average returns on his own half million dollar portfolio.[4] But in the form of a "Bayesian Machine" or "E-Filter," it generated another kind of profit.

Filtering many worlds

Although the E-Filter flopped at outsmarting the Dow Jones average, it proved useful in operations research. In 1973, Lambda physicist Gary Lucas wrote "Cassandra, A Prototype Non-Linear Bayesian Filter for Reentry Vehicle Tracking." Thirty-five years later, he enthusiastically diagramed his missile tracking system on a white board in his home office, which overlooks a lake in a forested area near Fairfax, Virginia. Talking to Lucas about operations research is the next best thing to talking to Everett himself, as the two men were intellectually in tune and very close.

Lucas spent part of his childhood at Los Alamos, where his father was a machinist working on the Manhattan Project. He came to Lambda in 1968 with a freshly pressed doctorate in theoretical physics from Yale. In one of their first conversations, Everett asked him to solve a series of tricky mathematical puzzles, and was delighted when he did so with alacrity. "Life was a game to Hugh," says Lucas, echoing so many others.

Lucas says that Everett ignored men with whom he could not compete intellectually; *and women were simply prey.* For example, on a business trip to Puerto Rico, he propositioned an employee with whom Lucas was having an affair. Frantic to escape her amorous boss, she telephoned Lucas, who flew overnight to Puerto Rico, joining her in time for breakfast. When Everett walked into the restaurant for his omelet, he did a double take and grinned at the lovers—win some, lose some.

On Fridays, after work, Lambda traditionally hosted a sherry hour for employees and clients that often lasted past midnight. Lucas spent many an hour sipping sherry and chatting with Everett about operations research. Although they did not talk about his many worlds theory, Lucas says that he has a good idea of how his friend felt about it:

> He thought that what he did was cool beyond belief, outrageously outside the nine dots. It tickled him to the core that it was at odds with what everyone had thought at the time. It challenged the authorities in physics, which gave him a quiet satisfaction. And it provided a simple, unarguable explanation for a fundamental problem in physics, which pleased him intellectually. He didn't know if it

---

[3] Series of Barry emails to Shikhovtsev, 2002.
[4] Everett had a bias for buying stock in local businesses, such as Washington Post, People's Drugs, Giant Food, and GEICO. He also invested heavily in companies whose products he enjoyed, such as Playboy, DeBeers, Tandy, and Control Data.

was true, and to a considerable degree he didn't care. It was all great fun. But when Bohr and friends strongly opposed the theory, he became invested in it and identified with it. He was not used to losing. He was also not used to living in the real world. It had all been a game for him. He couldn't handle that. He broke it off and started a new life.

Cassandra calling

During the Cold War, scientists were (and remain) daunted by the complexity of shooting at incoming ballistic missiles. Using radar and computers to accurately predict the continuous trajectory of a missile speeding thousands of miles an hour with enough accuracy to destroy it with another speeding missile seemed impossible. First, the incoming missiles did not fly linearly, i.e. they could be programmed to weave and duck and dodge and launch multiple warheads and scatter decoys. Second, in Everett's day computers did not possess enough memory and brain power to calculate all of the alternative trajectories—deviations from a smooth flight path—in time to launch an interceptor. Prior to Cassandra, predictive tracking filters were linear-thinking machines, hence they were practically useless when faced with maneuvering aircraft or flocks of reentry warheads. Assembling Cassandra from the corpse of Predictor, Lucas connected the problem of tracking average fluctuations in stock prices to the problem of predicting missile trajectories.

To simplify: Linear missile trackers continued to predict a straight line trajectory long after turns had occurred. But Cassandra created a "skyful" of all possible trajectories—branching paths—and randomly extracted a sample of the branching trajectories. The program then assigned to each sample a probability weight representing the likelihood or belief that the sample contained the true position of the missile through time. The tracker continually updated the sample with evidence from newly obtained radar data. Using Everett's algorithmic application of Bayes theory—his Bayesian Machine—Cassandra closed in on the actual trajectory by a process of deduction.

Unfortunately, it is a law of ABM research that the enemy can cheaply thwart just about any tracking mechanism with metallic decoys. The best defense is not to get into a fight with people that can throw nuclear missiles. Defeated by the ABM paradox, Lambda successfully applied the logic of the Bayesian Machine to designing air traffic control software for the Federal Aviation Commission, and, as passenger jetliners do not usually duck and weave, it worked well for years.

Many worlds machine

Lucas said there is a strong analogy between Everett's multiverse theory and the logic of the Bayesian tracker. The analogy is interesting because, although Everett did not use Bayes rule in the construction of his many worlds theory, some advocates of the many worlds theory at the University of Oxford approach the role of probability in that theory from a Bayesian point of view.[5]

Lucas explained:

> The measurement problem in quantum mechanics in which a measurement of an object, say a photon or an electron, is affected by the measurement itself has a parallel in the Bayesian tracker. The likelihood that a particular virtual vehicle is the real or, more precisely, close to the real vehicle is altered by an inherently imprecise radar measurement. Bayesian statistics are used to revise the probability of the virtual vehicle being the actual one. Since the distribution function for the real reentry vehicle is reflected by the density of virtual reentry vehicles, an entirely new set of virtual reentry vehicles must be constructed with each radar measurement.

> The entire process has a very quantum mechanical, multiverse-like feel to it. Very Hugh Everett like.[6]

---

[5] Wallace, D. (2003A).
[6] Lucas private communication, 2009; other Lucas quotes from interview, Oct. 2008.

31 The Death of Lambda

*Life, language, human beings, society, culture—all owe their
existence to the intrinsic ability of matter and energy to
process information. The computational capability of the
universe explains one of the great mysteries of nature: how
complex systems such as living creatures can arise from
fundamentally simple physical laws. These laws allow us to
predict the future, but only as a matter of probability, and
only on a large scale. The quantum-computational nature of
the universe dictates that the details of the future are
intrinsically unpredictable. They can be computed only by a
computer the size of the universe itself. Otherwise, the only
way to discover the future is to wait and see what happens.*

Seth Lloyd, 2006[1]

The future of software

In the late 1960s, the Secretary of Defense awarded Lambda a contract to
develop the CODE 50 Model, a new war gaming program that used Everett's
multipliers to model a variety of "force postures" during a nuclear exchange.
Complementing QUICK, one of CODE 50's jobs was to calculate how much of
the American nuclear force could survive a Soviet surprise attack and effectively retaliate.

As a young Army lieutenant, Jan M. Lodal worked in the Pentagon's Office
for Systems Analysis overseeing Lambda's work on CODE 50. He says that
QUICK and CODE 50 were used by the high command to double check what
the Strategic Air Command's programmers in Omaha were doing with the
SIOP. "The SAC war planners had a set rule of thumb: every target needed to
be hit by two warheads, unless it needed three. Prior to Hugh's work on CODE
50, every time the air force sent up a request for 10,000 nuclear weapons they
got them.

> McNamara had no way to definitively answer Air Force procurement arguments
> by asking SAC for their calculations. They would say, 'We can show you calcula-
> tions, but that cannot prove that the Soviets would not have lots of weapons left
> after the initial salvo.' But CODE 50 allowed us to determine what the real limits

---

[1] Lloyd, S. (2006). 3–4.

of a Russian attack were, so we could put in realistic numbers without making ridiculous assumptions about their capabilities. And that gave us a measure of assured destruction.[2]

Lodal has long been an influential voice in national security circles. He is a stickler for using the phrase "assured destruction," not "mutual assured destruction," because, he says, our own destruction is not part of the plan. He elaborates,

Especially after the deployment of submarines and MIRVs nobody knew how to define winning. Game theory did not make much sense given the high level of uncertainties, and the overwhelming capability of the weapons of both sides to wipe each other out. And nobody could come up with an escalation control scheme where the whole world would not be covered with fallout.

What Hugh proved to us with CODE 50 was that by 1970 there existed no totally disarming attack by the Soviets that could destroy our ability to retaliate.

Naturally, this worked both ways: a U.S. first strike was not likely to cripple the Soviet second strike force either, so, terminological caveats aside, mutual destruction was assured.

In the 1970s, Lodal was an aide to Secretary of State Kissinger. He attended summit meetings, sitting at the table with Soviet leaders, Andrei Gromyko and Leonid Brezhnev. His job was to provide Kissinger with technical expertise. He says that to the extent détente was successful, it was partially due to Kissinger's ability to marshal the technical details of assured destruction. And he credits Everett: "If I had not had Hugh as my mentor to teach me about force levels and nuclear parametrics, and if I had not had his computer models it might not have happened."

Lodal is still amazed by Everett's genius. "I find it astounding that he built both sides of CODE 50. He invented the mathematics and approximation techniques and then designed the FORTRAN program to run the calculations." He saw Everett almost as an physical extension of his computer. "The IBM 1604 had a loudspeaker that played tones as it was computing. Twenty minutes into a five hour run, just by listening to the tone, Hugh could tell where the computation was and say, 'Shut it off, it's making errors.'"

But what most astonished Lodal about Everett was his invention of "attribute value" programming. This path-breaking concept emerged from CODE 50 and QUICK. It was the first iteration of a "relational" database, a data sorting application of the sort later sold by Oracle and PeopleSoft. Lodal points out that the XML method now does for high speed computer systems what Everett's attribute value method accomplished for relatively low speed, small memory computers in the 1970s.

At the time, vast amounts of the Pentagon's war-gaming data was stored on spinning tapes or disks. Information was filed willy-nilly on the recording medium, so a search algorithm might have to parse the entire database to find

[2] Lodal interview, 2007.

a single byte. Lodal describes how Everett's attribute values revolutionized searching, and how the method reflected his many worlds theory.

> He realized that in data processing if you wanted to avoid entanglements in the code and preserve complete data sets without losing information content that you have to decide at the atomic level of the data base what items you most care about, what are the fundamental attributes of types of data, like names, ages, income in a social security database, or the various values defining destruction in a war gaming model.

Instead of mapping information by time or position on the tape, Everett assigned numerical tags to sets of data, categorizing them. Number 444, for example, would call up names of targets, 354 would call up the probability of a certain type of missile reaching a target, 666 would pop up its kill probabilities. Then, to further simplify and speed up the search process, says Lodal,

> Hugh invented probably the first efficient compression algorithm. It worked exactly like a three-dimensional spreadsheet with columns and rows, so you could pinpoint a certain value. Except Hugh's spreadsheet was N-dimensional. He could deal with an uncountable number of dimensions, just like his uncountable number of branching universes in his quantum theory, with no limit on the number of dimensions available.

> The stuff was amazing effective. Intellectual. Crisp. Precise. Like how he treated the Schrödinger equation: he said, 'Lets not get complicated, lets start with the simplest assumption, that it evolves linearly.'

Lodal's former boss, assistant secretary of defense, Ivan Selin, says Everett's innovation may seem trivial today, but at the time, it was mind-blowing.[3] In 1970, Lodal and Selin and another Pentagon official, Charles Rossotti, formed American Management Systems, a business consulting firm based upon Everett's attribute value concept. AMS soon became one of the largest, most successful information management consulting groups of the late 20th century.

AMS had its first big success in the mid 1970s when it used Everett's programming breakthrough to restructure the books of the nearly bankrupt City of New York. Lodal was featured on the front page of *The New York Times* as having saved the city from financial meltdown. But he gives much of the credit for that achievement to Everett's attribute value system:

> Hugh would say, 'We have to make the values independent of each other, orthogonal.' He conceptualized the data as consisting of glops of information in an n-dimensional Cartesian space where you could represent any collection of related data by a single point in that space. Instead of writing thousands of rules into the code, he came up with a general concept that applied to all of the data at all points.

And that was always Everett's genius: seeing ways to solve seemingly intractable statistical problems in computer science and quantum mechanics through simplification.

---

[3] Selin interview, 2007.

Unfortunately, he did not possess a similar eye for solving business problems.

Lambda sinks

Lambda's annual report for 1970 boasted that the company had 50 employees, most of them with PhDs, and annual sales over $1 million. In addition to designing nuclear wars, Lambda studied the best way to kill Vietnamese soldiers with conventional firepower. It had a contract to study "continuity of government," i.e. plans to institute martial law domestically in the aftermath of natural disaster, nuclear war, or insurrection. It designed data compression systems for the World-Wide Military Command and Control System, which was the Pentagon's global communications network. It wrote programs to train fighter pilots to maximize kill ratios in air combat. And at the same time it was tuning the war machine, it was designing an "arms control simulation study."

Although contracts with the Central Intelligence Agency and the National Security Agency were not listed in Lambda's official reports, they were listed in private spreadsheets kept by Everett, who personally controlled these sensitive jobs.[4]

Lambda also held a lucrative, politically sensitive contract with the Department of Health, Education and Welfare to assess busing as a school desegregation strategy. After Lambda produced an influential study showing that a large amount of desegregation could be obtained with a relatively small amount of busing, the company was contracted to write desegregation plans for most major metropolitan areas in the United States. But in 1972 the national press ridiculed Lambda as a bastion of clueless eggheads when it was revealed that its plan for the Washington D. C. area required students to walk across a busy interstate highway to get to the nearest bus stop.

The busing model was only a prototype and someone (perhaps an opponent of busing, Lucas suggests) leaked it to a *Washington Post* reporter. But the scandal came at a time when military think tanks were increasingly under attack for constructing idealized models of reality that were causing terrible bloodshed in southeast Asia and failing to work as advertised (e.g. carpet bombing Hanoi, napalming south Vietnamese families, assassinating political dissidents in Saigon). Lambda's government contracts were cut back after Congress questioned the wisdom of paying $585 million a year to 700 think tanks that could not figure out how to defeat an army of peasants wielding sharp sticks and second-hand rifles against the best-equipped military in human history.[5]

Reading the writing on the walls of the Pentagon, Lambda tried to wean itself from the military teat. Everett hoped to market his word processing

[4] Basement.
[5] Dickson, P. (1971). 14.

software, or to turn the revolutionary attribute value concept to commercial advantage. And he almost succeeded: In 1967, Lambda began a multi-year $600,000 study for Merck Corporation showing the pharmaceutical conglomerate how to transform one-drug factories into modular, just-in-time facilities producing many kinds of drugs. The computerized plan was elegant and workable—but, alas, far too expensive (a billion dollars) for Merck to adopt.

Everett's ideas were often far ahead of his time, but he was also stubborn and arrogant, and those traits hurt his business. For example, corporations were clamoring for software written in COBOL, a business-friendly programming language. But he absolutely refused to write in any format except FORTRAN, the scientific language. To prove its superiority, he wrote a version of COBOL in FORTRAN!

Desperate for income, Lambda tried to adopt its military software to commercial chores such as assigning train track routes for millions of empty freight cars, or improving the efficiency of open pit mining. But in the end, the company was rich in scientists and poor in marketers. Dean wanted to concentrate on getting consulting contracts from industry, but Everett was dead set on making a splash with proprietary software sales. Lambda's accountants were horrified that he spent $130,000 developing PROLOG, an anti-pirating system for software. He applied for a patent on this idea—which was to invisibly scratch a computer disk in such a way that its program would not decrypt itself and run unless the scratch was detected by an authorized user. But the Patent Office turned him down for lack of specificity in his design. He was unable to sell the product, and became convinced that IBM was trying to steal it from him.

Ironically, even as Lambda was declining, Everett's attribute value application was doing well for Selin and Lodal and Rossotti, the firm's former patrons inside the Pentagon. AMS knew how to sell advanced software in the corporate market, whereas Everett was adrift without a compass in the competitive world of business consulting. However, the AMS founders were so impressed by Everett's general brilliance, that they gave him and Lambda 14 percent of their initial stock offering in return for continued access to his brain. They subcontracted jobs to Lambda; and they made Everett a vice-president of AMS, although he had no administrative duties.

By late 1972, Lambda was drowning in red ink and laying off employees. A sad internal memo, probably authored by Galiano, related:

> We seem to me like a steamship, halted at sea, slowly drifting onto a reef. At first reluctantly, but eventually gaily, we begin to throw overboard the brass fittings, the cargo, and finally the crew, while hoping to float over the frightening coral. But, is the surviving derelict worth the price? Shouldn't we be below deck, stripped to the waist, stoking up the furnaces for all we're worth right now? Two years of drifting have left us precious little time.... We must model ourselves more on IBM and less on IDA memories.... The world, our current and

prospective customers, must be made aware of our superior performance, or it may well have been wasted.[6]

Concerned by Lambda's operating losses, its 50 percent owner, now called General Research Corporation, stepped in and folded the struggling company into its giant embrace, making it a division, and rewriting Everett's employment contract. The boilerplate listed as grounds for termination conviction of "a misdemeanor involving moral turpitude." Everett crossed out that phrase, scrawling next to it, "clarify meaning."

The ghost of Lambda struggled inside GRC for a few more years, but Everett left in 1973 to form a new company, DBS, with a young Lambda physicist, Don Reisler. DBS (not an acronym) specialized in the analysis of discriminatory patterns in the workplace, providing expert testimony on behalf of the government in federal lawsuits alleging bias. Rossotti was on the board; Reisler was president; and Everett wrote computer programs, when he bothered to go to the office.

He also parted ways with the AMS group. To this day, Lodal, Selin, and Rossotti maintain that the company—which was a billion dollar concern when they sold it in 2004—would not have gotten off the ground were it not for Everett's algorithms. The AMS founders all become wealthy, forging illustrious careers in both the public and private sectors. Lodal counseled several presidents on national security issues and is an expert on nuclear deterrence; Selin was an under secretary of state for President George H. W. Bush and later headed the Nuclear Regulatory Commission; Rossotti headed the Internal Revenue Service under Clinton and is now a partner in the Carlyle Group, which invests heavily in weapons systems and is run by former world leaders and defense department officials. They all remember Everett fondly, but sadly, as he did not fit into their world—he was not a team player.

Selin says, "Hugh modeled the world in his own mind and then forgot about the world and concentrated on the models, sometimes with brilliant insights, sometimes with ludicrous results. Only occasionally did he intersect with what the rest of us considered to be the real world."

The business of morality

Before coming to Lambda, Reisler worked with "some strange folks" at Research Analysis Corporation, the Army's non-profit think tank.[7] During the Vietnam War, RAC developed psychological warfare techniques under contract to the military.[8] "Psy-ops" deals in behavioral patterns: how people think, how they come to believe in political or religious ideologies, how they make decisions. Traditional game theory was useful to psy-ops researchers, until computerized

---

[6] Basement.
[7] Reisler interview, 4/7/06.
[8] Dickson, P. (1971). 150.

data-gathering took the art of social engineering to a whole new level. Still, the social engineers remained a strange breed within the already strange culture of op research with its custom designed morality.

At Lambda, for instance, Pugh started working on how moral values evolved biologically as algorithmic decision-making processes. After Lambda was folded into GRC, he wrote an important book, *The Biological Origin of Human Values.*[9] Everett and Wheeler were great fans of Pugh's book, published in 1977. It was praised by Harvard University's Edward O. Wilson, who was a prime mover in the new science of "sociobiology," and Pugh became tied to Wilson in the public eye.

Sociobiologists were using information theory and cybernetics to describe and explain the social practices of creatures, from ants to people. But some left-oriented academics attacked the work of Wilson and Pugh as reactionary and classist; unfairly, as it turned out, because theses critics did not grasp that sociobiology actually showed that evolution is wired to favor species-wide altruism and collective action over the "survival of the fittest" individuals,[10] which could mean that humans are currently trapped in an evolutionary cul-de sac by industrial and military technology. In his book, Pugh applied concepts underlying Lambda's targeting software to explain the evolution of morality-based decision-making. For example, targeting programs can be "taught" how best to assign values to targets, and to make decisions that optimize destruction. Taking human survival (not destruction of targets) as a basic biological value, Pugh showed how root emotions of love and fear affect the formation of higher moral values and choices. But he pointed out that biological evolution is relatively slow, and humans have been programmed by evolution to respond to perceived threats hormonally, in hyper-aggressive ways that are not advantageous to living in a modern civilization with its tremendous capacity for instant self-destruction. As a solution, Pugh appealed to "reason," hoping that humanity's leaders would (somehow) learn how to curb the inexorable excesses of the capitalist economic system and make political decisions based upon tolerance and love, not greed and fear.

Despite his insight into the nature of morality, and his concern about pre-programmed species-suicide, Pugh used his human values-driven decision-making model to design automated fighting programs for combat aircraft.[11]

After leaving Lambda, Pugh and Lucas partnered to form their own operations research firm, Decision Science Applications. They successfully used Everett's magic multipliers in a variety of military and corporate applications. Pugh retired quietly in 1992, and six years later the remaining owners sold the company for $42 million. Lucas walked away a multi-millionaire.

But back to the main story: Even as his Lambda world was collapsing, Everett's relative states theory, to his surprise and delight, was reborn as the Many Worlds Interpretation of Quantum Mechanics.

---

⁹ Pugh, G. (1977).
¹⁰ See Wilson, D. S. & Wilson, E. O. (2007).
¹¹ Pugh, G. E., Lucas, G., Gorman, G. (1978).

BOOK 10
MANY WORLDS REBORN

32 DeWitt to the Rescue

It would be the worse sort of folly to advocate that the study of classical physics be completely dropped in favor of newer theories…Nevertheless, we have a strong desire to construct a single all-embracing theory which would be applicable to the entire universe.

Hugh Everett III, 1956[1]

Many worlds in hibernation

After its publication in 1957, Everett's theory was officially ignored by mainstream physics for a decade; but his bright idea was not dead, merely hibernating. Science fiction writers set stories inside Everett's multiple universes. He made a cameo appearance in *The Scientist Speculates: An Anthology of Partly-Baked Ideas*.[2] The book's editor, a computer scientist named Irving Good, remarked that the notion of multiple universes was the stuff of science fiction until formalized by Everett. He crudely, but correctly summarized the theory as, "We all have innumerable identical twins with whom we very seldom communicate."[3]

But as the Xavier conference highlighted, the usefulness of a universal wave function held a powerful attraction for physicists who were dissatisfied with the standard interpretation of quantum mechanics. For instance, in 1963, MIT physics professor, Abner Shimony, published "Role of the Observer in Quantum Theory."[4] He slammed the von Neumann wave collapse postulate as unjustifiably dependent upon the agency of human consciousness. And he expressed disdain for Bohr's interpretation, saying it abandoned the possibility of understanding the quantum world on its own merits. He concluded that quantum mechanics itself was in need of correction. And in a footnote, he said that the

---

[1] DeWitt, B, and Graham, N, eds, (1973).

[2] Good, I. J. (1962). Wigner's famous "Wigner's friend" paper also appeared in this volume, which was packed with serious science.

[3] Ibid. 155. Good and Everett were acquainted: both did work for IDA's communications division in Princeton (probably cryptographic tasks for the National Security Agency). Among Everett's papers is a copy of a treatise by Good called "The Human Preserve." It speculates that our galaxy is secretly governed by telepathic "Chief Entities" who preserve inter-stellar law and order in what amounts to a galactic zoo. Good said that self-replicating intelligent machines would long ago have taken over the zoo were it not for intervention by the Chief Entities, whose beneficent occupation saves us from descending into anarchy. The analogy between guardianship by the Chief Entities and the American national security state's promotion of its role as a "global policeman" was intentional.

[4] Shimony, A. (1963).

relative state interpretation was deserving of more study. Shimony was motivated to treat the Schrödinger equation as real, but he saw Everett's model as a violation of Occam's Razor—the truism that the simplest explanation is the best explanation—by its multiplication of universes. And he left it at that.

In the late 1960s, Deborah van Vechten, a student at Brown University in Rhode Island told her physics professor that she wanted to write her undergraduate thesis on the measurement paradox in quantum mechanics. The professor, Leon N Cooper, was soon to win a Nobel Prize for pioneering work on superconductivity, which is highly quantum mechanical.[5] He told her there was "no such thing" as a measurement paradox; and then, "She pushed me on the subject and I realized what I had accepted as the Copenhagen interpretation or the reduction of the wave function was, indeed, paradoxical."[6]

Teaming up, van Vechten and Cooper wrote a paper for the *American Journal of Physics*, "On the Interpretation of Measurement Within the Quantum Theory." Referring to those who support the Copenhagen interpretation, the authors quoted Spinoza:

> They appear to conceive man to be situated in nature as a kingdom within a kingdom: for they believe he disturbs rather than follows nature's orders.[7]

Like Everett—of whose theory they were unaware—they attacked von Neumann's collapse postulate, concluding that it is an "anachronism because it divides mind and body." Like Everett, they allowed the Schrödinger equation to run its logical course, but they used Feynman's sum over histories method to support the idea of a non-collapsing wave function that includes the observer.

It was not until DeWitt and Graham asked them to permit their paper to be included in *The Many Worlds Interpretation of Quantum Mechanics* that Cooper and van Vechten learned of Everett's theory. They wrote a footnote acknowledging the primacy of his recognition of "the necessity of retaining all branches of the wave function," with the caveat that of all *their* possible worlds, only one is real. Cooper was disappointed when he learned that Everett had scooped him; he felt like Augustin Jean Fresnel when he learned in 1816 that his innovative wave theory of light had been previously discovered by Thomas Young.

Cooper remained interested in the measurement paradox,[8] writing two more papers on how a branching wave function works. In "How Possible Becomes Actual in the Quantum Theory," he asserted that the wave function as we perceive it does not seem to contain complete information about the present (as it has branches that we cannot perceive). This makes it difficult to reconstruct the history of the universe. However, the observer may ignore branches that he does not experience because universe-sized probability waves reinforce and cancel each other until one world emerges, "ours."

---

[5] Prof. Cooper says there is no period after his middle initial.
[6] Cooper interview, 2009; van Vechten interview, 2009.
[7] Cooper, L. and van Vechten, D. (1969).
[8] van Vechten went on to earn her PhD in super-conducting physics. She left her work in quantum foundation theory behind because it did not pay in the job market. She supervises the super-conducting electronics program at the Office of Naval Research.

We thus have a consistent but curious interpretation—a physical theory in which 'actuality' is an outcome of interference between 'possibilities'—a technical realization of the fantastic vision Jorge Luis Borges presents in *The Garden of Forking Paths*. Once more the poet has preceded the scientist.[9]

Enter DeWitt

In 1967, Bryce DeWitt awakened the relative state theory from its decade of hibernation. He made it acceptable to talk about multiple universes in public debates. And in doing this, DeWitt stuck his neck out, inviting intense criticism for championing Everett (who remained aloof from the tussle).

More than 30 years later, DeWitt talked about why he chose to promote many worlds. At a symposium, "Science and Ultimate Reality: Celebrating the Vision of John Archibald Wheeler," he explained that Everett had cut through years of fuzzy thinking about the measurement problem by simply assuming that quantum mechanics replaces classical mechanics as a *determinate* description of reality:

> This is a shocking idea, for it leads to a multiplicity of 'realities.' Few physicists in 1957 were prepared to accept it. And yet it can be shown to work.[10]

Shortly after he died of cancer in 2004, *Physics Today* published the ailing DeWitt's farewell to physics, "God's Rays." He wrote that if one accepts quantum theory as real, then,

> One is obliged to accept a stupendous number of simultaneous realities, namely, all the possible outcomes of quantum measurements as well as all the possible 'classical' worlds that emerge spontaneously from the wave function of the universe through the phenomenon of decoherence.[11]

DeWitt was an early proponent of decoherence theory, which, he thought, describes what happens to the information contained in a wave function as it *appears* to collapse. He said that the quantum decoherence effect is explained by Everett's interpretation, which is principally a theory about entanglement.

Bryce DeWitt was born Carl Bryce Seligman on January 8, 1923 in Dinuba, California. His father was a country doctor, and his mother taught high school Latin and mathematics. At age 12, DeWitt was sent off to Middlesex, an academically prestigious boarding school in far-away Massachusetts. On the east coast, he was shocked to encounter social prejudice based purely upon bearing a Jewish last name. Eventually, Seligman and his three younger brothers changed to DeWitt, a surname from their mother's lineage.

---

[9] Cooper, L. (1976). Cooper is still working on his interpretation. He says it is not necessary to assume that any world other than the one we are in is real. The wave function allows us to calculate the probability that we will have some particular future. But it does not tell us with certainty what the future will be (nor does it tell us by itself what the present is).

[10] DeWitt, B.S. (2004). 167.

[11] DeWitt, B. S. (2005). 33.

At Harvard University, he majored in physics. During the Second World War, he worked a short time on the Manhattan Project before enlisting in the Navy. After the war, he returned to Harvard and wrote his doctoral thesis on quantizing gravity under Julian Schwinger (a co-inventor of quantum electrodynamics, along with Feynman and Tomanga).

At Princeton's Institute for Advanced Study he met Cecile Morette, a French physicist, who was soon to found an internationally renowned physics institute in Les Houches, France. In 1951, Cecile and Bryce, now married, resided at the Tata Institute of Fundamental Research in Bombay, India, where the first of four daughters was born. Returning to the United States, DeWitt took a job designing nuclear artillery shells at Livermore National Laboratory in California. The mid 1950s found the couple working as research professors at the University of North Carolina, Chapel Hill where they organized the conference on gravitation, learning about Everett from Wheeler.[12]

DeWitt recalled his feeling when he first read Everett's short dissertation in 1957:

> First, I was tickled to death that someone had at long last, after so many years and so many tiresome articles, something new and refreshing to say about the interpretation of quantum mechanics. Second, I was deeply shocked.[13]

He wrote to Wheeler:

> It seems to me that the professional philosophers will have a greater appreciation of Everett's work than will the average physicist, at least for the present. [However] it has become increasingly clear that physicists themselves are obliged to be their own epistemologists, since no other persons have the necessary competence. Therefore Everett's work is to be praised.[14]

As previously recounted,[15] DeWitt balked at endorsing the full theory, because he could not "feel" himself "split." But Everett had quickly convinced him that this objection was actually a strength of the theory as it predicted that we will only experience one world, not many. Ultimately, DeWitt, as a cosmologist, was attracted to the universal wave function because a theory of quantum gravity must apply to the whole universe, precluding a role for an external observer.

Enter Graham

In the mid 1960s, one of DeWitt's graduate students, Neill Graham, wanted to write his PhD thesis on foundations of quantum mechanics. DeWitt tried to discourage him, saying that interpreting quantum mechanics was a topic better

[12] See Chapter 19.
[13] DeWitt, B. S. (2008A). 1.
[14] DeWitt to Wheeler, 5/7/57.
[15] See Chapter 18.

suited for old age.[16] But Graham persisted and, under DeWitt's guidance, he explicated the Everett interpretation from his own point of view.

Graham thought that Everett's derivation of a probability measure from the formalism was inadequate. So, he reworked Everett's formalism, intent on deriving a different type of probability measure, tantamount to a "principle of indifference." In doing so, he treated the Everett branches as countable (which Everett did not, he viewed them as uncountably infinite in number). In short, Graham's universe-counting approach to measuring probability was akin to proposing that as the sun will either explode or not explode tomorrow, the probability of the sun exploding tomorrow is 1/2.[17] Everett expressed his pique at Graham's measure by writing an exclamatory notation of disapproval in the margin of his copy of Graham's thesis.[18] And Graham's application did not catch on, although DeWitt liked it at first.

Graham also identified, but did not solve, the preferred basis problem.[19] This problem has a precise expression in the quantum formalism, but we can capture its philosophical angst in ordinary language. It arises in Everett's theory because he is not clear on how the classical world of our experience physically emerges from the universal ψ where all physically possible events occur. Why is *this* world *preferred* over all of the possible worlds I could find myself inhabiting? Graham asked. How does this moment in time connect to other moments in a causally consistent, finite history set in a coordinate system of infinite dimension? Considering the fuzziness of position and momentum delineated in the uncertainty principle: How does the set of precisely described positions of electrons in this universe connect to the set of precisely described momenta of these same electrons? Can a universe split along a position *basis* at the same time it splits along a momentum basis? How does a splitting universe choose which basis to split along?

Everett was not particularly troubled by the problem of preferred basis, believing that the selection of a measurement device (for recording position or momentum or spin) automatically determined the basis that describes a connected series of physical events. But it was not until decoherence theory evolved (thanks, in no small part, to Everett's influence) that the question of preferred basis became less controversial: A quantum system is said to decohere as it interacts with its environment, automatically selecting a causally connected, macroscopic history from the menu of possibilities inherent in its surroundings—while *decoupling* from those other possibilities. In this scheme, every Everett branch automatically selects a preferred basis as it correlates with an environment at light speed inside the universal superposition, the multiverse.

[16] DeWitt-Morette interview by Kenneth Ford, 2/28/95.
[17] Barrett, J. A. (1999). 166–172. Private communications with Barrett, 2008.
[18] Basement archive.
[19] Graham, N. (1973). See chapter 16 for a related explanation of the preferred basis problem. A basis is a vector in Hilbert space. See Albert, D. Z. (1992) for a readable, comprehensive explanation of basis.

Banner year: 1967

Graham's dissertation was not formally accepted until 1970, but DeWitt cited it as an unpublished dissertation in 1967. In the interim, it circulated like samizdat among physicists interested in cosmology. Inspired by working with Graham, DeWitt began publishing papers on the "Everett-Wheeler metatheory" in professional journals—breaking his profession's silence on the philosophically disturbing idea, while sanctioning it as a viable alternative to the Copenhagen interpretation.

DeWitt published two treatises in 1967 featuring Everett. First came a three-part opus in *Physical Review*, "Quantum Theory of Gravity,"[20] in which he summed up the state of search for a workable theory of quantum gravity. He wrote a section on Everett's universal wave function, explaining the formal logic of the theory of relative states based upon his reading of the short thesis, as he was not yet aware that there existed a longer, more complete treatise. He was, nonetheless, convinced that Everett had provided quantum mechanics with an internally consistent interpretation, while obviating the need for the wave collapse postulate.[21]

In *Physical Review*, DeWitt noted,

> Everett's view of the world is a very natural one to adopt in the quantum theory of gravity, where one is accustomed to speak without embarrassment of the 'wave function of the universe.'[22]

The concept of a universal wave function was a boon to cosmologists, but it carried philosophical baggage:

> Because of the size of the universe, we know that the 'Everett process' must be occurring on a lavish scale:...components of the universal state functional must be constantly splitting into a stupendous number of branches, all moving in parallel without interfering with one another...each branch corresponds to a possible world-as-we-actually-see-it.[23]

That summer, Wheeler and Cecile DeWitt organized a seven-week long conference in Seattle, sponsored by the military contractor, Battelle Memorial Institute. The conference was called Battelle Rencontres, or Battelle "encounter." The organizers brought together 33 of the world's leading physicists and mathematicians to ruminate about a wide range of problems, including the quantization of gravity.

DeWitt wrote a paper for the conference, "The Everett-Wheeler Interpretation of Quantum Mechanics." He defended the relative states model against the

---

[20]   DeWitt, B.S. (1967).
[21]   Around this time, Wheeler and DeWitt conceived a deceptively simple wave equation that symbolically represents a quantification of the energy of the whole universe. Known as the Wheeler-DeWitt equation, it does not tell one *how* to quantize gravity, but it is an equation that a successful theory of quantum gravity would have to satisfy, essentially, a universal wave function without time.
[22]   DeWitt, B.S. (1967). 1141.
[23]   Ibid. 1141–1142.

charge that the theory violated Occam's Razor with excessive complexity. He said that, contrary to Bohr's interpretation, Everett and Wheeler had *minimized* the concepts necessary to arrive at a workable model of quantum mechanics by treating the wave equation logically and by not resorting to metaphysics. He supported Graham's derivation of probability by counting universes, (although he was later to change his mind about that approach).

As a cosmologist interested in plumbing the initial conditions of the universe—where quantum mechanics and gravity ought to have been united—DeWitt was pleased to have a method of thinking about the whole universe from inside that universe, even if it had certain weaknesses. One weakness, he thought, was,

> As to why the good Lord chose to construct the world along these particular mathematical lines, Everett and Wheeler are silent.[24]

Another weakness:

> Although it is a beautifully self-consistent philosophy it can never receive operational support in the laboratory.[25]

These papers were but the opening salvos in DeWitt's life-long battle as Everett's champion. As quantum mechanics became more and more important to the study of conditions in the early universe, interest in the theory of the relative state soared.

DeWitt coins "many worlds"

DeWitt recalled that in 1969, he "had a visit in Chapel Hill from Max Jammer, who was writing a book on the foundations of quantum mechanics, and he had never heard of Everett. And I thought this was scandalous, because Everett had a brand new idea, it was the first fresh idea in quantum theory in decades, and he was being completely ignored. So I decided to write an article for *Physics Today*, which really put Everett on the map, and Wheeler promptly disowned Everett. The reason, as far as I can see, was that it was too revolutionary an idea, this idea of many worlds. It was anti-Copenhagen; Bohr was one of Wheeler's heroes, and he didn't want to be associated with it. He has refused to have anything to do with it in all the years since."[26]

DeWitt's September 1970 article in *Physics Today*, "Quantum Mechanics and Reality," was subtitled: "Could the solution to the dilemma of indeterminism be a universe in which all possible outcomes of an experiment occur?"

It began by criticizing Bohr:

> Bohr convinced Heisenberg and most other physicists that quantum mechanics has no meaning in the absence of a classical realm capable of unambiguously

[24] DeWitt, B.S. (1968). 320.
[25] Ibid. 326.
[26] DeWitt-Morette interview by Kenneth Ford.

recording the results of observations. The mixture of metaphysics with physics, which this notion entailed, led to the almost universal belief that the chief issues of interpretation are epistemological rather than ontological: The quantum realm must be viewed as a kind of ghostly world whose symbols, such as the wave function, represent potentiality rather than reality.[27]

Indeed, that statement sums up how many physicists and philosophers still feel about the quantum mysteries. DeWitt continued:

If a poll were conducted among physicists, the majority would profess membership in the conventionalist camp, just as most Americans would claim to believe in the Bill of Rights, whether they had read it or not. The great difficulty in dealing with activists in this camp is that they too change the rules of the game, but, unlike Wigner and Bohm, pretend that they don't.... The Copenhagen view promotes the impression that the collapse of the state vector [wave function], and even the state vector itself, is all in the mind. If this impression is correct, then what becomes of reality? How can one treat so cavalierly the objective world that obviously exists all around us?[28]

After castigating Wigner's "non-linear" solution to measurement, i.e. his theory that consciousness has a privileged role in the universe, DeWitt introduced the "Everett-Wheeler-Graham metatheorem." DeWitt later said, "The *Physics Today* article was deliberately written in a sensational style. I introduced terminology ('splitting,' multiple 'worlds,' etc.)...to which a number of people objected because, if nothing else, it lacked precision."[29] And in the article DeWitt was eloquent, not shying away from the "split" word:

This universe is constantly splitting into a stupendous number of branches, all resulting from the measurement like interactions between its myriad of components. Moreover, every quantum transition taking place on every star, in every galaxy, in every remote corner of the universe is splitting our local world on earth into myriads of copies of itself.... Here is schizophrenia with a vengeance.[30]

He was concerned, however, about how to tie any type of probability measure to physical reality:

The alert reader may now object that [Graham's probability] argument is circular, that in order to derive the *physical* probability interpretation of quantum mechanics, based on sequences of observations, we have introduced a nonphysical probability concept...alien to experimental physics because it involves many elements of the superposition at once, and hence many simultaneous worlds, that are supposed to be unaware of one another.[31]

DeWitt's confession summarized the probability problem in Everett: how can we prove a concrete, physical measure of the relative frequency of certain

[27] DeWitt, B. S. (1970). 160.
[28] Ibid. 159.
[29] DeWitt, B. S. (2008A). 3.
[30] DeWitt, B. S. (1970). 161.
[31] Ibid. 163.

occurrences over sets of branching universes when we have no hope of physically accessing those universes to make tests? He as much as admitted that the Born Rule was an assumption in the Everett interpretation, (a sentiment with which Everett disagreed). This admission was not in accord with his extravagant claim that the many worlds interpretation emerged from the quantum formalism without tinkering:

> Without drawing on any external metaphysics or mathematics other than the standard rules of logic, EWG are able...to prove the following metatheorem: *The mathematical formalism of the quantum theory is capable of yielding its own interpretation.*[32]

Shortly after the *Physics Today* article appeared, DeWitt made a more detailed, formalistic analysis of the Everett interpretation at a summer conference in Varenna, Italy. "The Many-Universes Interpretation of Quantum Mechanics" expanded on the problem of trying to derive the Born rule from the quantum formalism:

> This reality is not the reality we customarily think of, but is a reality composed of many worlds. [What we know about the meaning of quantum mechanics] is unfortunately not yet sufficient to tell us how to apply the formalism to practical problems. The symbols that describe a given [quantum] system...describe not only the system as it is observed in one of the many worlds comprising reality, but also the system as it is seen in all the other worlds. *We, who inhabit only one of these worlds, have no symbols to describe our world alone.* Because we have no access to the other worlds it follows that we are unable to make rigorous predictions about reality as we observe it. Although reality as a whole is completely deterministic, our own little corner of it suffers from indeterminism.[33]

DeWitt was not suggesting that we cannot use quantum mechanics to make sufficiently accurate predictions on "our" branch, but only that we can never know the exact composition of the universal wave function and, therefore, we cannot know or predict what happens in the universe as a whole, the multiverse.

Evoking Einstein

DeWitt concluded the *Physics Today* article asserting that the "EWG" interpretation would "breathe new life" into the philosophy of science. He posited: "Yet it is a completely causal view, which even Einstein might have accepted."

Letters to the editor streamed into *Physics Today*,[34] most of them taking furious exception to the evocation of Einstein. The journal gave substantial space to six prominent physicists to pick apart Everett and DeWitt, zeroing in

[32] Ibid. 160. "EWG" is his acronym for "Everett-Wheeler-Graham."
[33] DeWitt, B. S. (1971). 182. Italics added.
[34] DeWitt *et al.* (1971B).

on problems with preferred basis and the derivation of probability.[35] *But* they all treated Everett's interpretation as worthy of serious consideration, even when they were reluctant to endorse it as physics, or were ontologically repelled by it. And there were supporters: one scientist saw in many worlds an opportunity to apply parapsychology to quantum theory; another commented, wistfully, that people on a crashing airplane could take comfort in the certainty that they would survive in some branches of their personal universes.

DeWitt answered his critics point by point. Regarding the crashing plane, he said he *would* worry, because, "It's me I'm concerned about, not those other guys!"[36]

But he remained convinced that Einstein would have been

surprised and pleased at Everett's conception... for it is the only conception that appears capable of unifying general relativity in a profound way with the quantum theory, without changing either theory or adding any new formal elements. It is the only conception that, within the framework of the presently accepted formalism, permits quantum theory to play a role at the very foundation of cosmology.[37]

And then he *backtracked*:

I do confess to having somewhat overstated the case in my article in implying that the EWG metatheorem has been rigorously proven.[38]

He said that true rigor would require replacing all of the quantum mechanical symbols with new symbols stripped of formal rules for manipulating them. Rigor would require a proof

empty of an *a priori* meaning.... This remains a program for the future, to be carried out by some enterprising analytical philosopher.[39]

So, thanks to DeWitt and Graham, Everett became a minor celebrity in physics circles. Multiple universes were now a palatable topic for academic discussion. Philosophers started to pay attention. David Lewis, who was in the process of developing a theory of multiple worlds based purely upon philosophical concepts, signed Everett's dissertation out of the Princeton library.[40] Lewis' well-known "modal" construct of a plurality of worlds based

---

[35] One *Physics Today* correspondent, L. E. Ballentine, soon published a seminal paper in Everett studies. He explained Everett's failure to solve the preferred basis problem as arising because he gave "the measurement process a privileged position over other interactions [and this] seems contrary to the spirit of Everett's program, which was motivated in part by a reaction against the privileged status of measurement [as wave function collapse] in the orthodox interpretation." He also critiqued Everett's claimed derivation of the Born rule as only partially satisfactory. Ballentine, L. E. (1973).

[36] DeWitt, B. S. *et al.* (1971B).

[37] Ibid.

[38] Ibid.

[39] Ibid.

[40] Jeffrey Barrett, private communication, 2008.

on "counterfactuals" is purely philosophical, and not quantum mechanical,[41] but his work has influenced modern interpretations of how probability fits into the Everett worlds.

DeWitt and Graham proceeded to publish a resource letter in the *American Journal of Physics* analyzing 500 papers written on the interpretation of quantum mechanics; many were critical of Bohr and von Neumann. Clearly, they were not in the business of endearing themselves to colleagues:

> No development of modern science has had a more profound impact on human thinking than the advent of quantum theory. Wrenched out of centuries-old thought patterns, physicists of a generation ago found themselves compelled to embrace a new metaphysics. The distress which this reorientation caused continues to the present day. Basically physicists have suffered a severe loss; their hold on reality.[42]

Saying that the impulse to interpret quantum mechanics is a "malaise," the authors offered their bibliography so that successive generations of physicists and philosophers would not waste time serving up "half-forgotten 'solutions.'"

As Everett's controversial theory gained traction, Wheeler began distancing himself, asking colleagues to remove the "W" from "EWG." But DeWitt, who had not communicated with Everett since their exchange of letters in 1957, did not want to be left out on a limb as the "sole spokesperson" for the many worlds interpretation, as Everett declined to publicly participate in the spirited debate. DeWitt decided to produce a collection of commentary on Everett:

> I was also convinced that Everett's 'Relative State' paper could not have constituted a complete statement of his views.... With Wheeler's help, however, I was able to get Everett to send me a thick, faded, dog-eared manuscript entitled 'The Theory of the Universal Wave Function.'[43]

This was the original thesis (retitled) that Everett gave Wheeler in January, 1956. DeWitt convinced Princeton University Press to publish it, along with reprints of Everett's short thesis and Wheeler's companion piece from *Reviews of Modern Physics*. Also included in the collection were DeWitt's *Physics Today* article and his Varenna paper; Graham's paper on probability; and the Cooper-van Vechten paper.

Everett was happy to have the long thesis published, although he drew the line at copy-editing or reading proofs—a task left to Graham. But he did invest some time rewriting the section of the manuscript where he deduced the Born rule, i.e. as a probability measure over sets of branching universes. After the intense criticism of his derivation from his detractors *and* supporters, he

---

[41] "Counterfactuals" are considerations of that which goes on in possible worlds as indicated by such conditional terms as "if … then." Lewis writes, "It is only by bringing the other worlds into the story that we can say in any concise way what character it takes to make what counterfactuals true. The other worlds provide a frame of reference whereby we can characterize our world." Lewis, D. (1986). 22.

[42] DeWitt, B. S. and Graham, N. (1971A). 724.

[43] DeWitt, B. S. (2008A). 3–4.

was determined to improve his explanation, without spending *too* much time on it.[44]

In October 1973, the collection was published as *The Many Worlds Interpretation of Quantum Mechanics*, edited by DeWitt and Graham. In its first five months, the combined hardcover and paperback editions sold 811 copies netting Everett a royalty payment of $500. He purchased a dozen copies of the paperback and kept them on a shelf, occasionally giving one away. Most of them ended up in his son's basement.

Quantum corrections

Although Everett told one correspondent that he was pleased with the way DeWitt had presented his theory,[45] there is evidence that he was not pleased with DeWitt's and Graham's analysis of his derivation of probability.

George Wesley, M.D., works for the inspector general at the Veterans Administration; collecting physics books is his hobby. In the mid 1990s, he was making the rounds of second-hand bookshops in Washington D.C. He discovered that Everett (of whom he had never heard) had scrawled his name on the inside cover of the first (and only) edition of *The Many Worlds Interpretation of Quantum Mechanics*. He had also corrected two typographical errors and penciled several acerbic comments in margins. It was Everett's personal copy.[46]

Where De Witt wrote, in the Varenna paper:

> Everett's original derivation of this result invokes the formal equivalence of measure theory and probability theory, and is rather too brief to be entirely satisfying.

Everett noted: "!Only to you!"
Where Graham wrote:

> In short, we criticize Everett's interpretation on the grounds of insufficient motivation. Everett gives no connection between his measure and the actual operations involved in determining a relative frequency, no way in which the value of his measure can actually influence the reading of, say, a particle counter. Furthermore, it is extremely difficult to see what significance such a measure can have when its implications are completely contradicted by a simple count of the

---

[44] This manuscript was sent to the AIP in 1990 by Nancy Everett and appears to be the manuscript that was given to DeWitt and Graham and later returned by Princeton University Press. Although we cannot be 100 percent sure that the corrections were made for the 1973 publication, they are far too messy to have been handed in to Wheeler as a formal dissertation in January, 1956. The January manuscript was corrected for typos and small matters before being sent to Bohr in April, but, again, Everett had access to a typist, Nancy, and is not likely to have sent a scrawled mess to Bohr, so it is logical to presume that the changes were made for the 1973 book. See also pps 137–138.
[45] Everett to Harvey, 5/20/77. He also told Harvey that the ms. was his "last remaining copy."
[46] I have seen the book and identified the handwriting (not to mention the tone) as Everett's. Mark Everett has no idea of how it ended up in the used bookstore. It might have been sold by Liz, or someone who stole it from the Everett household: or, perhaps, by Everett himself in a fit of pique.

worlds involved, worlds that Everett's own work assures us must all be on the same footing.

Everett commented: "bullshit."

Everett also penciled remarks on a bound copy of DeWitt's Varenna paper, which the author must have sent to him. Next to the exact same paragraph about probability ("rather too brief …") he made a slashing down stroke with his pencil, scrawling: "Goddamit you don't see it."

But he was pleased on the next page when DeWitt wrote:

> All the worlds are there, even those in which everything goes wrong and all the statistical laws break down. The situation is similar to that which we face in ordinary statistical mechanics. If the initial conditions were right the universe-as-we-see-it could be a place in which heat sometimes flows from cold bodies to hot. We can perhaps argue that in those branches in which the universe makes a habit of misbehaving in this way, life fails to evolve, so no intelligent automata are around to be amazed by it.

Everett slashed downwards again to bracket these sentences, scrawling a satisfied "yes."

33 Records in Time

Most physicists end up as footnotes.

Susanne Misner (2007)[1]

Bell ringing

Although he never published another word of quantum mechanics after his dissertation was printed, Everett kept tabs on his theory as it matured and drew attention. He paid particular mind to a critique written by John Stewart Bell, a staff physicist at CERN, the particle accelerator complex in Geneva, Switzerland.

Bell was an experimentalist whose hobby was exploring quantum foundations. Attracted to Bohm's hidden variables theory, he published a famous paper in 1964 showing that—contrary to Einstein's EPR speculation in 1935—quantum mechanics acts non-locally, spookily.[2] In a mere six pages, he proved that the entanglement of quantum objects rules out the locally deterministic hidden variables that Einstein had searched to define, but *not* non-local hidden variables. This means that two entangled particles separated by light years may be treated as a single quantum system.[3]

Motivated by DeWitt's *Physics Today* and Varenna papers, Bell investigated the work of Everett and Louis DeBroglie. In 1927, DeBroglie had constructed a hidden variables theory. Decades later, Bohm, independently, had arrived at a similar formulation. The positions of particles in the deterministic quantum universe of DeBroglie and Bohm are governed by non-collapsing "pilot waves."

---

[1] Interview, 2007.
[2] Bell, J. S. (1964).
[3] Bell's theorem asserts that a change in one particle in an entangled two-particle system instantaneously impacts the other particle. Even for two particles separated by light years, the wave function of the whole two particle system changes without, apparently, violating the dynamical law prohibiting faster than light information transfers. Experimentally validated, Bell's "inequalities" raised deep questions, which Bell believed would only be resolved when the quantum mechanical description of reality is superseded by "an imaginative leap that will astonish us.... In this it is like all theories made by man. But to an unusual extent its ultimate fate is apparent in its internal structure. [Quantum mechanics] carries in itself the seeds of its own destruction." Bell, J. S. (1966).

Bell saw a link between pilot waves and the determinism of Everett's branching worlds.

Amongst Everett's effects, is a preprint of a paper written in 1971 by Bell, "On the Hypothesis that the Schrödinger Equation is Exact." (It was later revised and renamed, "Quantum Mechanics for Cosmologists.")[4] Bell had subtitled a section on the many worlds interpretation: "Everett (?)." Upon reading this, Everett penciled: "?why?"

It is too bad they never met, because Everett had a lot to say to Bell, who was treating his controversial theory seriously, as worthy of thoughtful criticism. Where Bell wrote,

> Now it seems to me that this multiplication of universes is extravagant, and serves no real purpose in the theory and can simply be dropped without repercussions. So I see no reason to insist on this particular difference between the Everett theory and the pilot-wave theory – where, although the wave is never reduced, only one set of values of the variables χ is realized at any instant.

Everett wrote: "not consistent."
Bell observed,

> Then there is the surprising contention of Everett and DeWitt that the theory 'yields it own interpretation.' The hard core of this seems to be the assertion that the probability interpretation emerges [as the Born rule] without being assumed....I am unable to see why, although of course it is a perfectly reasonable choice with several nice properties.

Everett penciled: "re-read Proof!"
Channeling Schrödinger's fear of jellyfishication (lack of preferred basis), Bell wrote,

> Thus the structure of the wave function is not fundamentally tree-like. It does not associate a particular branch at the present time with any particular branch in the past any more than with any particular branch in the future.

Everett replied that his branching wave functions were time-reversible: "tree both ways! branching only relative to choice of bases."
When Bell continued,

> And the essential claim is that this does not matter at all. For we have no access to the past. We have only our 'memories' and 'records.'

Everett wrote: "correct" and "still unique measure."
Dissatisfied, Bell concluded,

> The Everett theory provides a resting place for those who do not like the pilot wave trajectories but who would regard the Schrödinger equation as exact. But a heavy price has to be paid. We would live in a present which had no particular past, nor indeed any particular (even if predictable) future. If such a theory were

---

[4] Bell, J. S. (1981).

taken seriously it would hardly be possible to take anything else seriously. So much for the social implication.

In the margin, Everett scrawled "?Ha what difference from probabilistic? also no unique past!" (In other words, the efficacy of the Born rule does not require a non-branching past.)

Bell wrote another paper critiquing Everett in 1976. But, overall, he was intrigued by the no-collapse stance—he just could not come to terms with a multiplicity of worlds—and he frankly admitted to not being able to follow Everett's technical argument on how the classical world of our experience emerges from a superposition of all possible universes.

Bell died in 1990, even as the work of Everett-inspired decoherence theorists[5] was starting to gain currency. That such a profound thinker paid serious attention to the many worlds theory was testimony to its growing relevance.

Everett's nemesis, Leon Rosenfeld, was infuriated.

Rosenfeld's campaign

In the wake of DeWitt's high-profile endorsement of Everett in *Physics Today*, Bohr's philosophical legacy was in jeopardy. His amanuensis, Rosenfeld, viewed the ascendancy of Everett's theory as the tip of a poisoned spear aimed at the heart of Copenhagenism. After reading Bell's preprint on Everett, Rosenfeld grabbed for his pen, horrified that a physicist of Bell's stature was taking the idea of a universal wave function seriously.

My dear Bell,

Many thanks for the preprint of your last paper which I <u>did</u> read because you are one of the very few heretics from whom I always expect to learn something, and, indeed, I found this new paper of yours exceedingly instructive. To begin with, it is no mean achievement to have given Everett's damned nonsense an air of respectability by presenting it as a refurbishing of the idea of preestablished harmony.[6]...Is it not complacent of you to think that you can contemplate the world from the point of view of God?[7]

Bell's critical analysis of Everett was distressing to Rosenfeld, but he was utterly appalled when Frederick J. Belinfante of Purdue University wrote in partial favor of the relative state heresy.

In the basement archive, there are yellowing typescripts of two papers on quantum measurement written by Belinfante. Everett read them, penciling comments. These papers formed the core of a book published by Belinfante in

---

[5] Particularly Dieter Zeh, Wocjiech H. Zurek, James B. Hartle, Murray Gell-Mann, Robert B. Griffiths, Erich Joos.
[6] Rosenfeld was being sarcastic since "preestablished harmony" was a mystical, consciousness-centered view of physical reality proposed by Gottfried Leibniz in the 17th century. He typically labeled any idea that disagreed with Bohr's ideas as "theological," so now Bell, by dealing with Everett seriously, was a theologian, in Rosenfeld's eyes.
[7] Rosenfeld to Bell, 11/30/71.

Leon Rosenfeld, date unknown.

1975 on quantum measurement and time reversal.[8] "Measurements in Objective Quantum Theory" attempts to salvage the standard interpretation of wave function collapse in quantum mechanics by claiming to solve the measurement problem with help from Everett. To achieve this (supposed) feat, Belinfante employed a "re-interpretation" of Everett, which excluded branching universes and a universal wave function.

Using a "technically ingenious" argument,[9] Belinfante claimed that Everett eliminated the contradiction between the Schrödinger equation and wave function collapse by showing the irreversibility of memory records. For

[8] Belinfante, F. J. (1975).
[9] Brown, H. (1979).

Belinfante, Everett's relative states formulation was a "handy tool" for making predictions without having to call upon a postulate of wave function reduction.

But he cast aside Everett's universal wave function, reinterpreting his theory:

> The essential point is that the universe as we know it is just one arbitrary member of an entire ensemble of universes, and that the laws of physics are probability laws for this big ensemble, and therefore are not determinate for 'our' universe.[10]

Everett scrawled his view of Belinfante's reinterpretation: "Nonsense." But when Belinfante said it was *not* necessary to accept von Neumann's wave collapse postulate, as Everett obtained the same measurement result using his non-collapse model, he cheerfully inscribed: "right on." And where Belinfante critiqued Mott's explanation of a single track emerging from a spherical wave by a process of wave reduction, Everett made an approving check mark.

After crediting Everett for breaking ground on the measurement problem, Belinfante disparaged the core of the many worlds theory, "not claiming that it would be illogical, but stressing its unpracticability and its superfluousness." Echoing a sentiment evinced by many of those who have studied Everett, Belinfante remarked, "It is impossible to say that this point of view would be definitely wrong (i.e. illogical). Most people 'do not like it.'" His main objection was that "Everett's claim that his ψ would describe the universe is a hoax, and fortunately is a hoax," because it is impossible to calculate a universal wave function in real life.

In a series of semi-hysterical letters written to Belinfante during the summer of 1972 (after he read Belinfante's preprint), Rosenfeld labeled Everett's "heresy" a "muddle," commenting,

> With regard to Everett neither I nor even Niels Bohr could have any patience with him, when he visited us in Copenhagen more than 12 years ago in order to sell the hopelessly wrong ideas he had been encouraged, most unwisely, by Wheeler to develop. He was undescribably [sic] stupid and could not understand the simplest things in quantum mechanics.... I would suggest that Occam's Razor could be most profitably used to rid us of Everett or at least his writings.[11]

In a subsequent letter,[12] he railed against the "pitfall" of trying to "rescue" Everett's "wooly thinking." But a few months later, Rosenfeld reconsidered his opposition:

> As to the case of Everett, I see your logical point, and have therefore to grant you unrestricted right to adopt Everett's point of view for the sake of argument.[13]

---

[10] Belinfante, F. J. (1975). 51–52.
[11] Rosenfeld to Belinfante, 6/22/72.
[12] Rosenfeld to Belinfante, 8/24/72.
[13] Rosenfeld to Belinfante, 10/31/72.

He relented because he realized that Belinfante had modified Everett's theory to support an interpretation that it in no way supports. Of Belinfante's modifications, Everett scrawled, "baby with wash."

Ultimately, Belinfante's "translation" of Everett did not take off, not the least because he attempted to resolve the inconsistencies in his own approach by arguing that quantum indeterminacy is evidence for the existence of God.[14]

Pulp science fiction

By the mid 1970s, Everett's theory was becoming a touchstone for physicists exploring non-collapse approaches. By 1980, it was widely considered to be one of most important papers ever written on quantum measurement. Whether or not one agreed with the reality of multiple universes, the genie was uncorked: the monocracy of the Copenhagen interpretation had been rudely and successfully challenged, and branching universes were no longer figments of science fiction.

As physicist Wojciech Zurek, a pioneer in decoherence theory, says, "Everett gave us permission to think about the universe quantum mechanically."[15]

Even physicists who think that branching universes are bunk credit Everett with exposing the illogic of Bohr's ontological divide between the microscopic and the macroscopic—as making the quantum world real.[16]

Everett's small collection of annotated physics papers is a firm indication that he was proud of his theory. He seldom spoke of his achievement, but he was undoubtedly delighted by its rebirth. He remained perplexed that the best minds in physics could not follow his derivation of the Born Rule from the quantum formalism; but not perplexed enough, unfortunately, to work on improving the theory.

In his library he kept a book by British cosmologist P. C. W. Davies. He made the checkmark of approval by Davies' brief description of his theory as based on increasing and decreasing entropy.[17] And not long before he died, he bought a copy of *The Dancing WuLi Masters, An Overview of the New Physics*, by Gary Zukov. In it, he flagged the pages explaining the "Everett-Wheeler-Graham" theory of many worlds as mystical, although he had stated firmly that his theory was "unmystical."

His joy overflowed in December 1976 when the pulp science fiction magazine *Analog* ran a four-page article ("The Garden of the Forking Paths") explicating the "Everett-Wheeler" interpretation, largely drawn from DeWitt's *Physics Today* article.

Everett ordered multiple copies of the magazine, sending them to friends, including Wheeler, who had already begun to publicly dissociate himself from

[14] Brown, H. (1979). 189.
[15] Zurek interview, 2006.
[16] Conferences at University of Oxford (2007), Perimeter Institute of Theoretical Physics (2007), FQXi Azores (2009).
[17] Davies, P. C. W. (1974). A "complimentary" edition of book was sent to Everett, who read at least the pages on which his own name appeared.

the many worlds interpretation. Being praised in *Analog* probably added salt to Wheeler's festering wound. The article was in a section called "Quantum Physics and Reality." It ended with a quote from Jorge Luis Borge's story, *The Garden of Forking Paths*, the same quote that had inspired Cooper and, also, graced the preface of *The Many Worlds Interpretation of Quantum Mechanics*:

> ...a picture, incomplete yet not false, of the universe as Ts'ui Pên conceived it to be. Different from Newton and Schopenhauer, ... [he] did not think of time as absolute and uniform. He believed in an infinite series of times, in a dizzily growing, ever spreading network of diverging, converging and parallel times. This web of time—the strands of which approach one another, bifurcate, intersect or ignore each other through the centuries – embraces every possibility. We do not exist in most of them. In some you exist and not I, while in others I do, and you do not, and in yet others both of us exist. In this one, in which chance has favored me, you have come to my gate. In another, you, crossing the garden, have found me dead. In yet another, I say these very same words, but am an error, a phantom.[18]

[18]   Borges, J. (1962). 100.

34 Austin

*[My] theory is therefore capable of supplying us with a
complete conceptual model of the universe, consistent with
the assumption that it contains more than one observer.*

Hugh Everett III, 1956[1]

Spring, 1977

Shortly after Wheeler relocated to the University of Texas in Austin, he and
DeWitt invited Everett to give a seminar on the many worlds interpretation.
Even though he loathed public speaking, Everett was excited: In April, he
packed Nancy, Liz, and Mark into his second-hand Cadillac Seville and drove
from Virginia to a motel.

While their father in Austin lectured, Mark and Liz tooled about in a VW
Beetle drinking beer with a soldier friend of Liz's stationed nearby.[2]

The talk packed the university hall with curious students and teachers. An
exception to the school's no-smoking policy was made for Everett. Dressed in
his trademark black suit, he explained the basics of the theory, answering questions in a staccato voice as he paced back and forth, chain smoking and ges-
ticulating. At one point he went off on a tangent to promote Pugh's recently
published book on the biological origin of human values.

In an after-talk lunch at a beer-garden, a young, British graduate student
named David Deutsch sat next to Everett. He was researching quantum gravity
with Wheeler and DeWitt. DeWitt had introduced him to Everett's theory, and
after first disbelieving it, Deutsch became one of its firmest advocates. He is
now recognized as a pioneer in the field of quantum computation, which, he
argues, is made possible by the existence of many universes.[3]

Deutsch recalled that Everett was,

Full of nervous energy, high-strung, extremely smart, very much in tune with
the issues of the interpretation of quantum mechanics despite having left

---

[1] DeWitt, B. and Graham, N. eds. (1973). 109.
[2] Liz later moved to Hawaii to live with the soldier.
[3] In his popular book, *The Fabric of Reality* (1997), Deutsch constructs a candidate "theory of
everything" by interweaving Everett's many worlds scheme with the ideas of Karl Popper
(epistemology), Alan Turing (computation), and Richard Dawkins (evolutionary biology); Deutsch's
remembrances of Everett are from interviews with the author.

academic life many years before. I find that this is unusual: normally, people who leave become 'rusty' in time.

He was extremely enthusiastic about many universes, and very robust as well as subtle in its defense, and he did not speak in terms of 'relative states' or any other euphemism.

Deutsch asked him about the preferred basis problem. Everett said he did not think the question of how a preferred basis emerges was a problem; each universe, he said, is defined by the energetic structure of its system as a whole. In other words, the preferred basis is not something extra that you have add to the formalism, it emerges out of the theory naturalistically.[4] Deutsch is not alone in holding that the entanglement of a quantum system with the larger environment—the decoherence phenomenon—solves the preferred basis problem in the many worlds theory, not the least by prohibiting communication between branching—or decohering—universes.

Deutsch did not ask his lunch partner about the related problem of his derivation of probability; but he eventually agreed with DeWitt that Everett's probability measure did not work; he has several times tried to improve upon it. In recent years, Deutsch, who works with the University of Oxford's Center for Quantum Computation, has teamed up with a vibrant group of "Everettians" in the department of philosophy. Together, they are tackling the many worlds interpretation, searching to bridge gaps in the theory and, perhaps, to develop a new theory of quantum probability.

Only on Tuesdays

Returning to McLean, the Everetts invited the Misners over to celebrate the latter's 18th wedding anniversary. After numerous cocktails, Everett and Misner (pushed by Susanne Misner) made the rambling tape recording where Everett recollected the origins of his theory. He said that Wheeler had recently "confessed" to him that he reserved the right to disbelieve the many worlds theory once a month, on a Tuesday. Wheeler was being circumspect. He had already taken pains to distance himself from Everett, and he was about to publicly disown him in a series of papers called "Law Without Law."

Years later, Deutsch cut to the chase:

Wheeler told me that he was always implacably opposed to the theory—what he supported was Everett. He wanted Everett's idea to become known, not because he thought it was true, but because it was a jumping off point to look at the logic

---

[4] Deutsch's recollection of Everett's answer is worth quoting in its entirety, as there is no other record of him addressing this question after the Xavier conference. "When I asked him what defines the Hilbert space basis with respect to which one defines 'universes', in the general case (not just perfect measurements, where the answer is obvious), he said it was the structure of the system itself. I asked, which aspect of the structure the state itself, the Hamiltonian or what? He said the Hamiltonian, but he didn't think that this was an important issue." Deutsch email to Shikhovtsev, 2000.

of wave collapse, which Wheeler did not think was the right approach to the measurement problem.[5]

Shortly after the Austin trip, Paul Benioff, a researcher at the Argonne National Laboratory in Illinois sent Wheeler a paper critiquing the mathematics of the "Everett-Wheeler" derivation of probability as "vacuous." Wheeler forwarded the paper to Everett along with a copy of his reply to Benioff, which noted,

> The point you make is obviously important…I should add that Everett's Princeton PhD thesis was on a topic entirely conceived by him and ought to be called the Everett Interpretation not the Everett Wheeler Interpretation. Though I have difficulty subscribing to it today, I still feel it is one of the most important contributions made to quantum mechanics in recent decades.[6]

Everett circled the word "difficulty," scrawling "only on Tuesdays." A few weeks later, Wheeler copied Everett on another letter to Benioff[7] (who had written a second paper on the many worlds theory) asking him to *please* stop referring to the Everett-*Wheeler* theory. For Benioff's edification, he enclosed a reprint of his new paper (dismissing Everett's theory), "Include the Observer in the Wave Function?"

Everett was incensed; he slash-penciled corrections all over Benioff's first paper, which examined the many worlds probability measure from Graham's point of view, not from Everett's. Everett wrote: "I made identification with Mathematics of Probability, and short-circuited all such nonsense. Only understanding of isomorphism is required!" Once again, his derivation of the Born rule was not being understood. Asked to referee Benioff's paper for *Foundations of Physics*, Everett drafted a reply letter rejecting it "in toto." Then he (diplomatically) revised the letter, saying that he did not have time to review it.[8]

A year after the Austin trip, Nancy drafted a letter to Wheeler, thanking him for forwarding the Benioff papers and letters, and apologizing for her husband's inability to correspond with him. She never mailed this letter.[9] Another year passed, and one evening the Everetts watched Wheeler host a PBS television show on Einstein. Inspired, Nancy picked up the old draft, slightly rewrote it, had Everett sign it, and mailed it off.

It was an odd letter, written in the third person. It read, in part,

> There are two things about Hugh that perhaps need clearing up. One is, tho' it appears he plays hard-to-get by refusing to correspond, the truth is, he feels the written word is totally inadequate in comparison to a one-to-one conversation. This is why the meeting in Austin two years ago with Bryce DeWitt, [and Deutsch] was such a great thing for him to participate in….
>
> Far from being totally unconcerned, Hugh may even feel some gratification to be receiving a small measure of recognition for his work done under your

---

[5] Deutsch interview, 2004.
[6] Wheeler to Benioff, 7/7/77.
[7] Wheeler to Benioff, 9/7/77.
[8] Documents in basement.
[9] Basement.

counsel....Now we read in <u>Physics Today</u> that even more is being done to expedite the flow and exchange of ideas what with the Institute forming in Santa Barbara. (Hugh always thought Santa Barbara a lovely spot).[10]

In July 1979, in what may very well have been their last communication, Wheeler wrote to Everett:

Thank you for your letter of too many weeks back. I think you got a great subject going and I am overjoyed at the thought of your getting back and going to bat for it![11]

On that same day, Wheeler copied Everett and DeWitt on a letter to the director of the Institute for Theoretical Physics at University of California, Santa Barbara:

Hugh Everett who did that fascinating Everett interpretation of quantum mechanics and who ought to be got back into it to go on with it has written to me indicating that he might conceivably get free to spend a period at the Institute.

I have written Bryce DeWitt about this and believe that it has real possibilities for quite fruitful interactions.[12]

DeWitt replied to Wheeler,

There is nothing in Bohr's description that is not contained in Everett. Everett, however, would amplify Bohr's statement. When Bohr says that an elementary act of measurement (or quantum interaction or what have you) is 'brought to a close by an irreversible act of amplification,' Everett would add that the process of one 'world' splitting into many is simultaneously brought to completion.

Everett in fact delves deeper by pointing out that nowhere does Bohr give a rigorous definition of 'irreversible' or make a distinction between 'irreversible in practice' and 'irreversible in principle.' Some might say that therein lies Bohr's strength, that the distinction has no operational significance.

Everett suggests (and I believe) that it is a mistake to transform the wonderful lessons that Bohr has taught us into points of dogma. The history of physics has taught 1. that one should never be dogmatic, 2. that one should never hesitate to push a formalism to its ultimate logical conclusions however absurd. In the case of the formalism of quantum mechanics one cannot say that the interferences are there at one moment but gone the next.

All that Everett is really trying to say is that the interferences are in principle always there. As David Deutsch so aptly puts it: 'Quantum theory is the Everett interpretation.' The theory may ultimately be proved wrong, but at the present time you cannot have one without the other.[13]

---

[10] Everett to Wheeler, 3/21/79.
[11] Wheeler to Mr. & Mrs. Hugh Everett, 7/12/79.
[12] Wheeler to Scalapino, 7/12/79.
[13] DeWitt to Wheeler, 9/25/79.

Despite Wheeler's enthusiasm, Everett made no effort to re-ignite his theo-retical career. In 1980, two years before his death, he wrote a letter to physics enthusiast L. David Raub:

I certainly still support all of the conclusions of my thesis.... Dr. Wheeler's posi-tion on these matters has never been completely clear to me (perhaps not to John either). He is, of course, heavily influenced by Bohr's position... It is equally clear to me that, at least sometimes, he wonders very much about that mysteri-ous process, 'the collapse of the wave function.' The last time we discussed such subjects at a meeting in Austin several years ago he was even wondering if some-how human consciousness was a distinguished process and played some sort of critical role in the laws of physics.

I, of course, do not believe any such special processes are necessary, and that my formulation is satisfactory in all respects. The difficulties in finding wider accept-ance, I believe, are purely psychological. It is abhorrent to many individuals that there should not be a single unique state for them (in the world view), even though my interpretation explains all subjective feelings quite adequately and is consistent with all observations.[14]

Decades later, Deutsch summed up why the many worlds model is so impor-tant in the history of physics and philosophy:

Everett was before his time, not in sense that his theory was not timely—every-body should have adopted it in 1957, but they did not. Above all, the refusal to accept Everett is a retreat from scientific explanation. Throughout the 20th cen-tury a great deal of harm was done in both physics and philosophy by the abdi-cation of the original purpose of those fields: to explain the world. We got irretrievably bogged down in formalism, and things were regarded as progress which are not explanatory, and the vacuum was filled by mysticism and religion and every kind of rubbish. Everett is important because he stood out against it, albeit unsuccessfully; but theories do not die and his theory will become the prevailing theory. With modifications.[15]

[14] Everett to Raub, 4/7/80.
[15] Deutsch interview, 2006.

35 Wheeler Recants

Given for one instant an intelligence which could comprehend all the forces by which nature is animated and the respective situation of the beings who compose it—an intelligence sufficiently vast to submit these data to analysis—it would embrace in the same formula the movements of the greatest bodies of the universe and those of the lightest atom; for it, nothing would be uncertain and the future like the past would be present to its eyes.

Pierre Simon Laplace, 1814[1]

In 1972, Max Jammer was researching the final chapters of *The Philosophy of Quantum Mechanics*. He wrote to Wheeler asking for Everett's address. The professor had lost track of his whereabouts, replying, "Bohr did not take to [Everett's] way of describing q. mechanics, as he also earlier had not accepted Feynman's way of ascribing q. mechanics."[2]

Was he still carrying Everett's torch?

Jammer later learned the McLean address from DeWitt, who also told him about the existence of the long thesis. He wrote to Everett,[3] who replied at length, recounting the influences on this thinking when he wrote the original thesis. He said his theory was the simplest explanation of quantum mechanics available, but accepting it was "a matter of taste." He said he had sent his last copy of the long thesis to DeWitt for publication. And that the only other copy was locked in a file drawer along with the doctoral thesis on EPR of his business partner, Reisler. When they had started DBS, the two physicists had locked up their respective dissertations, agreeing not to talk about quantum mechanics for ten years, "when presumably we could afford the luxury of such a diversion."[4]

Everett's spirits must have been lifted by the attention being paid to his theory. But the failure of Lambda weighed heavily upon him; he was sexually obsessed; and alcoholism was slowly degrading his ability to perform intellectually. Then, Wheeler, his mentor, and once his strongest advocate, publicly disowned him. Why? Let's revisit 1957 and take it from there.

[1] Laplace, P.S. (1814). "Philosophical Essay on Probability." Quoted in Wheeler, J. A. (1977). 11.
[2] Wheeler to Jammer, 3/19/72.
[3] Jammer to Everett, 8/28/73.
[4] Everett to Jammer, 9/19/73.

The power of Bohr

In the spring of 1957 Bohr and his inner circle categorically rejected Everett's "heretical" quantum model because—above all other considerations—it violated Bohr's prohibition on talking about reality as quantum mechanical, as fundamentally non-classical. Wheeler went on to publish his enthusiastic endorsement of Everett's work in *Reviews of Modern Physics*, but he was distressed about having unsuccessfully challenged Bohr, and he soon found an opportunity to reaffirm his loyalty.

In October 1957, Bohr received the first Atoms For Peace award in Washington D.C. The $100,000 award was sponsored by the Ford Motor Company to honor scientists working on peaceful uses for atomic energy. Wheeler's and Bohr's work on fission had made the atomic bomb possible, and Wheeler was a prime architect of the hydrogen fusion bomb, but these credentials did not hurt them with Atoms for Peace, which asked Wheeler to make the keynote speech.[5] Delivered in best-courtier fashion, it cast Bohr as the epitome of civilized man. According to Wheeler, Bohr's interpretation of quantum mechanics was "the most revolutionary scientific concept of this century," presumably surpassing Einstein's contributions.

Wheeler had invited Everett to the Atoms for Peace award dinner. But he declined to attend; if he had gone, he probably would have gagged when Wheeler lauded Bohr's institute as a place where, "young men come from all over the world to work in an atmosphere free of envy and jealousy, one that could be described in the words of Sir William Rowan Hamilton's poem, 'Yet with an equal joy let me behold, Thy truth's chariot o'er that way by others rolled.'" In Copenhagen, said Wheeler, "the climate of ideas opens the way to new advances in physics." This was, at best, wishful thinking on Wheeler's part. After the Second World War, Bohr's institute had become more of a temple or ideological fortress than a laboratory for new ideas.

As long as the world of physics talked of Everett behind closed doors, Wheeler could live with being associated to the upstart theory. But, after DeWitt started up his Everett bandwagon, the world at large started paying attention to the Everett-*Wheeler* theory of multiple universes. Wheeler was torn. He did not deny that the theory had merit, but the attachment of his good name to it harmed his standing with committed Copenhagenists, such as Rosenfeld and Heisenberg. Many prominent physicists, including Feynman, thought many worlds was a ludicrous idea. And espousing a belief in multiple universes was not necessarily consistent with the image of rationality that Cold War scientists typically strove to project.

But if Wheeler was to repudiate Everett, he might end up looking the fool if an experimental proof of the theory ever panned out.

So, he vacillated.

---

[5] Wheeler. J. A. (1957A).

Trieste, Italy, 1972

The "Symposium on the Physicists Conception of Nature in the Twentieth Century" in Trieste was attended by two score of the finest quantum thinkers of the day, including, Wigner, Heisenberg, Dirac, Bell, Lamb, Rosenfeld, Schwinger, Cooper, and Chen Ning Yang. Wheeler came to the cosmopolitan seaport having just passed the magic age of 60, when physicists often celebrate a *festschrift*, call it a day, and become *emeritus*. But he was hyperenergetic, joyously churning out strange ideas in the service of uniting gravity and the quantum. He had recently named the gravitational singularities at the heart of quantum cosmology "black holes." He continued to obsess about "quantum foam" seething with electromagnetic geons and gravitational wormholes. He was fascinated by the relationship of the probabilistic chaos of the microscopic universe to the macroscopic order of our classical world. And, quixotically, he desired for all of these new ideas to fit into Bohr's complementary mould.

Wheeler arrived in Trieste on the heels of having published a massive textbook, *Gravitation*, written with Misner and Kip Thorne. He referred repeatedly to *Gravitation* in a lengthy, provocative paper he delivered on the intersections of general relativity and the quantum in which he catalogued milestones in the evolution of quantum mechanics from Planck to Bohr to Heisenberg, Dirac, von Neumann, Feynman, and, finally, to *Everett*, whose relative states theory allowed physicists to treat the whole universe as quantum mechanical.

After equating Everett with quantum giants, Wheeler suggested that entire universes can float in "superspace," a foamy space-time geometry without the dimension of time. Probability waves representing histories of whole universes propagate through superspace scattering off black holes. Within this mathematical framework, Wheeler postulated the existence of "leaves of history," i.e. alternative histories of the universe: multiple, co-existing universes. Each universe exists on a "classical leaf in history in superspace" and has "quantum spread." Due to the nature of gravitational collapse, "There is also not the slightest possibility to travel to another leaf of history."[6]

He was describing a type of many worlds theory:

> No one can deny the 'co-existence of alternative histories of the universe' who accepts the existence of quantum fluctuations in the geometry of space.[7]

It was the superposition principle writ large.

> One has only to recall the famous double slit electron interference experiment to see the same principle in a simpler context. The 'co-existence of two histories' of the electron is the very heart of the observed interference. No one has ever successfully contested it.[8]

---

[6] Wheeler, J. A. (1973).
[7] Ibid. 232.
[8] Ibid. 232.

Philosophically, Wheeler did not care for the notion of a Big Crunch, the prediction that our universe will ultimately collapse into a black hole, squashing the laws of physics as we know them (he was writing long before it was discovered that we probably live in an outwardly accelerating universe). Confronted by the prospect of oblivion and anarchy in the foundations of physics, Wheeler harkened back to ancient Grecian themes of eternal change as the ultimate physical law. He postulated that the deepest laws of physics will survive the gravitational collapse of our universe:

> It is difficult to find any other way to summarize the situation as it now appears than this: 'There is no law except the law that there is no law;' or more briefly, 'Ultimate MUTABILITY is the central feature of physics.'[9]

Searching for a philosophical tether in a multiversian reality roiled by chaos, uncertainty, the end of time, Wheeler was becoming more and more attached to Bohr's principle that the act of observation changes the object observed. In Trieste, he called for the concept of "observer" to be replaced by "participator." He then abandoned the concept of the wave function as physically real. He reached back 2,500 years, to quote Parmenides of Elea, a poet who spoke of possible worlds brought into being by observation: "What is,…is identical with the thought that recognizes it."[10]

And Leibniz:

> Although the whole of this life were said to be nothing but a dream and the physical world nothing but a phantasm, I should call this dream or phantasm real enough if, using reason well, we were never deceived by it.[11]

This was basically a restatement of Wigner's idealist philosophy that human consciousness creates physical reality. Everett believed the reverse: that physical reality creates human consciousness, which is why his theory of quantum entanglement does not require human participation as a *sine qua non*.

Strasbourg, France, 1974

Two years later, Wheeler gingerly removed Everett from his list of quantum greats. The occasion was "A Colloquium on Fifty Years of Quantum Mechanics," held at the University of Louis Pasteur, Strasbourg. It was attended by more than a dozen physicists, including a young French man, Jean Marc Lévy-Leblond. Wheeler's presentation was called, "Include the Observer in the Wave Function?"[12]

The most striking feature of Everett's universal wave function is that it includes the observer. In Strasbourg, Wheeler attacked the core of the many worlds model by *excluding* the observer from the wave function. In doing so he referred to Wigner's argument that wave mechanics *only* describes correlations between

⁹ Ibid. 242.
¹⁰ Ibid. 244.
¹¹ Ibid. 244.
¹² Wheeler, J. A. (1977).

conscious observations. According to Wigner, and now Wheeler, observers of quantum systems cannot stand outside a wave function, but neither can they be part of a wave function as they must observe it. If this seems confusing: it is. Wheeler summed up: "There has to be a wave function for the universe but there can't be a wave function for the universe: that is the dilemma."[13]

Wheeler was concerned that if one adopts the use of a deterministic universal wave function, "predictability perishes." And yet we experience the world probabilistically. Clearly, he no longer believed his claim of two decades previous that Everett's universal wave function explained the "appearance" of probability in a branching universe. He now called Everett's claim "extreme."[14] Desiring to keep probability, he was ineluctably drawn toward an amalgam of the Wigner-von Neumann postulate of conscious intervention as creating physical reality, and Bohr's admonition that, Wheeler paraphrased, "an observation is only complete when there is an [external] observer."[15]

Hence, the observer cannot be included in the wave function, Wheeler concluded, attacking a pillar of Everett's theory: that reality does not require the participation of an observer. Wheeler cast away Everett's main physical idea: that the observer is included in the wave function, i.e. that the whole universe is quantum mechanical. Rather, he said, the universal wave function is "beyond the reach of the laws of physics itself."

Wigner seized gladly upon Wheeler's quasi-spiritual conversion, writing, probably with some relief, that his friend no longer supported Everett.[16] Wigner himself rejected the many worlds theory on the basis that "it can neither be confirmed nor refuted."[17] Of course, the same could be said for his own theory that conscious impressions form the "primitive reality."[18]

Hand washing contest

Several years later, Lévy-Leblond sent Everett a copy of a paper he had presented at Strasbourg, "Towards a Proper Quantum Theory."[19] In his cover letter, Lévy-Leblond said that, at the symposium, Wheeler had

> suggested that I directly ask your opinion on what I believe to be a crucial question concerning the 'Everett & no-longer-Wheeler' (if I understand correctly!) interpretation of Qu. Mech.[20]

Lévy-Leblond had his own interpretation of Everett's interpretation:

> The question is one of terminology: to my opinion there is but a single (quantum) world, with its universal wave function. There are not 'many worlds,' no

---

[13] Ibid. 2.
[14] Ibid. 13.
[15] Ibid. 14.
[16] Wigner, E. (1981). 289.
[17] Ibid. 294.
[18] Wigner, E. (1973). 382.
[19] Lévy-Leblond, J. (1976).
[20] Lévy-Leblond to Everett, 9/17/78.

'branching', etc., except as an artifact due to insisting once more on a classical picture of the world.

Uncharacteristically, Everett answered Lévy-Leblond, writing that his offering "is one of the more meaningful papers I have seen on this subject, and therefore deserving of a reply."[21] Specifically Referring to of Lévy-Leblond's analysis of the many worlds theory, Everett wrote, "In this case, your observations seem entirely accurate (as far as I have read)."

On those pages, Lévy-Leblond had written:

> To me, the deep meaning of Everett's ideas is not the coexistence of many worlds, but, on the contrary, the existence of a single *quantum* one. The main drawback of the 'many-worlds' terminology is that it leads one to ask the question of 'what branch we are on', since it looks as if our consciousness definitely belonged to one world at a time. But this question only makes sense from a classical point of view...It becomes entirely irrelevant as soon as one commits oneself to a consistent quantum view, exactly as the question of the ether was deprived of meaning, rather than answered, by a consistent interpretation of relativity theory.[22]

That the multiverse is a giant superposition embracing branching worlds, some classical, some not, was a key feature of Everett's theory—and this important feature was not clearly explained in either version of his dissertation. It is relevant to why he always viewed the uncountably infinite branches as "equally real."

Everett explained to Lévy-Leblond:

> I have not done further work in this area since the original paper in 1955 (not published in its entirety until 1973, as the 'Many-Worlds Interpretation etc.'). This, of course, was not my title as I was pleased to have the paper published in any form anyone chose to do it in! I, in effect, had washed my hands of the whole affair in 1956.

The hand-washing claim was, as we have seen, an exaggeration. Subsequent to 1956, he had made some substantial efforts to explain his theory, and he had monitored its trajectory. He apparently edited the section on probability in the long thesis prior to publication. But even Everett's closest friends were unaware how much he cared about the fate of his theory because he cultivated the appearance of not caring about the most significant achievement of his life.

Wheeler, on the other hand, was scrubbing his hands like mad.

Great smoky dragons

After Wheeler retired from Princeton and moved to Austin, Texas in 1976 to teach and research physics at the University of Texas alongside DeWitt, one of the first papers he wrote was "Bohr's 'Phenomenon' and 'Law Without Law.'"

---

[21] Everett to Lévy-Leblond, 11/15/77. Either Everett or Lévy-Leblond had the year wrong.
[22] Lévy-Leblond, J. (1976). 194.

In successive years, he wrote several papers called "Law Without Law," as he shed his Everettian baggage and reaffirmed his allegiance to the Copenhagen interpretation, adding a dose of Wignerian idealism. In his pantheon of quantum heroes, he replaced Everett with *Rosenfeld*.[23] And he waxed poetic about his late mentor's philosophical commitment to indeterminism:

> The great smoky dragon, Bohr's phenomenon, has its tail sharply localized at the point of entry to the apparatus. Its teeth are sharply localized where it bites the grain of photographic emulsion. In between it is utterly cloud-like, localized neither in space nor in time.... The central point of quantum theory can be put into a single, simple sentence. 'No elementary phenomenon is a phenomenon until it is a registered ('observed,' 'indelibly recorded') phenomenon.'... We are inescapably involved in bringing about that which appears to be happening.[24]

Everett, of course, thought that quantum interactions left indelible traces in the environments (including, but not limited to observer brain states) of respective branches; and that physical change was *not* dependent upon consciousness and human participation, quite the opposite.

In 1979, Wheeler explicitly disposed of the many worlds interpretation, as well as the "branching histories" approach he had favored a few years before.[25] He questioned any and all assumptions of a deterministic universe. Those who agree with Laplace, said Wheeler, may claim: "The Universe is a machine. No, we have to tell him: that is a cracked paradigm."

And no paradigm was as cracked as Everett's:

> Imaginative Everett's thesis is, and instructive, we agree. We once subscribed to it. In retrospect, however, it looks like the wrong track. First, this formulation of quantum mechanics denigrates the quantum. It denies from the start that the quantum character of Nature is any clue to the plan of physics. Take this Hamiltonian[26] for the world, that Hamiltonian, or any other Hamiltonian, this formulation says. I am in principle too lordly to care which, or why there should be any Hamiltonian at all. You give me whatever world you please, and in return I give you back many worlds. Don't look to me for help in understanding this universe.
>
> Second, its infinitely many unobservable worlds make a heavy load of metaphysical baggage.[27]

In 1983, a year after Everett's death, Wheeler and Wojciech Zurek published a comprehensive collection of the all-time great papers in quantum measurement theory, among them Everett's short thesis. The "Law Without Law" essay, in which Wheeler had dismissed the many worlds theory as "metaphysical baggage," was also included: but he excised his negative references to Everett,

---

[23] Wheeler, J. A. (1985). 363.
[24] Ibid. 365.
[25] Wheeler, J. A. (1979A).
[26] A Hamiltonian is a function used to express the energy of a system.
[27] Ibid. 396.

replacing them with eliding dots.[28] It might have been hard to explain, after all, why he was including Everett's paper in a collection of the best foundational ideas in the history of quantum theory, when he totally disbelieved it. Or, perhaps, he had simply changed his mind again.[29]

---

[28] Wheeler, J. A. and W. H. Zurek (1983). 201.
[29] In 2001, Wheeler co-authored a paper with cosmologist (and Everett-fan) Max Tegmark, "100 Years of the Quantum." The paper explains how decoherence may be construed to solve the preferred basis problem in the many worlds interpretation, and it credits Everett with causing a shift in how physicists interpret quantum mechanics as fundamentally unitary (non-collapsing). Tegmark, M. and Wheeler, J. A. (2001).

BOOK 11
AMERICAN TRAGEDY

36 The Final Years

*However unreasonable or immoral an action may be, man
has an insuperable urge to rationalize it, that is, to prove to
himself and to others that his action is determined by reason,
common sense, or at least conventional morality. He has little
difficulty in acting irrationally, but it is almost impossible for
him not to give his action the appearance of reasonable
motivation.*

Erich Fromm, 1955[1]

Everettian blues[2]

Listening to Everett's sing-songy voice on the only known tape recording of
him gives one a feel for his crackling energy when inebriated. He sounds like a
man who wants to enjoy the good life with friends; but he also he sounds a bit
fey, a bit self-deprecating, a bit sad. In truth, he was not wont to show his sad-
ness, or any other emotion, to friends and colleagues. He projected the attitude
of a cynical man of rare intelligence who viewed life as a game in which smart
people minimized the worst that could happen. He seemed largely indifferent
to the feelings of others. Indeed, he seemed indifferent to the fate of his many
worlds theory, as he seldom spoke of it.

Susanne Misner was more observant. She remembers her friend of a quarter
century as a clown laughing to hide tears.

His wife and children knew *that* Everett: the sad, uncommunicative alcoholic napping on the living room couch after dinner, snoring loudly as the
television blared, waking to write computer code at the dinner table, chain
smoking and drinking into the wee hours while the family slept.

Absorbed by her inner life, Nancy was an indifferent house-keeper. The
house reeked of tobacco smoke; newspapers and magazines were stacked all
over. The "fallout shelter" in the basement was gradually taken over by batch-
numbered bottles of homemade wine jokingly labeled "DP" for Dom Perignon
(it was barely drinkable). After Everett harvested his backyard grapes in the

---

[1] Fromm, E. (1955). 65.
[2] This narrative is based on interviews from 2006 to 2009 with Mark Everett and many of Everett's
colleagues and friends.

Everett making wine in kitchen, circa 1976.

autumn, he squashed them in garbage cans in the kitchen—permeating the house with the pungent smell of fermentation.

For a while his hobby was CB-radio. He kept a unit in the kitchen, and another in his used Cadillac, regularly chatting with redneck trucker friends, occasionally inviting them over for beer and barbecue. His handle was Mad Scientist, and he looked the part, with his goatee and his freakishly long, curled fingernails. In a nod to self-preservation, he smoked his Kent cigarettes through a long plastic filter.

Nancy was allowed into his world to the extent she joined his pursuit of fine, fatty, French food, washed down with copious amounts of premium alcohol, followed by occasional bouts of sport sex. She was widely viewed by his associates as self-effacing, childlike, an odd match for the scientist. But she was loyal to him, and he to her, in his own way. And she was in charge of the children.

Unlike the stereotypical neighborhood Dad, Everett did not coach sports or take the kids on outings—except *once* he took Mark to a circus. In vain, the children craved attention from their remote, self-absorbed father. Nancy was a constant presence, ready to give them rides and feed them. But Liz and Mark were largely unsupervised—allowed to drink, smoke dope, and have sex with friends at home. They were not chastised or put on restriction or given the limits that children need to learn *who* they are. They were often in hot water with teachers, neighbors, cops. By her late teens, Liz was a full-blown mess. Manic-depressive, she turned to sex, alcohol, and a variety of drugs—from LSD and pot to cocaine and heroin—to kill the pain of being alive. It was not uncommon for her to drink a beer for breakfast. Mark, six years younger, revered his sister, even as she spun out of control. Downstairs in the basement, he constantly banged his drum set, filling the house with percussive demands, and, even then, his parents left him to his own devices.

Reisler met Everett when he applied for a job at Lambda in 1970. He shyly inquired of Reisler, who holds a PhD from Yale University in foundations of quantum mechanics, if he'd ever heard of the relative states theory? Reisler thought, "Oh my God, you are *that* Everett, the crazy one." He quietly acknowledged that he was aware of the interpretation. And that was the last time they spoke about quantum mechanics. Even when Everett returned from Austin he did not mention the trip to Reisler.

Reisler sums up: "Hugh was a Renaissance man. It is a mistake to view him only as a physicist or operations researcher; he was more than that, he was a problem solver. And a hedonist." Reisler, a non-smoker, non-drinker and an outdoorsman, was, in terms of healthy lifestyle, a world apart from his pallid-faced partner.[3]

When Everett and Reisler started DBS, they agreed that Everett would tend to the science, and Reisler would manage the business. To Everett's dismay, Reisler insisted that they *not* procure prostitutes for clients. Their clients were mostly officials at federal agencies, including the departments of Health, Education, and Welfare, and Justice. They also had a lucrative subcontract with AMS to optimize ship maintenance for the Navy. For a while, DBS was closely aligned with AMS, which owned 25 percent of the company. The AMS founders were reluctant to let Everett stray too far out of sight, so they kept him on a financial tether, and lunched with him regularly, picking his brain.

DBS's bread and butter was overseeing affirmative action laws for federal agencies. Everett designed computer programs to ferret out patterns of discrimination. His forte was inventing algorithms to sort huge quantities of data while checking for computing errors. Reisler delivered expert testimony on behalf of agencies suing corporations and local governments for race, disability, and gender discrimination. At its height, the privately owned company reached $1.25 million in sales and kept a score of employees. But the bulk of DBS's work was nowhere near as interesting or as demanding as military operations research, and Everett quickly became bored. With a girlfriend, Georgia Bailey, he started a travel agency, giving her a $10,000 ownership stake in the firm, which she was to manage.

---

[3] In late 2007, Reisler was diagnosed with lung cancer, which his doctors attributed to being around Everett's second-hand smoke. Reisler private communication, 2007.

Everett and Donald Reisler, circa 1979.

On a yellow legal pad he tested a succession of names for the company: Many Worlds, Black Hole, Crossroads, Trantor-Terminus, Quantum Quality, Green Knight, Stoned Stranger, Mighty Mouse, Virgin Territory. He settled on Key Travel.

It was never a money-maker, quite the contrary, but it did enable him to get discounts on the week-long ocean cruises he took at every opportunity. He loved to travel first-class, whether by air or water. But family vacations were often spent at a modest beach house designed and built by his father in Kitty Hawk, North Carolina.

He also bought a condo on St. Thomas in the Virgin Islands. Friends recount that Nancy did not mind when her husband invited "other" women along on their Caribbean adventures. He was always on the make, shamelessly slipping his hand under the dinner table onto the thighs of wives of his friends; but he did not hold a grudge when his advances were rejected (and they were not always rejected).

He loved playing the big spender, picking up dinner tabs with $100 bills, flipping Krugerrands in the air to startle friends with the flash of gold (he kept about $20,000 worth of gold medallions in a safety deposit box).

He played poker for modest stakes and enjoyed the art of bluffing.

In the late 1970s, Everett became enamored of mini-computers. He bought a Tandy-Radio Shack TRS-80 and programmed it to run spreadsheets for Key Travel. His software designs were cutting edge, testing the calculational limits of the desktop computers. Reisler says Everett's programs might have made bundles of money if he had taken the time to develop and market them, but he was always moving on to the next iteration. He fed his programming addiction by buying $35,000 in personal computer equipment, an expense, which, as it turned out, he could ill afford.

In the late 1970s, Everett financed the start-up of an artificial intelligence software development company with a former DBS data-programmer, Elaine Tsiang. They named it Monowave (after Mark's garage band of that day). Tsiang had earned her doctorate in physics at North Carolina, studying with DeWitt. After graduation, she had trouble finding a job in physics—being female, and of Chinese descent. Serendipitously, she applied for a programmer position at DBS, only to learn, from Everett's careful question, that he was *that* Hugh Everett III. Subsequently, she became very close to him and Nancy, once going on a Caribbean cruise with them, (and being mistaken by a passenger for their daughter).[4]

Tsiang recalls, "Hugh liked to espouse an extreme form of solipsism as part of his constant, light-hearted one-upmanship. Although he took pains to distance his theory proper from any theory of mind or consciousness, obviously we all owed our existence relative to the world he had brought into being."

Being Everett's business partner was not easy: A furiously penciled note that he left behind castigates Tsiang for not taking advantage of tax loopholes. It shows how contemptuous and angry he could be with colleagues who made minor mistakes.

Everett considered himself a libertarian, says Tsiang: "Everything should be allowed, except physical force. People should be allowed to sell themselves into slavery, if they so chose, or can be persuaded." He admired Machiavelli, she

---

[4] Tsiang to Nancy Everett, 7/22/82. Tsiang's relationship to Everett was purely Platonic.

says, for his "penetrating theory of political behavior, and evolutionary biologist Richard Dawkins for his selfish gene theory."[5]

Another former DBS programmer, Keith Lynch, says Everett liked to talk about libertarianism and, also, logical paradoxes. Lynch was an amateur physicist, so Everett gave him a paperback copy of his many worlds book. Lynch says that he and Everett once talked about "quantum suicide," the proposition that it makes sense for a believer in the many worlds interpretation to play high stakes Russian Roulette as, in some universe, some version of you is bound to win.[6] It is unlikely, however, that Everett subscribed to this view, as the only sure thing it guarantees is that the majority of your copies will die, hardly a rational goal.

Everett kept his life compartmentalized—his family knew next to nothing about his work, his colleagues knew little or nothing about his family. His friends were largely unaware that he was drinking and smoking himself to death. But Nancy was keenly aware that both her daughter and husband were afflicted by alcoholism. She researched the disease, giving them both Alcoholic Anonymous books and pamphlets. But, between Everett's drinking and smoking and eating raw meat and fatty foods and refusing to see a doctor for any reason, "he kind of orchestrated his demise," she later wrote to a friend. Everett often joked that layering itself with cholesterol was the heart's way of protecting itself,[7] and there is some truth in that, but slathering on too much of the wrong kind of cholesterol proved to be fatal to him.

End of days

Pretty, blonde, outwardly vivacious, inwardly despairing, Liz bounced in and out of the McLean residence over the years. She partied from dusk to dawn. When she was 16, she shacked up for a while with a man twice her age. She gained and lost boyfriends, she gained and lost jobs. In 1979, she moved to Hawaii to live with her army friend from Austin, a loser who lived off her and her family. Worried about their unstable daughter, Hugh and Nancy shipped her a used Mazda. They paid her credit card charges. They sent a constant stream of checks to cover the newest emergency: crashed vehicles, dental work, lost apartment rent, no money for food. They flew her home for a Christmas celebration and she partied in a drunken, stoned blackout for days, not even remembering how she eventually got back to Honolulu. Eventually, her

---

[5] Quotes from Tsiang interview and Tsiang private communications; Dawkin's selfish gene theory asserts that humans are vehicles used by genes to reproduce themselves, although the theory is not reducible to that observation.

[6] Lynch interview. Everett obviously liked Lynch, who describes himself as "anarcho-capitalist." He refuses to allow the government to take his photograph, so he cannot drive a car, nor fly on airplanes. He is, however, one of the few people that Everett talked to about his theory in his later years. Lynch has his own strange tale of how he got his job at DBS after being unfairly imprisoned for burglarizing the firm: http://www.keithlynch.net/prison.html.

[7] Reisler interview, 2006, Lynch interview.

Liz, circa 1977.

boyfriend became a fundamentalist Christian and they broke up. She returned to McLean and partied harder, because that was all she knew how to do.

After Colonel Everett retired from active duty in the army with a sizeable pension in 1958, he and Sara embarked upon a world tour for more than a year. Returning, they purchased a mansion called The Knoll in rural Berryville, Virginia, filling it with such rococo furnishings as dining room chairs with Cupid heads. The colonel earned a credential as fallout shelter engineer, and is thought by family members to have worked on Site R inside Raven Rock Mountain in Pennsylvania, the hardened command bunker into which U.S. government leaders would retreat in time of nuclear war.[8] He enjoyed a long retirement, and died in 1980 after a struggle with lung cancer and diabetes. In many ways, Everett emulated his father—the compulsive photography, the excessive drinking, the love of smoking and women, the service to militarism. Hugh Jr. had been his son's anchor ever since Katharine walked away. And when he died, so did Everett's model for living.[9]

There were some happy moments in Mclean: Mark remembers that for his father's 51st birthday, Liz cooked up a surprise.

> She asked him to talk to a friend of hers who was interested in learning about computers. He put wine and cheese out on the dining room table. The friend came over and he was very charming and outgoing with her, and then she suddenly ripped off her dress revealing a Wonder Woman outfit and sang Happy Birthday to him. Liz had hired her! I can still see the look on his face when he realized he'd been had. His face turned red, and then he laughed and had a great time with it. Of course, the shock might not of helped his heart.[10]

The summer of 1982 was the nadir of the worst economic recession since the Great Depression. To control inflation brought about by high energy costs, and the Reagan administration's military spending spree, the Federal Reserve contracted the money supply, causing interest rates to shoot upwards of 20 percent. The recession was long and deep and lethal to the luxury travel business.

As the principal owner of Key Travel, Everett could not walk away from the firm, which had not done well even in better times. Strapped for ready money, in 1979, he had refinanced the McLean house, taking out $57,000 in cash at 11.5 percent interest. During the next year, he loaned Key Travel $140,000. Nonetheless, the company fell behind on payroll withholding taxes, sinking ever deeper into debt.

---

[8] Basement files, Everett family interviews.

[9] Colonel Everett was a dogged advocate for his own interests. In the basement is a file of letters documenting his tussle with the Internal Revenue Service over an audit of his 1960 tax return (he was also audited in 1957 and 1959). He wrote to high ranking IRS officials that the IRS auditor had been "motivated by malice or sheer stupidity [and] I am concerned that something must be done to...remove the incompetent from the public payroll." He complained that his tax dollars were being "frittered away in the Congo." And he buried the IRS in details about the cost of every item in his home, including light bulbs. In the end, the IRS capitulated. The colonel also received tax-free disability payments from the Army, although there is no record of a service-related disability, and he was in fine health when he retired.

[10] Mark Everett, private communication, June 2009.

Hugh Jr., Sara, Nancy, Everett, circa 1979.

Everett was in serious financial trouble: he had first and second mortgages on his McLean house. He had to meet substantial monthly payments on his condominium in the Virgin Islands. He had car loans for his collection of used luxury cars: a Lincoln Continental, Cadillac Seville, and Mercedes Benz. Plus, he had to pay a high rate of interest on substantial loans he had taken from DBS to finance Key Travel. And he had his hedonistic lifestyle to maintain.

He had considerable assets: In mid 1981, counting his AMS stock, he held over a half million dollars in securities. He owned a $75,000 insurance policy, and a $135,000 annuity for Nancy. But he did not want to liquidate the family's nest-egg, so, beset by Key Travel's bottomless deficit, he borrowed another $80,000 from DBS. When he failed to pay back DBS in June, as agreed, the debt rolled over to December at 16 percent interest.

Reisler says that Everett became obsessed by his problems with Key Travel; seldom showing up for work at DBS. As lost hours could not be billed, the company was on the verge of stopping his salary. Family financial records show that by early 1982 he was liquidating the AMS stock, pumping the cash into Key.

Trying to calculate a way out of the financial mess, he invented complicated accounting programs for his Radio-Shack computer, to no avail. As of June 1979, his computerized spreadsheets showed that he owned $651,000 in stocks ($495,000 in AMS stock alone). But, three years later, his stock holdings totaled $160,000, due to a falling market and selling off almost all of the AMS stock even as it was plummeting in price. By July, 1982 his once-sizeable AMS holding had shrunk to a mere $36,750. His net worth was diminishing at a fantastic pace, and his expenses were rising during the unrelenting recession. Key Travel became a black hole, valued at zero.[11]

---

[11] Spreadsheets; IRS Estate Tax form; probate filing for IRS.

Everett at home, circa 1981.

One night in early May, Everett sat at the dining room table, furiously writing code for a financial software program he called "Winning Mortgage." It was designed to automatically calculate mortgage payments after varying the principal and interest amounts and length of payment periods. He thought it would appeal to people losing their jobs who were in danger of foreclosure, such as the air traffic controllers who had recently been fired en masse by Reagan. He was, according to Reisler, inventing a framework for a user-friendly interface of a type that appeared in the marketplace many years later. In the dead of night, the near-bankrupt scientist laid out an advertising campaign for the software to appear in the *Washington Post*, "the business section not the LIBERAL rest of the paper." In addition to helping people in danger of foreclosure, the program would facilitate investors preying on foreclosures. He wrote in large letters on a yellow sheet:

WANT TO FORECLOSE.!!
THINK OF THE FUN!
Remember SNIDELY WHIPLASH
Remember UNCLE TOM
Remember J. R. !!! ?
Don't you wish you were ONE OF THEM? Join in the FUN OF FORECLOSURE!!
Think of what you are MISSING!

Want to get Promoted? Catch your <u>BOSS</u> in a <u>Flagrant error!</u>

The <u>WORLD</u> will be using <u>WINNING MORTGAGE</u>
CAN YOU AFFORD TO BE WITHOUT IT

One afternoon a few days later, Nancy told Mark that Liz was "sleeping" on the bathroom floor. Mark found her unconscious, clutching an empty bottle of sleeping pills. Paramedics rushed her to the hospital; she was revived by doctors moments after her heart stopped. When Mark returned home from the hospital that evening, Everett was reading his favorite magazine, *Newsweek*. He looked up and said, "I didn't know she was so sad."

On the evening of July 18, Mark was doing dishes in the kitchen. Nancy and Liz were out of town visiting cousins. Hugh took the trouble to chat pleasantly with his son about music and poker, which was unusual as they generally passed each other by like wraiths. Leaving the house a bit later, Mark glimpsed his father lying in exactly the opposite of his usual position on the couch as he watched the television news. It seemed odd to him, but he was in a hurry and continued on his way out the door.

The next morning, when Mark arose, the house seemed quiet, too quiet. Glancing into his parent's bedroom, he saw Everett fully clothed in his dark suit lying sideways on the bed, feet almost touching the floor. Mark shook him and yelled at him to wake up. Frantic, he called 911. The operator talked him through CPR. But it was far too late—Everett's body was cold and stiff.

He'd had a heart attack and passed out on the bed, dying soon after. His right coronary artery was stuffed with ruptured plaque, which was the immediate cause of death. His heart was abnormally enlarged, probably due to high blood pressure and stress. He suffered from severe arterial sclerosis. His prostate gland was chronically inflamed. And, at the time of death, he was legally drunk.[12]

As the paramedics carted Everett away in a black body bag, Mark sat at the living room table leafing through his dad's *Newsweek*, thinking that he did not remember ever having touched his father in life.

That night, after Nancy and Liz returned, shocked and sad, the three of them slept together in the parents' bed, seeking comfort.

There was a cremation. There was a memorial service and a eulogy delivered by Neil Killilea, but his closest friends, Reisler, Pugh, and Lucas were all out of town, vacationing. The *Washington Post* ran an obituary praising his work in operations research and physics. Nancy kept his ashes in an urn inside a filing cabinet in the dining room for a few years. Then, one day, in accordance with his express wishes, she tossed the cremains into a garbage can. He was gone, but a record of his life—his achievements and his failures and his secrets—remained behind.

---

[12] Report of autopsy, Chief Medical Examiner, No. Va. District, 7/30/82; interview with cardiologist Dr. Patrick Devlin.

Everett, in his foyer, circa 1982.

Among Everett's basement papers were notes that he had made after the American Institute of Physics asked him to prioritize his top five scientific capabilities.[13] At the bottom of the list, he put "servomechanisms." Followed by "operations research." Skill number three was "relativity and gravity." Two was "decision game theory." And at the top of the list, in pride of first place: "quantum mechanics."

---

[13] American Institute of Physics, National Registry of Scientific and Technical Personnel, 5/14/57.

37 Aftermath

Self-awareness, reason and imagination disrupt the 'harmony' which characterizes animal existence. Their emergence has made man into an anomaly, into the freak of the universe. He is part of nature, subject to her physical laws and unable to change them, yet he transcends the rest of nature. He is set apart while being a part; he is homeless, yet chained to the home he shares with all creatures. Cast into this world at an accidental place and time, he is forced out of it, again accidentally. Being aware of himself, he realizes his powerlessness and the limitations of his existence.

Erich Fromm, 1955[1]

Not long after Everett died, Nancy wrote a note to Mark explaining that he could not have saved his father. She and Liz had recently watched a PBS program, *Heart Attack*, and learned that within minutes irreparable brain damage occurs: "You wouldn't want to save anyone after the 1st 4 minutes any way," she wrote.[2] Everett was a "victim of his addictions." Of his death: "I think he knew it was inevitable, but he also wanted to spare us any worry over it. I hope these thoughts may set your mind at rest about some aspects of Hugh's way of leaving."

In the days and weeks and months following her husband's sudden passing, Nancy had her hands full figuring out his convoluted financial affairs. Reisler forgave the large debt to DBS in return for most of Everett's stock in the firm. Between the annuity, income from what was left of the stock portfolio, sale of family land in Vermont, and monthly social security checks, Nancy did not have to work. She was able to pay the mortgage, donate small sums to charity, and take bird-watching vacations with her friends. She met regularly for tea and bridge with a social circle of widows and ex-wives of Lambda men, who found her to be bubbly and fun-loving. She hiked, gardened, went to music concerts, puttered around the house, and did her best to monitor and preserve Everett's scientific legacy.

Six months after he died, she bought P. C. W. Davies' *The Accidental Universe* (1982). Davies has long been a subtle proponent of a quasi-religious

---

[1] Fromm, E. (1955). 23–24.
[2] Everett, Nancy to Everett, Mark, 9/27/82.

"anthropic" principle in physics, i.e. the concept that the concatenation of physical constants shaping our universe are so improbably precise and finely-tuned that reality must have been designed by an intelligent being. Nancy marked where Davies explained that her husband's theory, if true, destroyed the anthropic argument for the existence of a God. Everett, a committed atheist, would have been delighted to have performed this service to philosophy.[3]

Nancy enjoyed watching science shows on PBS. Upon learning the measure for entropy—$S=k \log W$—she scribbled it constantly in her journals, accompanied by such comments as, "order <u>tends</u> to vanish, but not all – Probability too low – evolution = simple to complex step by step or stratified complexity – biology dependent on $S=k \log W$." Like her husband, she looked for hidden information in the law that all things decay.

Over the years, she slowly went through Everett's papers and belongings. She transcribed the post-Austin tape recording and showed it to Misner and Wheeler for technical corrections. In 1991, *Scientific American* ran an article on quantum cosmology, featuring pioneers in the field, including Schrödinger, Wheeler, DeWitt, George Gamow, Stephen Hawking, and Everett. She found a photograph of her husband as an unbearded, perky young man for the

Misner, Nancy, Wheeler review transcript of Everett-Misner tape, 1991.

[3] Davies, P. C. W. (1982). 122–126. In the many worlds interpretation, the anthropic principle can be pulled inside out to skewer the intelligent design hypothesis: If our universe is but one of many worlds, then its fine-tuned improbability is unremarkable and a type of evidence for the existence of multiple universes, since our world would not be likely to exist *unless* there was a spectrum of worlds.

magazine. She also corresponded with several science writers who approached her with biographical proposals—but none of their projects panned out.

In the early 1990s, Nancy shipped a small, incomplete collection of papers and letters to the American Institute of Physics for archiving. But most of Everett's papers remained boxed and unexamined until 2007.

Becoming close to an Episcopalian congregation, she dated a church-going man, 20 years her senior. When the relationship soured, she sought to dispel her pain and depression through stream of consciousness journaling. Fortunately, she found another older man to love, and he was a kind person, solicitous of her well-being.

Liz was gang-raped one night by a group of black men she met at an ATM machine while she was drunk. Traumatized, she started talking like a racist, Southern cracker. Then, in 1987, she married a convicted drug-dealer while he was in prison. Alcoholism and drug use pulled her toward the abyss. She could not hold a job. Nancy regularly wrote checks to cover her health insurance and to buy groceries.

Liz and the bad-news husband moved to Hawaii. They hopped from island to island, apartment to apartment, bad scene to bad scene. She did temporary clerical work, when she wasn't too drunk or stoned to type or file. Sadly, changing geography was not enough to alter the course of her disease. She was constantly in and out of mental hospitals and rehab centers and 12-step groups. She'd put together a few months of sobriety, and then relapse, each relapse more painful and harder to dig out of than the one preceding—she lived in a hellish feedback loop of addiction, anxiety, depression, and self loathing.

After knocking around Northern Virginia for several years, taking college classes, playing music, waiting tables, substitute teaching, pumping gas, Mark packed up everything he owned and drove to Los Angeles, hoping to make a new life for himself as a songwriter and rock singer. He called himself "E," shorthand for "M.E.," Liz's affectionate nickname for him. With his tiny band, Eels, he began churning out albums. Shy and reclusive in his private life, on stage E is for Entertainer. By the early 1990s, he was touring the world, performing in major concert halls. His scratchy, melancholy voice conveys the wry humor of his original songs, thrilling fans who hear reflected in his music their own fears and insecurities and smallish hopes.

Liz was E's most enthusiastic fan. Like many a troubled American teenager Liz had spent countless hours listening to Neil Young's album, "After the Gold rush." But M.E. was her soul mate; his music soothed her.

She wrote to him,

> People just don't understand that it has nothing to do with whether things are going good or bad in your life. You just feel like shit, regardless. It's a real drag that I guess you should be able to relate to, having such a great career and all, and feeling like shit too. I don't think even Mom understands it.... But, like you, I'm willing to check out anything that will shed some light on things, like childhood therapy stuff.

Another thing I've been thinking about is when you were home and Mom said, 'I wouldn't want to be anybody but myself,' and we both said that we'd rather be ANYbody than ourselves. I don't know anyone else that feels that way, short of the people I met in the booby hatch who all think that everybody's got their shit together but them. Anyhow I've had that 'feeling of worthlessness' shit for ages.[4]

As adults, Liz and Mark each spent a lot of time in therapy, trying to find and fill the holes in the emotional life of their childhood.

Nancy explained to Mark in the spring of 1996,

Hugh III – as a family man – was very very very reserved – with all of us – He had a great sense of humor and enjoyed people, but I would not say he was warm and giving in a relaxed way. He didn't hate or dislike <u>any</u> of us! He tried to show support in his way – He was supportive of your career – the bands et al.

It may have seemed that he didn't care about you in a disciplining sort of way – He did not agree about setting limits et al – He was dead set against prohibitions of one's freedoms.

I heard a definition of a liberal the other day as one who was so wishy-washy they didn't know how to say NO to their kids. I cringed! I knew how – but no one was listening. So I didn't do it right.

Anyway, — I don't know any better than you – but its my belief that Hugh was happy to have you aboard. But in those years he was overly wound up in running companies, or getting someone else to run it for him, or setting up a company for his girlfriend, just sort of enjoying his good fortune so early on – that's why 'success' is scary. He felt he deserved to dine out and eat overly rich food, wine, etc.[5]

Nancy explained why she had stopped going to church for many years,

so it couldn't be used as an excuse for our breaking up. Sundays he'd 'work' many a wkend at the office . . . Anyway – one has to work at making a family work – and we didn't cooperate or communicate too well along those lines . . . we had a strange family situation – tho seemingly normal on the outside. We were a mismatch in that we each needed someone stronger to bring the other out. But we shared many interests (Well – there must be some)![6]

Ten weeks later, on July 11, in Hawaii, a few days after her 39th birthday, Liz succeeded in killing herself with an overdose of sleeping pills.

She left a note, that read, in part:

Funeral requests: I prefer no church stuff. Please burn me and DON'T FILE ME☺. Please sprinkle me in some nice body of water . . . or the garbage, maybe that way I'll end up in the correct parallel universe to meet up w/ Daddy.[7]

Nancy penned a eulogy for her daughter:

---

[4] Everett, Liz to Everett, Mark, 10/15/91.
[5] Everett, Nancy to Everett, Mark, 4/26/96.
[6] Ibid.
[7] Recall that Nancy had kept Everett's ashes in a filing cabinet before throwing them in the trash, as per his request.

Liz, Jan. 1996.

Liz, thank you for sharing with us your sense of humor (some say it could be wicked), your style, your wit, your steadfast loyalty, your caring for others and your indomitable spirit, sometimes while suffering cruel affliction. …

Your father having been a theoretical physicist who wrote of a parallel universe, and your grandmother – a poet who wrote of interstellar space, perhaps it was natural for you to think of other worlds more peaceful.

Liz, you have fought so bravely with such patience. We regret we could not help you more.

Peace be with you, til we meet again.

Two years later, Nancy was diagnosed with lung cancer, attributed to breathing her husband's second-hand smoke. Mark nursed her at home in McLean during her final days of pain. He asked her a lot of questions, including if she and Everett had ever considering divorcing? She said, "Oh, no. Never. He was so unique and such an original thinker. There was something about him I knew I'd never find in anyone else." Star-crossed as lovers, they were always friends.

Nancy and Mark, Feb. 1997.

Nancy died on Armistice Day, Everett's birthday.

Crushed, Mark organized her funeral, making a remembrance book of photographs celebrating her love of the outdoors. The last photograph in the album was of a teenage Liz, looking sweet and sane. He sold the house, and moved the boxes stuffed with the record of his parent's and grandparent's and sister's lives into the basement of his modest home in the hills of Los Angeles.

And then he recorded and released several albums of songs grieving for his lost family. Engineering the albums in a small recording studio on the other side of a concrete wall from the family archive, he came to terms with sole survivorship by making his story into a work of art for the world. And his career thrived. Eels' albums were critically acclaimed. He was asked to score songs for "Shrek," "American Beauty," and other movies, and his international following grew.

As Mark aged, he began finding small pieces of his father in his own behaviors, and, finally, some peace. His 2005 album, "Blinking Lights and Other Revelations," ended with a song of recognition and forgiveness, "Things the Grandchildren Should Know."

> I'm turning out just like my father
> Though I swore I never would
> Now I can only say that I have love for him
> As I never really understood

What it must have been like for him
Living inside his head.
I feel like he's here with me now
Even though he's dead.

As part of healing, Mark wrote an autobiography, *Things the Grandchildren Should Know*. It is mostly about his "long and whacky road surviving the life that led from being his father's son."[8] As the book was being readied for

Everett on cover of *Nature*, July 2007.

---

[8] Mark Everett, personal communication, 2009.

publication, *Scientific American* interviewed him for a profile of his father. Mark grinned, "When my rock singer friends tell me they are being interviewed by *Rolling Stone*, or *People*, I say, so what, I'm being interviewed by *Scientific American*!"

July 2007 marked the 50th anniversary of the publication of Everett's many worlds theory. *Nature* put him on the cover, as did several other science magazines. At the University of Oxford, and elsewhere, physicists and philosophers gathered to dissect the influential idea. The BBC filmed Mark traveling around the country talking to his late father's friends and colleagues—Reisler, Pugh, Misner, Trotter and others—while trying to understand the gist of the many worlds theory. *Parallel Worlds, Parallel Lives* premiered to great reviews in the United Kingdom. It was picked up in the United States by NOVA and received accolades in publications ranging from *The New York Times* to *Hollywood Reporter*. People were captivated by the image of the rock singer searching to forgive his strange, brilliant father.

Eels used the hour-long film as an opening act during their international tour in 2008. It was a spunky move. Concert halls full of unsuspecting rock fans suddenly found themselves watching a movie about quantum paradoxes and multiple universes punctuated by atomic mushroom clouds. Night after night, when the film ended, E strolled on stage to ask the audience, "What do you want? physics? or rock and roll?" And, despite what you might think, "physics" got a lot of applause (most nights).

And then he sat down at the piano and sang his heart out—having done his best to give his father a day in the sun.

Mark Everett performing at Royal Festival Hall.

BOOK 12
EVERETT'S LEGACY

38　Modern Everett

> *Today, physics is in crisis. Physical theory is unbelievably*
> *successful; it constantly produces new problems, and it solves*
> *the old ones as well as the new ones. And part of the present*
> *crisis—the almost permanent revolution of its fundamental*
> *theories—is, in my opinion, a normal state of any mature*
> *science. But there is another aspect of the present crisis: it is*
> *also a crisis of understanding. This crisis of understanding is*
> *roughly as old as the Copenhagen interpretation of quantum*
> *mechanics.*
>
> Karl Popper, 1982[1]

Fiction and science

Probability formulas measure uncertainty and entropy, quantify ignorance and information, reduce the need to make wild guesses, and inform our beliefs about reality. Could it be that intelligent decision-making is reducible to the calculation of odds, to *betting*? The puzzle of probability has flummoxed scientists, philosophers, and novelists ever since humans consciously confronted the nature of choice—the existence of alternatives. The trick, it is thought, to making a *rational* choice is being able to predict the consequences of a decision before making it by measuring probabilities. Experience provides us with information about possible consequences. Measuring the relative frequency with which an event has occurred in the past allows us to predict the likelihood of it recurring. And the generalization of experiential data allows us to predict chances for events we have yet to experience. For example, the probability that we will pick a white marble out of an urn filled with 99 black marbles, and one white marble, is 0.01 or one percent each time we insert a hand into the urn. Of course, it is not *certain* that in 100 or 100 million attempts we will snag the white marble!

Despite having used it on a daily basis for eons, we still do not know what probability *is*—why the mathematics of probability tracks change in the physical world. What does it *mean* to say that an event is possible? And, for that matter, what happens to unrealized possibilities? One of the first Western scientists, the Pre-Socratic philosopher, Anaxagoras of Clazomenae, wondered

---

[1] Popper, K. (1982). 1.

how the ingredients of everything that can possibly happen can all be contained in the same space, the same universe, so to speak. His puzzlement about the operation of probability foreshadowed our modern puzzlement over quantum superposition: how to come to terms with points in an abstract "space" which documents the coincidence of all physical possibilities. Anaxagoras' solution was to model a universe in which all possibilities, all possible worlds, are somehow manifest, although he could not say how this could be physically so. But he did question why we should assume that there is only one universe, and not many. Twenty-five centuries later, it is even more reasonable to ask whether or not multiple universes exist. Remarkably, the stuff of science is stranger than fiction.[2]

Science fiction

In 1922, H. G. Wells wrote one of the first stories about a parallel universe. "Men Like Gods" treats of a parallel world with "no parliament, no politics, no private wealth, no business competition, no police, no prison, no lunatics."[3] It was a Utopia defined by its differentiation from our world. The Utopians had shared our bloody past until history (inexplicably) branched. Given the contrast between the two worlds, there was little doubt about which world a rational being would choose to live in, although Wells' characters could not see all possible worlds.

In the 1934 story *Sideways in Time*, Murray Leinster's universe-traversing protagonists must choose,

> between the forks in the road... there is more than one future we can encounter, and with more or less absence of deliberation we choose among them. But the futures we fail to encounter, upon the roads we do not take, are just as real [although] we never see them.[4]

Everett was addicted to reading science fiction. The possible existence of parallel universes would have intrigued him long before he wrote his dissertation. And his theory certainly intrigued the world of science fiction. Perhaps the first science fiction story to base itself directly on Everett was "Store," written by Robert Sheckley in 1959.

> Yes, my friend, though you might not have suspected it, from the moment this battered earth was born out of the sun's fiery womb, it cast off its alternate-probability worlds. Worlds without end, emanating from events large and small; every

---

[2] Writing for *Scientific American*, Max Tegmark identifies four different kinds of parallel universes: 1. an infinite universe in which all events would occur repeatedly outside our cosmic horizon; 2. bubble universes of the eternally inflating string theory "Landscape" type; 3. the Everettian many worlds model; 4. Platonic mathematical structures that do not distinguish between the physical and the abstract, giving form to all possible worlds. Tegmark, M. (2003).
[3] Wells, H. G. (1922).
[4] Leinster, M. (1978). 20.

Alexander and every amoeba creating worlds, just as a ripple will spread in a pond no matter how big or small the stone you throw.[5]

Scheckley's post-World War III tale of multiple universes appeared in an anthology of such science fiction, *Beyond Armageddon*. The collection's editors claimed that it had been inspired by the publication of Everett's theory two years previous. They probably had no idea what Everett did for a living, but they would have been appalled to find out. They wrote,

> When human extinction (as a result of a decision) is assigned a probability, however small... the statistician and his employer should be detected, apprehended, and led away in straitjackets to the nearest lobotomy ward. Such people are not just running loose in Washington and Moscow, they're running things.[6]

A few years after Everett died, Frederik Pohl tapped into the many worlds idea to fictionalize a government research program to travel between parallel worlds. In *The Coming of the Quantum Cats*, a physicist is testifying to a congressional panel:

> 'Those parallel universes, created at the rate of millions every microsecond, are just as "real" as the one I'm testifying before you in.'... A congressman from New Jersey leaned over to whisper in my ear: 'Do you see any military application in this, Dom?'
>
> 'Ask him, Jim,' I whispered back, and when the congressman did the physicist looked astonished.
>
> 'Oh, I do beg your pardon gentlemen,' he said. 'And ladies, too, I mean,' he added, nodding toward Senator Byrne. 'I thought all that had been made clear. Well. Suppose you want to H-bomb a city, or a military installation, or anything at all, anywhere in the world. You build your bomb. You take it into one of the parallel universes. You fly it to the longitude of Tokyo—I mean of whatever the place is—and you push it back into our world and detonate it. Boom. Whatever it is, it's gone. If you have ten thousand targets—say, the entire missile capacity of another country—you just build ten thousand bombs and push them all through at once. It can't be defended against. The other people can't see it coming. Because, in their world, it isn't coming... until it's there.'[7]

As the specter of nuclear war insinuated itself into the nightmares of Baby Boomers, a meme of multiple universes began to reproduce itself in Western literature and cinema. It permeates the novels of Andre Norton, Philip K. Dick, Greg Egan, and Neal Stephenson. Jet Li starred in a martial arts film doing kung-fu leaps between parallel worlds. Harry Turtledove makes a living writing alternate history sagas. Philipp Pullman's trilogy of children's novels, *His Dark Materials*, roams through a series of Everett-type worlds.[8]

---

[5] Miller, W. M. and Greenberg, M. H. (1983). 39.
[6] Ibid. 5.
[7] Pohl, F. (1986). 58. Pohl. © 1986.
[8] Pullman says he was inspired to write about multiple universes after attending a lecture on the many worlds interpretation by University of Oxford physicist, David Deutsch.

Typical consumers of many worlds fantasia may not realize that their entertainment is mirrored by serious scientific theory. But if you pick up a contemporary book on quantum mechanics—scientific or philosophical—it is likely that "Everett, Hugh III" will be listed in the index.

Popper speaks

Philosopher of science, Karl Popper, was drawn toward Everett's "simple and ingenious" argument.[9] In 1982, he published *Quantum Theory and the Schism in Physics*, a collection of essays attacking the "anti-realism" of positivism, instrumentalism, and the Copenhagen interpretation. Popper found Bohr's division of the universe into quantum and classical realms to be a "mistaken and even a vicious doctrine."[10] He concocted his own theory of probability to bridge the mysterious gap, using film strips as a metaphor to explain how superpositions decompose into single worlds:

> Since each 'still' in the filmstrip attached to a real time slice consists of a whole catalogue of weighted possible states, my proposal really involves that the predictive filmstrips split at any interaction into as many filmstrips as there are possibilities in the catalogue. In this my picture...greatly resembles Everett's; only that the many worlds remain mere possibilities instead of becoming real.[11]

Popper liked Everett's theory because it,

> is a completely objective discussion of quantum mechanics. In Everett's approach (as opposed to the Copenhagen interpretation) there is no need, and no occasion, to distinguish between 'classical' physical systems, like the measuring apparatus, and quantum mechanical systems, like elementary particles...Instead, all physical systems are regarded as quantum mechanical systems, especially the apparatus used in measurements; and, indeed, the universe.[12]

Popper found a use for the universal wave function in his "propensity" interpretation of quantum mechanics, but his main critique of Everett was that, according to Schrödinger's wave mechanics, the branching universes ought to *fuse* as well as *split*. Popper found that prospect to be "clearly absurd," as the branches would have "no interaction between them before their fusion."[13] Be that as it may, Everett had considered the question of fusion (time reversibility). He did not view it as a problem, as the microscopic superpositions expressed in his universal wave function were time-reversible, whereas the macroscopic splitting process appears to be irreversible in accord with entropy

[9] Popper, K. (1982). 90.
[10] Ibid. 42.
[11] Ibid. 187.
[12] Ibid. 89.
[13] Ibid. 94.

and our experience of the arrow of time.[14] And Everett's argument that the Schrödinger equation applies to the *entire* universe—micro and macro— gained in credibility after quickly decaying superpositions of molecules were observed in the laboratory.[15] (Which is not to say that fusion is impossible.)

Everett claimed that macroscopic objects—amoebas, cannonballs—can exist in a superposition of states, but we (who are also macroscopic objects) do not see such superpositions because we entangle with each version of the superposed object and our copies embark on different histories. Where he saw branching histories, modern physicists now see an (essentially) irreversible correlation between objects and their environs known as decoherence—a phenomenon amenable to both uniworld and multiworld interpretations.

Decoherence in a nutshell

Decoherence theory has emerged from decades of foundational research in quantum mechanics. In the 1920s, Paul Ehrenfest theorized about how quantum superpositions map the world described by classical mechanics. And we have witnessed Mott puzzling over how the environment inside a Wilson cloud chamber caused spherical probability waves to collapse into single tracks. But it is only in recent times that the quantum mechanical formalism has been able to reasonably explain the dynamics of what happens when an object in superposition interacts with its surroundings. Decoherence theory may be more of a heterogeneous collection of techniques than a systematic theory,[16] but it goes some distance toward solving the measurement problem. And all of the pioneers in decoherence theory credit Everett with making it possible to even talk about the *coherent* quantum mechanical space inside which objects and, indeed, entire universes *decohere*.

After Bohm and Everett openly rebelled against the monocracy of the Copenhagen interpretation (and the collapse postulate, which many physicists, rightly or wrongly, believed to be a part of Bohr's complementarity), theorists began looking at the formalism of quantum mechanics in a new light: searching for a way that the wave function could *contain the observer* without falling into the infinite regression of always having to place the (classical) observer outside the (quantum) system observed. John Bell had strong words for Bohr's unbridgeable partition:

> From some popular presentations the general public could get the impression that the very existence of the cosmos depends on our being here to observe... So I think it is not right to tell the public that a central role for conscious mind is

---

[14] DeWitt, B. and Graham, N. eds. (1973). 94–100; Everett did not have the last word on this difficult problem, of course, and Popper's worry remains real as we do not see large objects moving backwards in time, but given the reversibility of the Schrödinger equation, that is not unthinkable.

[15] Mesoscopic superpositions have been experimentally witnessed in double slit-type interferences of fullerenes. Superpositions of quantum objects have been recorded in experiments using Josephson junctions, atom interferometry, ion traps, microwave cavities. See Zurek, W. H. (2002). 6.

[16] Saunders, S., "Many Worlds: An Introduction," in Saunders, S, Barrett, J., Kent, A., and Wallace, D. eds. (2010).

integrated into modern atomic physics....The founding fathers of quantum theory decided...that no concepts could possibly be found which would permit direct description of the quantum world....The 'Problem' then is this: how exactly is the world to be divided into speakable apparatus...that we can talk about...and unspeakable quantum system that we cannot talk about?...It is not essential to introduce a vague division of the world of this kind.[17]

Decoherence theorists seek to speak the unspeakable by including the apparatus and the observer in the wave function. Decoherence accounts for the loss of interference effects between overlapping wave functions; the loss of coherence between different elements of a superposition in a quantum system; the seeming loss of quantum information as a superposition decomposes into a single state or a collection of non-interfering states.

For example, the ability of experimentalists to manipulate decoherence rates—the speed at which an atomic superposition decomposes—is part of the technical basis of quantum computation. The longer that a quantum transistor or "qubit" can be sustained in a "coherent," superposed state without disintegrating (decohering), the longer a quantum computation can run using that qubit.[18]

Decoherence occurs when a complex quantum system entangles with its surroundings, thereby appearing to lose information content and becoming "solid," i.e., behaving classically, macroscopically. But decoherence is *not* an interpretation of quantum mechanics: It is a mathematical technique for describing how microscopic systems of a certain size or complexity segue into macroscopic systems. In it, there is no inexplicable collapse or quantum jump and the transition is governed by the non-collapsing Schrödinger equation. The structure of decoherence does not exclude the measuring apparatus nor the observer from the wave function that includes the object observed. And as a physical phenomenon, it is amenable to explanation by the Everett interpretation, which encompasses—but is not ruled by—the externality central to the Copenhagen interpretation. In this sense, the Everett model is thought to supersede, but not to destroy the Bohr and von Neumann models (which are self-limiting explanations when considering the universe as wholly quantum mechanical). The concept of branching universes can model decoherence; but there is no consensus among those who do so on whether or not the branches are physically real.

Zeh and Zurek

In the late 1960s, Dieter Zeh was a young lecturer in nuclear physics at the University of Heidelberg. While exploring the foundational questions in

---

[17] Bell, J. S. (1987). 171.
[18] Because of the superposition property, it is thought that a quantum computer will be able to process information exponentially faster than its classical counterpart. Classical transistors are limited to holding values 0 *or* 1. A qubit can hold a superposition of values 0 *and* 1! When linked in arrays of entangled qubits, the processors will operate exponentially faster than non-quantum processors. The trick is to extract the answer from the superposition without destroying the answer.

quantum mechanics, he drafted a short paper arguing that quantum mechanics is universally valid, i.e. that the wave function need not collapse.[19] Speaking the supposedly unspeakable, Zeh proposed that macroscopic systems exist in superpositions that are not isolated from their surroundings, rather, they are constantly correlating or entangling with the remainder of the universe (which includes observers). He wrote, "It must be concluded that macroscopic systems are always strongly correlated in their microscopic states."

Zeh is widely credited with being the first physicist to analyze decoherence as a possible solution to the measurement problem.[20] In Zeh's scheme, each element in a superposition over a certain size or level of complexity "decouples" from the others as the physical process of entanglement with the environment destroys the interferences encoded in their shared wave function. Each element of the superposition delinks from the composite wave function describing the erstwhile superposition and assumes its own, separate wave function. Zeh cited Everett's relative states formulation as analogous to his decoupling theory, which also employs the concept of a universal wave function.

Echoing Everett, Zeh claimed that the EPR paradox and the mystery of non-locality are solved by his theory, as each measurement result happens in a different "world component," and no superluminal contradictions occur. Zeh recalls that he did not know about Everett's theory until shortly after he wrote his now-famous paper, "On the interpretation of measurement in quantum theory," which was, in effect, "an argument for the Everett interpretation."

Decades later, after reviewing the mini-papers (from the basement) that Everett had submitted to Wheeler in 1956, Zeh commented that Everett's "derivation" of the Born Rule from the quantum formalism was more of a plausibility argument than a precise derivation, but that it made sense overall as being the only reasonable choice for a probability measure if objective reality is represented by the universal wave function. He said the metaphor of the splitting amoeba shows that Everett viewed the branching as physically real. And he decided that, "Everett seems to come very close to decoherence. One may assume that Everett would some day have discovered the full implications of decoherence, but when he refers to the 'environment,' he seems to mean classical correlations, not microscopic entanglement."[21]

Everett did not realize the importance of solving the preferred basis problem, says Zeh, but this problem, according to many (but certainly not the majority of) physicists and philosophers who think about these matters, has since been resolved by decoherence theory, which treats the universe as completely quantum mechanical and models how classical worlds emerge (decohere) from within the universal superposition of all physically possible events.

[19] Zeh, H. D. (1970).
[20] Camilleri, K. (2008).
[21] Zeh, private communications, 2008. See Chapter 15 for Everett's view of how microscopic systems interact with their environments.

Unfortunately for Zeh, it turned out that associating one's career with Everett's theory was not a wise move at the time. Zeh's (soon to be classic) paper was turned down by several prominent journals; one editor called it "senseless." And a Nobel Prize winning heavyweight at the University of Heidelberg, J. Hans D. Jensen, sent a copy of Zeh's paper to Rosenfeld, who replied acerbically, urging Jensen to make sure that "such a concentrate of wildest nonsense is not being distributed around the world with your blessing." Jensen told Zeh to "let sleeping dogs lie," and Zeh's academic career was negatively impacted by his inability to stop kicking the dog.[22]

In 1970, the paper was published in a new journal, *Foundations of Physics*, refereed by Wigner, Margenau, Bohm, de Broglie and other dissenters from the Copenhagen interpretation. DeWitt and Zeh hooked up at the Varenna conference that same year and each championed Everett's work in his own way. Zeh's decoupling analysis evolved into the now widely accepted theory of how quantum systems transition into classical systems through decoherence—a process which is constantly taking place and does not require the causal agency of consciousness or the presence of an external observer. But until the early 1980s, decoherence theory was ignored or maligned by many mainstream physicists; Zeh calls that period "the dark ages."[23] But the maverick from Heidelberg blazed a trail soon widened by others, including Wojciech Zurek, Erich Joos, Murray Gell-Mann, Stephen Hawking, and James B. Hartle: all of whom credit Everett with inspiring them to think beyond the limitations on understanding imposed by the Copenhagen interpretation and the collapse postulate.

In the late 1970s, Wojciech Zurek was a graduate student at the University of Texas. After hearing Wheeler and Deutsch talk about the preferred basis problem in the Everett interpretation, Zurek was inspired to tackle it. By 1981, he had mathematically modeled a method of predicting how the continuous information transfer caused by the entangling of objects with their environments automatically selects the preferred basis—the menu of possible choices that includes the classical world of our experience.[24] This addressed the central problem with Everett's formalism, which did not satisfactorily demonstrate *how* specific classical systems (branch histories) emerge out of superpositions of all physically possible events. Preferred basis[25] is the measurement problem writ large: why does our world (or a collection of worlds like ours) emerge from an infinity of alternatives?

---

[22] Interview with Zeh recorded by Fabio Freitas 7/25/08. See also: Camilleri, K. (2009) and Freire, O. (2009).
[23] Camilleri, K. (2008). The term "decoherence" was coined in the late 1980s.
[24] Zurek, W. H. (1981). Zurek adds that "the central idea was to apply the insight of Bohr (that measurement alters the quantum state) with Everett's insistence that everything (including information transfer that occurs in measurements) can be modeled using quantum theory." Zurek, W. H., private communication, June 2009.
[25] Also called "pointer basis," as the needle or pointer of a measurement apparatus only points to one number on a dial at a time, not a superposition of numbers. See Chapter 16 for a related discussion of preferred basis and decoherence.

Zurek, who works at the Los Alamos National Laboratory, is concerned with local dynamics, with information transfers. He explains decoherence in terms of "quantum Darwinism." Making an analogy to biological natural selection, Zurek shows how certain possible branches in the multiverse become more lasting and robust than others by repeatedly copying themselves into the quantum environment. Obviating the need for postulating wave function collapse, Zurek says that macroscopic observers perceive the classicality of these foliating branches without disturbing their smooth evolution through time as per the Schrödinger equation. As a metaphor, consider how differently positioned readers can access the information in this sentence by intercepting select copies of the text that proliferate throughout the photon environment at different angles. You cannot read this sentence from behind the book, but there are a vast number of angles from which you can view it and incorporate its information into your brain. Meanwhile, endless cascades of photons ricochet off the printed page in various directions holding similar information and recording it (redundantly)in the environment.

Physicists still talk about how Zurek put decoherence on the map in October 1991 with his article "Decoherence and the Transition from Quantum to Classical" in *Physics Today*.[26] He showed how the local environment dynamically "sucks information out" of the superposed quantum system with which it is correlating, entangling—leaving behind our classical world. The residual information is not destroyed—it is hidden from our sight, dispersed into *other* correlations between the system and the environment.

This is heady stuff! Zurek says it "makes the branching analogy in Everett's writings more literal." Inspired by the role that entanglement plays in decoherence, and the relativity of Everett's "relative states," Zurek went on to derive a probability measure for quantum mechanics that does not require assuming the Born rule (as Everett seems to have done, if only unconsciously).[27]

Zurek does not know if Everett's worlds are real or not, but he credits him for breaking up an intellectual log jam in physics. The many worlds interpretation gave physicists "permission" to treat the environment as quantum mechanical by viewing the universe as a closed system describable by a non-collapsing universal wave function. Permission was granted to move beyond the limitations imposed on inquiry by Bohr and von Neumann. But "quantum Darwinism" does not contradict Bohr, nor von Neumann, says Zurek. Without collapsing the wave function, per se, decoherence, in effect, smoothly inserts a

---

[26] Zurek had previously published two papers laying out the basic idea of decoherence as a loss of information as a superposition decomposes and entangles with the environment. See Zurek, W. H. (1981).

[27] Zurek, W. H. (2003); Zurek, W. H. (2005). In analyzing Zurek's derivation, Arthur Fine and Maximilian Schlosshauer point out that, in general, the problem of assumptions of probability in proposed derivations "of the Born rule can be traced back to a fundamental statement about any probabilistic theory: We cannot derive probabilities from a theory that does not already contain some probabilistic concept; at some stage we need to 'put probabilities in to get probabilities out.'" They praise Zurek for advancing our understanding of this problem by a "quantum leap." Schlosshauer, M. and A. Fine (2005).

quantum-classical partition at the moment of measurement of a branching or non-branching quantum system. Keeping intact the linearity of the Schrödinger equation, it is conducive to single world and many worlds interpretations, without ruling out either.[28]

Decoherence theory does not resolve the most vexing question about the Everett worlds, as articulated by Zurek:

> One might regard [quantum] states as purely epistemic (as did Bohr) or attribute to them 'existence'. Technical results...suggest that truth lies somewhere between these two extremes. It is therefore not clear whether one is forced is attribute 'reality' to all branches of the universal state vector [i.e., universal wave function].[29]

In the realm of interpretation, Zeh has long been opposed to Bohr's ontology. In a letter to Wheeler in 1980, he complained:

> I expect the Copenhagen interpretation will sometime be called the greatest sophism in the history of science, but I would consider it a terrible injustice if – when someday a solution should be found – some people claim that this is of course what Bohr always meant, only because he was sufficiently vague.[30]

Zeh focuses on the non-local aspect of a decoherent event as the effect of local entanglement propagating at light speed like a giant, relativistic "zipper." Within the giant superposition of the universal wave function, Zeh views *only* the subjective consciousness of each observer as "real." He takes what he calls a "many minds" approach (although the minds are virtual copies of brains, not some supra-physical consciousness or Mind). In his model, all branches, including the particular decohered branch of each conscious observer can only be known as "a heuristic picture—not a physical process."[31] To some extent, this singles out the concept of consciousness as a special process—a burden which Everett did not impose upon his physical theory which considers all the branches as equally real.[32]

In sum, both Zurek and Zeh address the problem of describing how macroscopic systems decohere from coherent superpositions without changing the Schrödinger equation. They model how quantum systems behave as they entangle with their immediate surroundings and the remainder of the quantum universe. In decoherence theory, the universe—including measured systems, measuring apparatuses, and their respective environments—evolves according to the time-dependent Schrödinger equation. And a branching

---

[28]   Zurek, W. H. interview with author, 2006; private communications, 2009.

[29]   Zurek, W. H. (2007).

[30]   Zeh to Wheeler, 10/30/80.

[31]   Zeh, H. D. (2009). 15. See also: Zeh, H. D. (2000).

[32]   According to Kristian Camilleri, Zeh's interpretation of decoherence reflects the philosophy of the neo-Kantian Hans Vaihinger who argues "that the underlying reality of the world remains unknowable, but we behave 'as if' the constructions of physics such as electrons, protons and electromagnetic waves exist, and to this extent such 'heuristic fiction' constitutes our reality. In this way, Zeh argued that the universal wave function may be employed as a 'heuristic fiction,' but it is no less 'real' than the entities posited by other physical theories (e.g. quarks), the existence of which is routinely taken for granted." Camilleri, K. (2008).

observer can be included in the universal wave function: Unlike the more restrictive Copenhagen interpretation, which defines the observer as classical and forever external to the observed microsystem.

Convinced by experiments showing mesoscopic objects decohering from coherent superpositions, many theoretical physicists and philosophers accept the decoherence model as explanatory of the quantum to classical transition without having to decide whether or not it is best interpreted in terms of an Everettian multiverse, or a single universe.[33] In this sense, decoherence theory subsumes the standard conception of measurement with its postulated discontinuity of wave function collapse. As cosmologist James B. Hartle observes, "In short, classical physics is an approximate emergent feature of the kind of entirely quantum universe that Everett talked about."[34]

Nous

In November 1984, the philosophical journal, *Nous*, published a special issue on foundational questions in quantum mechanics. It was dominated by discussions about the meaning of entanglement and the many worlds interpretation. Echoing Wheeler's fluctuations on the many worlds question, some philosophers were simultaneously attracted by the notion of a universal wave function as a solution to the measurement problem and repelled and disconcerted by its logical implications.

Howard Stein of the University of Chicago found Everett's branching worlds to be "a bizarre notion with no compensating gain," because the many worlds theory was, in his opinion (as well as Everett's) not falsifiable by experiment:

> On the other hand, the view that [wave function collapse] never occurs, and that all processes are governed by quantum mechanics, is one that deserves to remain in the field ... It is worth remembering that great advances in physics have sometimes resulted from the discovery of effects, predicted by theories, that had long seemed unlikely to occur and that many physicists regarded as in principle impossible to detect.... The problem may not be ripe for solution.[35]

Stein's colleague, physicist Robert Geroch, upheld the core of Everett's theory in a way that avoided using the concept of "splitting" observers or branching universes. He found in Everett a new way of conceptualizing probability in a totally quantum universe; he was attracted by its compatibility with relativity and its implications for a theory of quantum gravity.[36] He proposed a kind of "one-world version of Everett."[37]

---

[33] The question is often posed: how can a universe decohere when it does not have an environment? There are several ways to answer this, including that only local pieces of it need decohere, so the whole universe does not have to decohere, or the environment of a universe decohering non-locally is the multiverse, which, by definition, never wholly decoheres.

[34] Hartle, private communication, July 2009.

[35] Stein, H. (1984).

[36] Geroch, R. (1984).

[37] Hellman, G. (2009). 221.

Philosopher Richard A. Healey of the University of California, Los Angeles raised the question of how an observer maintains a sense of identity while constantly dividing into beings with shared pasts and divergent futures. This question particularly bothers philosophers who are attracted by the non-collapse model, but are unwilling to let go of a notion of an indivisible personal identity. Some influential philosophers and physicists are not at all comfortable with the idea that "mind" is a purely physical phenomenon.[38] Healy was not of this mentalist school, but his understanding of Everett reflected a common attitude among physicists and philosophers who wished to extirpate the collapse postulate via a universal wave function, but declined to accept the consequence that their bodies—and/or their immortal souls—split.

Healey remarked that Wheeler had recently withdrawn his support of Everett, even though the relative states theory "fit naturally with Wheeler's conception of super space, considered as an approach to quantizing space-time as well as its contents." In the end, Healy bemoaned the lack of consensus on what Everett was actually saying: "The interpretation needs interpreting."[39]

Complexity and information

In 1989, the Santa Fe Institute in Santa Fe, New Mexico hosted a remarkable conference, "Complexity, Entropy and the Physics of Information." Zurek, Zeh, and Wheeler made presentations on the relation of information theory to quantum mechanics, as did a score of other prominent physicists. Everett was widely credited as bringing information theory to bear on quantum mechanics, and the debate over whether or not his branching universes are physically real, or simply a conceptual tool, continued.

The conference issued a manifesto—"The specter of information is haunting science." It affirmed the importance of understanding thermodynamics, the arrow of time, and the measurement problem as "transfers of information." The usefulness of Everett's universal wave function was acknowledged:

> The distinction between what is and what is known to be, so clear in classical physics, is blurred, and perhaps does not exist at all on a quantum level. For instance, energetically insignificant interactions of an object with its quantum environment suffice to destroy its quantum nature. It is as if the 'watchful eye' of the environment 'monitoring' the state of the quantum system forced it to behave in an effectively classical manner. Yet, even phenomena involving gravity, which happen on the most macroscopic of all the scales, bear the imprint of quantum mechanics.

---

[38] As we shall learn in the next chapter, several philosophers have proposed pseudo-Everettian theories that claim it is not the physical observer that splits as universes branch, but only human consciousness that branches into many "minds" governed by "Mind," echoing the religio-idealism of Katharine Kennedy Everett's poem, "Unified Field."
[39] Healey, R. A. (1984).

In fact, it was recently suggested that the whole Universe—including configurations of its gravitational field—may and should be described by means of quantum theory.

Interpreting results of the calculations performed in such a 'Wave function of the Universe' is difficult, as the rules of thumb usually involved in discussions of experiments on atoms, photons, and electrons assume that the 'measuring apparatus' as well as the 'observer' are much larger than the quantum system. This is clearly not the case when the quantum system is the whole Universe.[40]

Decoherence theory was one of the main themes of the conference—as was the role of algorithmic randomness, which defines the information content of an object based on the theory of computation rather than on probabilities.[41] Importantly, Hartle and Murray Gell-Mann presented "Quantum Mechanics in the Light of Quantum Cosmology," a theory of decoherence known as "consistent histories."[42] Hartle had been thinking about Everett since the late 1960s, when he derived a probability measure from the quantum formalism without postulating the Born rule or wave function collapse.[43]

Hartle and Gell-Mann credited Everett with suggesting how to apply quantum mechanics to cosmology. They considered their "decohering sets of histories" theory as an "extension" of his work. Using the Feynman path integrals, they painted a picture of the initial conditions of the universe when it was completely quantum mechanical. Their method treats the Everett "worlds" as "histories" giving "definite meaning to Everett's 'branches.'"[44] They assign probability weights to possible histories of the universe, and, importantly, include observers in the wave function.

Hartle declines to state whether or not he considers the branching histories outside the one we experience to be physically real, or purely computational. And he says that "predictions and tests of the theory are not affected by whether or not you take one view or the other."[45]

Everett, of course, settled for describing all of the branches as "equally real," which, given that our branch is real, would mean that all of the branches are real.

---

[40] Zurek, W. H. (ed.) (1990). vii–ix.

[41] In a nutshell, algorithmic information content (AIC) measures the entropy (information) of a message. One presenter perceived a version of AIC in Everett's correlation equations.

[42] Zurek, W. H. (ed.) (1990). 425–458. Cal-Tech's Gell-Mann was awarded a Nobel Prize in 1969 for work on particle physics. University of California Santa Barbara's Hartle has done important work on cosmology and quantum gravity with Stephen Hawking and others.

[43] Hartle, J. B. (1968). Hartle's probability derivation from the quantum formalism was similar to Graham's method of deriving probability in the Everett scheme, although each arrived at their conclusion independently. However, Hartle's derivation has proven to be more robust and influential.

[44] Zurek, W. H. (ed.) (1990). 440.

[45] Hartle, private communication, 2009; Hartle says that the consistent histories approach (which is also due to work by Robert Griffiths and Roland Omnes) does not contradict Bohr: "Copenhagen quantum mechanics is not an alternative to decoherent histories, but rather contained within it as an approximation appropriate for idealized measurement situations." Hartle, J. B. (2008). 17. As we learned in Chapter 15, Everett said that his model "generalized" Bohr, by which he meant that it superceded Bohr's self-limiting explanation by treating the universe as whole.

Conference participant Jonathan J. Halliwell of MIT later wrote an article for *Scientific American*, "Quantum Cosmology and the Creation of the Universe." He explained that cosmologists owe Everett a debt for opening the door to a completely quantum universe. The magazine ran photographs of the most important figures in the history of quantum cosmology: Schrödinger, Gamow, Wheeler, DeWitt, Hawking and "Hugh Everett III, a student of Wheeler in the 1950s at Princeton [who] solved the observer-observed problem with his 'many worlds' interpretation."[46]

And so, even as his ashes sank into a Virginia landfill, Everett's intellectual progeny were hard at work taking his theory to new levels.

[46] Halliwell, J. (1991). 76.

39 Everett goes to Oxford

> Cosmologists, even more than laboratory physicists, must
> find the usual interpretive rules of quantum mechanics a bit
> frustrating.... It would seem to me that the theory is
> exclusively concerned with 'results of measurement' and has
> nothing to say about anything else. When the 'system' in
> question is the whole world where is the 'measurer' to be
> found? Inside, rather than outside, presumably. What
> exactly qualifies some subsystems to play this role? Was the
> world wave function waiting to jump for thousands of
> millions of years until a single-celled living creature
> appeared? Or did it have to wait a little longer for some
> more highly qualified measurer - with a PhD? If the theory
> is to apply to anything but idealized laboratory operations,
> are we not obligated to admit that more or less
> 'measurement-like' processes are going on more or
> less all the time more or less everywhere?
>
> John Bell, 1981.[1]

The rise of mentalism

A century has passed since Max Planck and Albert Einstein discovered the
quantum world. The basic quantum paradoxes—uncertainty, non-locality, and
the measurement problem—are either unsolved or remain highly contentious.
The most influential proposals for adjudicating the measurement paradox are
the Copenhagen Interpretation (which declares the problem not a problem),
Bohm's non-collapsing hidden variables theory (which supplements the
Schrödinger equation), the "GRW" spontaneous collapse theory[2] (which mod-
ifies the Schrödinger equation), and the Everett interpretation, which makes
do with the Schrödinger equation alone.

Apart from the Copenhagen interpretation, these theories are "realist," i.e.
they assume that the wave function is physically real, that the human mind is
a natural object, and that consciousness does not play any role in shaping a

---

[1] Bell, J. S. (1987). 117.
[2] Formulated in 1986 by Italian physicists G. C. Ghiradi, A. Rimini, and T. Weber.

reality that exists independently of human agency.[3] It is an interesting phenomenon that, faced with the measurement problem, resort is sometimes made to theories of consciousness, as if the mind is an all-purpose repository of the inexplicable. For instance, in 1988, even as Everett's fundamentally realist theory was gathering steam in scientific and philosophical circles, an attempt was made to transform it into a theory called "many minds" by philosophers David Albert and Barry Loewer, (Albert later disavowed this model).[4]

The Albert-Loewer interpretation set out to keep the Schrödinger equation intact *and* to solve the preferred basis and probability problems in Everett's formulation. It postulated that a continuous infinity of *non-physical* minds exist and operate—not according to a physical dynamics, not according to quantum mechanics, but according to *mental dynamics* that exist *apart* from our physical reality (there are no superpositions in this model). In this scheme, subjective *beliefs* about nature are considered as physically determinate in nature, i.e. every physical outcome in a probabilistic universe is correlated to a single mental state drawn from the vast array of independently evolving minds that collectively define the identity of the observer.[5] Everett, on the other hand, deliberately replaced human minds with record-keeping machines in his theory, obviating the need for a special theory of consciousness or a mentalist approach. There are, unremarkably, many minds or rather many brains in Everett's theory because each splitting observer-brain embarks on its own future. But after branching, these purely physical minds are no longer linked, nor part of an over-arching Mind.

It is worth recalling what Everett wrote in 1980 about Wheeler:

> [He was] wondering if somehow human consciousness was a distinguished process and played some sort of critical role in the laws of physics.
>
> I, of course, do not believe any such special processes are necessary, and that my formulation is satisfactory in all respects.[6]

Inserting non-physical processes into Everett's theory is antithetical to his intention and what it says.

---

[3] Bohr opposed privileging the role of human consciousness, but the Copenhagen Interpretation can be construed as anti-realist because it places the measuring device forever outside the wave function. Few modern physicists or philosophers are satisfied with Bohr's emphasis on the primacy of classical physics, although it remains influential, probably due to cultural inertia and the lack of a commonly accepted alternative. Eminent physicist, N. David Mermin, for example, is attracted to the Copenhagen viewpoint, noting that "quantum states are calculational devices and not real properties of a system." Mermin, N. D. (2009). 8.

[4] Albert, a proponent of Bohm's hidden variables theory, is the author of a widely read and accessible text on quantum mechanics, *Quantum Mechanics and Experience* (1992). In 2004, he was deceived by the makers of a New Age film called *What the Bleep?* into talking about physics on camera. His comments were then edited out of context to make it appear as if he supported anti-realist, thoroughly ridiculous and inaccurate notions linking quantum mechanics to religion. He disavowed the scientifically dishonest film.

[5] See Barrett (1999) for an explication of the Albert-Loewer theory and other many minds models that claim Everett as a source of inspiration.

[6] Everett to Raub, 4/7/80.

Bohm again

In 1957, Everett tried to find an address for the ex-patriated physicist, David Bohm.[7] Perhaps he wished to commiserate with him about Bohr's resistance to new ideas. There is no evidence that the two mavericks ever communicated, but Bohm was fascinated by Everett's theory. He died in London shortly before his book on quantum ontology (with Basil Hiley), *The Undivided Universe*, was published in 1993. In it, Bohm argued that the wave function does not collapse, and that human consciousness is not a causal force in physics. He considered the many worlds theory as vital and worthy of

> special attention because it has come to be fairly widely accepted, especially by those who work in general relativity and cosmology and who therefore feel a need to regard the universe as existing in itself whether observed or not.[8]

But Bohm did not understand Everett. He misinterpreted Everett's use of memory records to register splitting events as a statement that only consciousness branches. Bohm demarcated what he called Everett's "speculative theory of mind" from DeWitt's account of many worlds as a physical theory. Specifically, he stated that Everett's theory did not support DeWitt's statement that observers and universes physically split, or branch.[9]

Regarding Dewitt's "version" of Everett, Bohm criticized it for assuming, not deriving the Born rule, and requiring, he thought, additions to the quantum formalism. More to the point:

> It defies the imagination to grasp intuitively how the universe could split, and even more as to how it could be doing so in a stupendous number of ways.[10]

But Everett and DeWitt *were* in accord on the existence of splitting (as we have seen in previous chapters). It is a red herring (and an all-too-common mistake) to divide Everett's view from DeWitt's view in regard to splitting: They both asserted the "reality" of branching universes. DeWitt and Everett differed regarding the method of deriving a probability measure, but they agreed that the branches were "equally real." This agreement is substantiated by the published and unpublished writings of both men, including Everett's "split" footnote, rebelliously inserted into the published thesis of 1957. Nonetheless, Bohm and Hiley insisted that

> Everett did not contemplate the splitting of the universe.... Indeed Everett's view should not even be called the many-worlds interpretation but rather, as Albert and Loewer have suggested, the many minds interpretation.[11]

This error was amplified by philosopher David Chalmers in *The Conscious Mind* (1996). Chalmers was overtly promoting mind-body dualism, i.e. the

[7] Everett to American Institute of Physics, 6/1/57.
[8] Bohm, D. and Hiley, B. J. (1993). 296.
[9] Ibid. 301.
[10] Ibid. 315.
[11] Ibid. 303.

hypothesis that human consciousness is not subject to physical laws. And he looked to Everett in support of his non-Everettian notion. Attracted by Everett's argument that the Schrödinger equation does not break down in the face of measurement, Chalmers echoed Bohm-Hiley by insisting that Everett had constructed an Albert-Loewer type of many minds theory. For Chalmers, Everett's splitting is "only in the minds of the observers."[12]

Other philosophers of physics have construed Everett as a many-minder, but those arguments consistently fail in the light of his long thesis, and the basement papers that speak to his intention. Philosophers and physicists may create Everett-type many minds theories, but it is not correct to attribute them to Everett.

It was up to an iconoclastic physicist allied with a small group of philosophers at the University of Oxford to protect Everett's materialism.

The Oxford connection

David Deutsch talked to Everett only that one time at lunch in Austin in 1977, but he has devoted much of his career to making a case that quantum mechanics itself is the best proof of the existence of multiple universes, or the multiverse, as he calls the set of all physically possible worlds. Born in 1953, Deutsch is slightly built, pale of pallor, long-haired, and somewhat reclusive. A night owl, he seldom ventures out of the dreamy confines of Oxford town. In the early 1980s he introduced the idea of a universal quantum computer.[13] And in the early 1990s, he co-invented the Deutsch–Jozsa algorithm, demonstrating the potentially amazing speed and parallel processing capacity of a quantum computer. The insight that is possible to use superposed microscopic systems as "qubits" running computations exponentially faster than a classical processor was the shot that set off a billion-dollar research race.[14] He is a Visiting Professor of Physics at Oxford. And in 2008, he was elected a fellow of the prestigious Royal Society, which has 60 Nobel laureates on its roster.

Deutsch recounts that DeWitt convinced him of the truth of many worlds. But, DeWitt used to tell him there was a terrible problem with probability in the otherwise great theory: "And I would understand the problem only for as long as it took me to walk out of DeWitt's office. After several years, I finally got it and realized it was important and started working on it."[15] In 1984, Deutsch proposed a method of considering quantum mechanics as universal and the Everett worlds as "continuously infinite."[16] But this particular method added to the equations of quantum mechanics, defeating the point of the Everett interpretation, and he later turned to another method of introducing a measure of probability in the

[12] Chalmers, D. J. (1996). 346–357.
[13] Deutsch, D. (1985).
[14] Deutsch told science journalist Julian Brown that he first thought about quantum parallelism as a way to test the existence of the Everett worlds. Brown, J. (2000). 22.
[15] Deutsch interview, 2006.
[16] Deutsch, D. (1985A).

multiverse.[17] Nonetheless, Oxford philosopher Michael Lockwood used Deutsch's method to construct a many minds interpretation of Everett:

> Instead of postulating a continuous infinity of worlds, we could credit every sentient being with a continuous infinity of simultaneous minds or conscious points of view, which differentiate over time.[18]

An intense debate over Lockwood's proposal ensued in the pages of the *British Journal of Philosophy* in 1996. Deutsch wrote an essay congratulating the philosopher on taking physics seriously, but he noted,

> Lockwood's preference for the term 'many minds' over 'parallel universes' risks giving the impression that it is *only* minds that are multiple, and not the rest of reality.... Admittedly, [we can] detect other universes only indirectly. But then, we can detect pterodactyls and quarks only indirectly too. The evidence that other universes exist is at least as strong as the evidence for pterodactyls or quarks.[19]

Deutsch expressed amazement that physicists and philosophers are reluctant to consider the Schrödinger equation as universally valid:

> Despite the unrivaled empirical success of quantum theory, the very suggestion that it may be *literally true as a description of nature* is still greeted with cynicism, incomprehension and even anger.[20]

In the same journal issue, Oxford philosophy professor Simon Sanders said it was a mistake to focus physics debates on "mentality and the nature of consciousness." Saunders, a realist, is a pivotal player in many worlds research. Intellectually, he is a hybrid, typical of philosophers of science: he is learned in both the higher mathematics of physics and the syllogisms of philosophy. And he is a bit of a rabble-rouser inside his rarified circle of colleagues.

After a year as an undergraduate in the early 1970s in the newly founded Physics and Philosophy degree school at Oxford, Saunders eschewed the socially elite, all-male "champagne glass twirling" atmosphere of his ancient, walled-in college to live communally on a farm while puzzling out the quantum paradoxes and earning his degree, and later his doctorate. By 1990, he was working as a professor of philosopher at Harvard University in Cambridge, Massachusetts, where he met his no-nonsense wife, Kalypso Nicolaidis, an expert in international relations. Kalypso recalls that Saunders talked about multiple universes on their first date. So, she married him.

Saunders returned to Oxford in the mid 1990s as a philosophy professor. Soft-spoken and unassuming, the professor is the polar opposite of Everett's persona.

---

[17] Deutsch, private communication, July 2009. In his 1984 paper Deutsch made the important observation: "Unlike the C.I.[Copenhagen interpretation], the Everett interpretation can therefore be applied at all instants, not just after measurements. It gives a picture of a world (i.e. everything that exists) consisting of many coexisting universes (i.e. maximal sets [of] observables with values) evolving approximately independently on large scales, but in intimate interaction, through interference effects, on small scales." Deutsch, D. (1985A). 19.

[18] Lockwood, M. (1996). 173. Lockwood's argument is interesting, but too complex to fully describe here.

[19] Deutsch, D. (1996).

[20] Ibid.

But for two decades, he has been grappling with the problem of defining probability in many worlds theory, picking up where the original theorist abruptly left off. Saunders was one of the first philosophers of science to recognize the importance of decoherence to the many worlds interpretation. He has also developed distinct parallels between Everett's relative state idea and Einstein's theory of relativity according to which the difference between the past, present and future is purely *relational*. Just as it seems that there is no "flow of time" in Einstein's relativistic view of the whole of space and time, so there is no "transition from the possible to the actual" in Everett's wave function of the universe. Put simply, "relational fact" theory speaks to the problem of identity by asserting that one can locate oneself as an object inside the multiverse of all actualities by identifying the fact or sets of facts to which the self-object is related, e.g. I am *that* Simon that exists in a universe in which London has a population of 7,556,900 at noon sharp on Tuesday, not the *other* Simon where the population is 7,556,901!

In the late 1990s, Deutsch dropped a theoretical bombshell that astonished Saunders: he proposed that the Born rule can be derived in quantum mechanics from considerations of pure rationality.[21] In a move that would have delighted Everett, Deutsch turned to decision theory—the modern incarnation of game theory—to define probability in the multiverse. Deutsch recalls

> I eventually got around to publishing a decision-theoretic version of probability, and there was a welter of criticism, almost entirely missing the point, from people objecting to the Everett interpretation itself, rather than to this particular derivation of probability. They got rather hot under the collar about it.[22]

His technical argument concluded that rational decisions are expected to satisfy personal preferences with a certain probability. Taunting one-worlders, Deutsch argued that people are rationally compelled to make quantum mechanical decisions in a branching universe *as if* the Born rule applied. This addressed one of the main objections to the many worlds interpretation: that if everything physically possible happens, then assessing outcomes probabilistically is not necessarily a reasonable criterion for deciding what to do. He concluded:

> Thus we see that quantum theory permits what philosophy would hitherto have regarded as a formal impossibility, akin to "deriving an ought from an is," namely deriving a probability statement from a factual statement. This could be called deriving a "tends to" from a "does."[23]

Savage axioms

As the millennium turned, a graduate student in physics at Oxford, David Wallace, improved the rigor of Deutsch's argument, knocking down some

---

[21] Deutsch, D. (1999).
[22] Deutsch interview, 2006.
[23] Deutsch, D. (1999).

of the objections to it.[24] Saunders quickly endorsed the newly christened "Deutsch-Wallace" theorem's consideration of the "practical" problem in a theory of branching universes: How should we act as to realize our goals when every outcome actually occurs? According to Everett's basic theory, the quantum world appears to be indeterminist—even though the multiverse is objectively determinist because everything occurs in some measure.

Wallace observes,

> The co-existence of determinism with the in-principle-unknowability of the future is from a philosophical point of view perhaps the Everett interpretation's most intriguing feature.[25]

The name of the game is to derive probabilistic judgments from constraints of rationality without postulating a probability rule or changing the essential formalism of quantum mechanics in which the Schrödinger equation evolves deterministically. This research is guided by the statement: "Whatever (objective) probability *is*, our empirical access to it is via considerations of rationality and behavior: it must be the case that it is rational to use probability as a guide to action."[26]

In defining rationality, Wallace referred to a set of logical axioms developed in 1954 by Leonard J. Savage.[27] Savage noted:

> It is one of my fundamental tenets that any satisfactory account of probability must deal with the problem of action in the face of uncertainty.[28] ... From the *personalistic* point of view, statistics proper can perhaps be defined as the art of dealing with vagueness and with interpersonal difference in decision situations.[29]

Interpreting quantum mechanics with decision theory is not as far-fetched as it may sound. Wallace constructed (and continues to refine) a technical proof that it is rational to use the Born rule (squaring the wave function of a quantum event to determine its classical probability) to make decisions in a multiverse—without postulating that rule.

One weakness in the original Deutsch-Wallace approach (acknowledged as such by its proponents) was that it required making a different kind of assumption: that the Everett interpretation is true. If one does not believe in branching universes, then the decision-theoretic argument loses meaning because it is an argument that one should accept the Born rule *if* one believes in branching universes. This limitation was notably addressed by Hilary Greaves, a fellow in philosophy at Oxford's Summerville College. Despite the fact that everything

[24] Wallace, D. (2003). Wallace subsequently switched his profession to philosopher of science, a discipline he now teaches at Oxford. He is writing a book on Everettian quantum mechanics for Oxford University Press.
[25] Wallace, D. (2007). 11.
[26] Wallace, D. (2002). 3.
[27] Savage, L. J. (1954). Savage was a game theorist and a Bayesian (as was Everett).
[28] Ibid. 60.
[29] Ibid. 154. Italics added.

happens, the multiverse theory is still accountable to empirical test on the basis of measuring relative frequencies, says Greaves. Rigorously applying Savage's axioms, which are independent of quantum mechanics, she argued that it is rational to use the Born rule to make personal decisions in both single universe and multiverse models. And her argument did not require believing that the Everett worlds exist.[30]

In sum, the Oxford philosophers want to "operationalise" probability in quantum mechanics by reducing it—not to relative frequencies—but to quantifiable "preferences of rational agents."[31] For Wallace, "Rationality is the only way in which the concept of 'probability' makes contact with the physical world."[32] And a rational being (or machine-driven algorithm) ranks preferences among decision options according to principles of utility, i.e. usefulness to the ranker.[33] Utilities are here treated as values—prices—placed upon optimizing certain goals. Rational people compare the value they have assigned to the achievement of a goal with its perceived probability of occurrence before they choose to act. But in the Everettian multiverse, of course, this method is questionable: how can rational decisions be based upon relative frequencies when there appears to be no such thing as probability?[34]

Yet, probabilistic quantum mechanics works.

Savage's basic tenet is that probability is personal. So, say the Oxford Everettians, in a multiverse, it is functionally logical to make decisions based on personal preferences and beliefs about possible outcomes in one branch even though "you" do not know which post-decision branch "you" will end up on.[35] The Born rule serves to quantify this uncertainty, and becomes a rational

---

[30] Greaves, H. (2004); Greaves, H. and Myrvold, W. (2007). Greaves appeals to Bayesian updating and the evidential nature of "confirmation theory" to arrive at the Born rule.

[31] Wallace, D. (2007).

[32] Wallace, D. (2002). 62. For Saunders, "branching should be understood as objective probability, and Deutsch-Wallace should be taken as showing that subjective probability should track those objective probabilities." Saunders, private communication, July 2009; see also Saunders, S. (2005). Everett, himself, tended to view probability as a purely subjective phenomenon for the single observer, even though he laid the foundation for Saunder's view that there is an objective probability (from the bird's eye perspective). Unfortunately, the definition of "single" observer gets increasingly hazy the more you think about it in these terms.

[33] It is worth noting that a half century ago, a RAND researcher named Daniel Ellsberg (later of Pentagon Papers fame) wrote a critique of the rationality of using the Savage axioms to make risky decisions. In the face of *ambiguous* information, Ellsberg explained, in technical language, it is not irrational for decision-makers to *not* use "their best estimates of probability," i.e. people who do not use the Savage axioms are not necessarily irrational. Ellsberg, D. (1957).

[34] Philosopher Jeffrey Barrett (University of California, Irvine), an Everett expert who has substantial disagreements with Oxford Everettians, notes: "If every consequence is in fact realized, then there cannot possibly be constraints on rational action since action must be determined as rational or not by differing consequences contingent on the action, but all rewards and punishments are typically in fact fully realized regardless of what action an agent performs." Barrett, private communication, 2009.

[35] Another way of looking at the situation, from a bird's eye view, is that the possible consequences of one action are correlated with the possible consequences of the many other actions with which it is entangled as described by the universal wave function. Therefore, a particular outcome occurs with relative frequency across the set of all post-event branches that include all possible outcomes of that action in their subsequent history. That is, from the bird's eye view, outcome A is recorded in 20 percent of the branch histories, B in 30 percent, C in 50 percent. And in a single branch, a frog-observer records a sequence of squared amplitudes: A at 0.2, B at 0.3, C at 0.5.

guide for decision-making even when all consequences appear to be possible. The argument gives Everettians permission to keep using probability in quantum mechanics, making a case that it is rational to accept the Everett interpretation.[36]

Probability derivations aside, why do we only experience one world?

Splitting tigers

Building on Saunders' reliance on decoherence theory to explain the emergence of a classical reality, Wallace appealed to philosopher Daniel Dennett's concept of real objects defined as persistent patterns.[37] The insight goes something like this: Think of a tiger and the jungle and the branching universe in which it hunts. The tiger and its environment (including us, the observers) are composed of entangled microscopic systems constantly exchanging photons and electrons and waves of energy. The tiger, and everything it interacts and correlates with, constantly branches into copies that embark upon different histories. Quantum mechanically, the tiger is a smear of patterns shape-shifting through a quantum jungle spread over multiple worlds and times: but for macroscopic observers (and our copies), the tiger (and its copies) persists as a pattern, an *approximation* of a sharply defined object at an atomically coarse-grained level of nature, but an entity, nonetheless: a living, breathing regularity emerging from a "swirl of molecules."[38]

Wallace elaborates:

> It makes sense to consider a superposition *of* patterns, but it is just meaningless to speak of a given pattern as being *in* a superposition.... Patterns are not superposed, but duplicated by the measurement event.... It is worth remembering the crucial role that decoherence is playing in this account: without it, we would not have the sort of branching structure which allows the existence of non-interacting multiple near-copies of a given process. Multiplicity occurs at the microscopic level, thus macroscopic objects... are genuinely multiplied in number.[39]

Wallace speaks to the problem of understanding the definiteness of individual universes inside the giant superposition of the multiverse in which interference terms between branching universes are not wholly lost.

> Are consistent histories, and worlds which persist over time, real? Yes, in the sense that rivers, or animals, or persisting objects, are real: like worlds or instants they are not directly present in the formalism, and unlike worlds or instants they are only approximately definable, but that is no reason why they should not be seen as legitimate entities or used in our explanations (any more than we should

---

[36] Wallace's work is more sophisticated than this summary. The point is that modern Everettians are addressing the probability problem by expanding the terms of the debate and, like Everett, questioning probability (as we think we know it).

[37] See: Dennett, D. C. (1991).

[38] Wallace, D. (2003A). 6.

[39] Ibid. 12–17.

expect to be able to describe zoology in any useful or explanatory way using only the language of quantum field theory).[40]

Following Saunders, Wallace also links general relativity and Everett's relative states:

> We are undoubtedly more at home with Minkowski space-time than with the universal state. Partly this may be because we have worked with the [relativity] concept in physics for rather longer, but more importantly we have long been used to the idea that multiple times exist (in some sense)—the innovation in relativity theory is the unification of these instants into a whole, and the identification of the instants as secondary concepts. Everett asks us to take both steps at once: to accept that there exist many worlds, and then to fuse them together into a whole and accept that the worlds are only secondary. Clearly this is a significantly larger conceptual jump; still, if we are prepared to accept the existence of many worlds and if we are happy with the step from many times to space-time, there seems no reason to avoid a similar step in the case of quantum theory.[41]

It's all about information flow, says Deutsch, who likens the structure of the multiverse—with its classically separated, but quantum mechanically linked realities—to a computation run by a computer straddling parallel worlds.[42] Taking a hard line, he says that talking about many worlds as an interpretation of quantum mechanics "is like talking about dinosaurs as an 'interpretation' of fossil records."[43]

The burning question is: are the Everett worlds *real*? For Deutsch, Saunders, Wallace and their supporters, yes, according to quantum mechanics. They have yet to convince the world, but Saunders makes the excellent point that nothing similar to their logical arguments has been made for sustaining a one-world realist interpretation of quantum mechanics—or any one-world realist theory of probability, for that matter. The idea that indeterminism rules the quantum world in one-world theories rests on familiarity and tradition, rather than an understanding of what physical chance really is, say the Oxford Everettians.[44]

Everett@Fifty

Even as the July 2007 issue of *Nature* featured Everett's many worlds theory on its cover in celebration of its half century anniversary, the Oxford philosophers organized an international conference, Everett@Fifty. About 30 philosophers and physicists gathered to hotly debate the pros and cons of the many worlds theory at Oxford's faculty of philosophy just off Logic Lane. In October, the

[40] Wallace, D. (2001). 23.
[41] Ibid. 23–24.
[42] Deutsch, D. (2002).
[43] Quoted in Saunders, S. *et al.* (2010).
[44] Ibid.

debate continued at a conference at the Perimeter Institute for Theoretical Physics in Waterloo, Canada.

The ground rules for the Everett conferences mandated a realist, physicalist approach, no idealists need apply:

> Ask not if quantum mechanics is true, ask rather what the theory implies. Specifically, ask what quantum-state *realism* implies, with the state as given by the Schrödinger equation. What follows, in particular, when quantum theory is applied without restriction to the whole universe?[45]

But within realism, there is plenty of room for disagreement. Especially when it comes to critiquing what amounts to a new theory of probability being worked out by the Oxford Everettians. For instance, although the Savage axiom-based approach to deriving probability does not do any better than von Neumann and Morgenstern did in explaining *what probability is* (or how utilities are set in the real world), it, arguably, surpasses military-style game theory as Everett practiced it by validating the use of an *ethically neutral* probability measure—the Born rule.[46] It explains why it is not necessarily irrational (in theory) to use this measure when making self-interested decisions in a deterministic universe; (Everett, in his own way, was aiming at the same target: explaining why using the Born rule is rational).

At the conferences, critics of the decision theory approach questioned the criteria used to define utility and rationality in a deterministic multiverse. Huw Price, professor of philosophy at the University of Sydney, made a case that if a bottom line of maximizing self-interest is replaced by a form of "distributive justice" (which might weight the concerns of poor people greater than the rich, or the lame as greater than the fleet) then the appeal to decision theory might lose its rational connection to the Born rule. Price also insisted that metaphysical questions of personal identity be kept separate from decision theory since it is difficult to define "self-interest" for multiple "selves".[47]

As usual in Everettian conversations about splitting observers, the question of personal identity came to the fore. At the center of several conference debates was the question of how and to what extent should a decision-maker in an Everettian multiverse care about the fate of her or his "successors." What if she takes an action that will impoverish all of her successors save one billionaire–

---

[45] Ibid.

[46] In military game theory, probability measures were inextricably linked to maximizing utility values placed upon a limited range of supposedly rational choices. As Anatol Rapoport pointed out in his critique of game theory: "For the most part, decisions depend on the ethical orientations of the decision-makers themselves. The rationales of choices so determined may be obvious to those with similar ethical orientations but may appear to be only rationalizations to others. Therefore, in most contexts, decisions cannot be defended on purely rational grounds. A normative theory of decision which claims to be 'realistic,' i.e. purports to derive its prescripts exclusively from 'objective reality,' is likely to lead to delusion." Rapoport, A. (1964). 75. But in quantum mechanics, nature itself selects the menu of choice, and assigns probability weights, and the Born rule discovers the weights. It is rational to use the Born rule to construct a mini-max solution, (even if, as Everett said, probability is, ultimately, an illusion).

[47] Saunders, S. *et al.* (2010).

but that's the one that she identifies with (she only cares about winners). Is that irrational? Or what if it is the other way, and whatever else happens, she is concerned to defend the worst off—to alleviate their sufferings—at some expense to the great majority? If these matters seem esoteric, think about the question of mutual assured destruction that Everett contemplated: What would it mean to launch (or not launch) a nuclear first strike in a branching universe in which all physically possible events occur with some frequency?

Bohmians and metaphysicians chimed in with a range of objections to the arguments of the Everettians. Leading the charge, David Albert worried that if the Everett interpretation is true, there would be no such thing as probability, and we would not be able to do statistical physics. He strongly objected to the Deutsch-Wallace-Greaves argument from decision theory:

> The fission [observer splitting] hypothesis (since it is committed to the claim that all such experiments have all possible outcomes with all possible frequencies) is structurally incapable of explaining anything like that. The decision-theoretic program seems to act as if what primarily and in the first instance stands in the way of need of being explained about the world is why we *bet* the way we do. But this is sheer madness!...a bait and switch...an argument to the effect that if we held an altogether different set of convictions about the world [i.e. believing the Everett interpretation] than the ones we actually hold, we would bet the same way as we actually do.[48]

Love it or hate it, the influence of many worlds grows with age as science and philosophy continue to contemplate the strangely beautiful idea proposed by Everett a half century ago.

---

[48] Ibid. See also Kent, A. (2009). In which physicist Adrian Kent dissects the Oxford Everettian arguments.

Epilogue Beyond Many Worlds

Now, at the beginning of the 21st century, faced with scientific claims like neo-Darwinism and the multiverse hypothesis in cosmology invented to avoid the overwhelming evidence for purpose and design found in modern science, the Catholic Church will again defend human nature by proclaiming that the immanent design evident in nature is real. Scientific theories that try to explain away the appearance of design as the result of 'chance and necessity' are not scientific at all, but, as John Paul put it, an abdication of human intelligence.

Christoph Schönborn, Cardinal Archbishop of Vienna, 2005.[1]

Vatican attacks on a scientific theory usually are a sign that it is intelligent and correct—just ask Copernicus, Galileo, Bruno, Newton, Darwin, and Einstein. Their revolutionary insights survived the scorn of theologians and provided foundations for brave new experimental discoveries and increasingly explanatory theories. Assuming that the concept of a multiverse is here to stay: In addition to Everett's relative states formulation, how many kinds of multiverses *are* there? Is any multiverse experimentally verifiable? Even in principle?[2]

This was the subject of my conversation with an Albanian-born cosmologist, Laura Mersini-Houghton, as we traipsed a path along the rocky shore of San Miguel, an island in the Azores. It was the summer of 2009. We were attending a conference on foundational questions in physics, and Everett's theory was on many minds.[3] Mersini-Houghton believes there is newly discovered cosmological evidence that points to the possible existence of an Everett type multiverse. Popular science magazines are lionizing her theory because if there is even a smidgeon of proof of a multiverse: that is *news*.[4]

The history of physics is littered with strange ideas—but the fact that Everett's theory is a tool in the kit of contemporary physics underscores the observation by Wheeler with which this book began: "We can believe that we will first understand how simple the universe is when we recognize how strange it is."[5] Most new ideas are strange—at first.

[1] Op-ed *The New York Times*, 7/7/2005, A23.
[2] Bernard Carr's *Universe or Multiverse?* (2007) is a collection of 28 articles by top physicists ruminating on how to test for the existence of a variety of multiverses. Many of the essays discuss Everett's considerable influence on cosmology.
[3] The biannual meeting of the Foundational Questions Institute, www.fqxi.org
[4] For example, Chown, M., (2007). Geftner, A. (2009). Bignami, L. (2008).
[5] Wheeler, J. A. (1973). 245.

The Landscape

As the 21st century dawned, the marriage of string theory and cosmology derived a "Landscape" of 10^{500} universes. That is a 10 followed by 500 zeros: a rather large number considering that there are only 10^{80} atoms in "our" universe. Each of the universes in the (abstract, unseen, unverified) Landscape—think of it as a multiverse—differs in the arrangement of its physical constants (vacuum energy, gravity, fine-structure, etc.). Most of these universes would be inhospitable to life as we know it, because the configuration of constants that permitted life to evolve from collections of hydrogen atoms needed to be so precisely calibrated that our presence seems altogether improbable. Until the advent of the Landscape, scientifically inclined theologians eagerly fastened upon this cosmological improbability. Wielding the anthropic principle that an explanation of the structure of the universe must be compatible with our presence as observers, they claimed that our finely tuned existence is so unlikely to have occurred that nature must have been designed by a super intelligent Being!

In defense of secular reason, physicist and author Leonard Susskind has explained that the Landscape sets the emergence of our seemingly improbable world inside such a vast distribution of possible worlds that our puny place in the scheme of all things seems unremarkable, inevitable, mundane.[6] It is hubris to assert that the universe is anthropocentric, focused on humans. If the Landscape (or any type of multiverse) exists, we are probabilistic accidents, not the children of angels, nor the toys of a Grand Inquisitor.[7]

The Landscape is related to "eternal inflation," the constant bubbling into being of new universes. As a metaphor for inflation, think of two pencil marks on a balloon expanding away from each other as the balloon is blown up. In a forever-expanding balloon, Big Bangs are the result of local braking actions. In a highly metaphorical sense, an area of this cosmically inflating balloon underwent a probabilistic shift in the value of its constants, "nucleating" a bubble, a balloon within the balloon, in a shower of quantum sparks followed by a powerful burst of temporary inflation that shaped our particular universe. The configuration of the cooling sparks (post-inflation) is preserved as Cosmic Microwave Background Radiation (CMB), an observable record of the early history of our bubble—one of many bubble universes in the Landscape of eternal inflation.

Susskind reflects,

> The many-worlds of Everett seems, at first sight, to be quite a different conception than the eternally inflating megaverse. However, I think the two may really

---

[6] Susskind, L. (2005).
[7] There are several sorts of multiverse—including an incredibly tedious one. If we consider the universe to be infinite in size, it should, according to the laws of probability, reproduce an infinite number of recurrences of everyone and everything in the vastness beyond our cosmic horizon: an infinitely repetitive Groundhog Day (à la the Bill Murray movie). It would include the Everettian splits.

be the same thing.... There are branches (as well as real worlds) for every location on the Landscape.[8]

Physicist Juan Maldacena of the Institute for Advanced Study also draws a parallel between the Everett worlds and eternal inflation: "In principle they are different, however it is important in the notion of inflation that there are some quantum fluctuations that change the parameters along the spatial slice of this universe. So perhaps the many worlds of eternal inflation that occur at different points in space should be viewed as the many worlds of Everett where you think about these things but occurring at the same point in space."[9] In other words, nucleating bubbles and branching worlds may be—somewhere, somehow—linked by the rules of the quantum mechanical game.

Naturally, scientists need more than mathematical models to validate ideas; they need evidence that physical reality is congruent with ideas.

Matthew Kleban at New York University thinks about what might happen when two bubble universes collide. He says that a signature of a past collision could be recorded in the CMB. One sign of a collision would be the existence of a record that the otherwise uniform directional flow of our expanding universe has been marred by a contrary flow—what Kleban calls, tongue-in-cheek, an "axis of evil." ("Evil" because a significant collision with a neighboring bubble would vaporize us.)[10]

Enter the Dark Flow, a directional anomaly in the CMB—8 billion light years distant—discovered in late 2008 by astronomers. The Dark Flow could signify a movement of galaxies flowing in a direction opposite to the rest of our universe. That possibility has some cosmologists in a tizzy—wondering if it is evidence of the existence of a space-time outside our space-time? Mersini-Houghton thinks it might be a gravitational pull on our universe by a neighboring universe (a "brane" in the parlance of string theory).[11]

It could also be a blip in the data.

Cold spot

In 2007, NASA astronomers announced that they had discovered a giant void in the CMB, nicknaming it the Cold Spot. Mersini-Houghton's take on it was intriguing from the Everett point of view. She speculated that the Cold Spot recorded a moment during the Big Bang before inflation really took off, in which a quantum-gravitational superposition of possible universes existed in a very tiny patch. A moment later, an unfathomable number of universes—proto-bubbles in the Landscape—decohered from the patch, leaving behind the tiny, hot void, which gradually grew to a large, cold void as the CMB

---

[8] Susskind, L. (2005). 317–323.
[9] Private communication, Maldacena, July 2009.
[10] Chang, S., M. Kleban, *et al.* (2008); Kleban interview, August 2009.
[11] Mersini-Houghton, L. and R. Holman. (2009).

expanded. When writing about the emergence of this quantum multiverse, Mersini and her colleagues used Everett's idea of a universal wave function:

> By proposing to place the wave function of the universe on the landscape and use its many-worlds interpretation, the quantum mechanical universe was thus embedded onto the string landscape thereby making the Everett and the Landscape multiverse equivalent.[12]

More than 50 years after Everett aired his controversial idea, it remains alive and kicking and deeply disturbing. Maldacena remarks, "When I think about the Everett interpretation in everyday life, I do not believe it. But when I think about it in quantum mechanics, it is the most reasonable thing to believe."[13]

In 1730, Isaac Newton—who knew nothing of Planck's constant or the Born rule or string theory Landscapes—wrote, in *Opticks*:

> And since Space is divisible in infinitum, and Matter is not necessarily in all places, it may also be allow'd that God is able to create Particles of Matter of several sizes, and in several proportions to Space, and perhaps of different Densities and Forces, and thereby to vary the Laws of Nature, and make Worlds of several sorts in several parts of the Universe. A least, I see nothing of contradiction in this.[14]

And when all is said and done, why should there be but *one* universe? Is not the notion that there is only one world just as strange as that there might be many? And what if there are many worlds? What then?

> Let sea-discoverers to new worlds have gone;
> Let maps to other, worlds on worlds have shown;
> Let us possess one world; each hath one, and is one.[15]

The End.

---

[12] Kobakhidze, A. and L. Mersini-Houghton (2006). See also: Aguirre, A., *et al.* (2007). On the other hand, researchers at the University of Michigan assert that the Cold Spot may be a statistical anomaly. See: Zhang, R. and Huterer, D. (2009); Merali, Z. (2009). Regardless of media-driven bubbles of excitement, there is a distinct prospect of finding evidence of other universes in the CMB. See Kashlinsky *et al.* (2009).
[13] Maldacena interview, 2006.
[14] Newton, I. (1730). 379–380.
[15] Donne, J. (circa 1621). "The Good Morrow." 3.

Glossary

ABM Acronym for anti-ballistic missile systems, which are expensive to build and cheap to fool.

Alcoholism, drug addiction Substance abuse is a mental illness with definable symptoms.

Altruism Evolutionary biologists now regard group altruism as a significant force behind evolutionary success.

Anthropic principle The idea that whatever explanation one has of the structure of the universe, it must be compatible with the fact that we exist as observers.

Amplitude, wave function The wave-like attributes of a quantum system expressed as a complex number, i.e. a number that uses "i," the "imaginary" number that is the square root of minus 1. An amplitude (or wave function) is constructed from a series of measurement experiments, although, arguably, it exists regardless of human agency.

Born rule The postulate that squaring the value of an amplitude obtained from a measurement interaction gives the classical probability that the same amplitude will result from a repetition of the measurement experiment. This predictive rule is the fundamental tool of quantum mechanics.

Bell's theorem John S. Bell's mathematical proof that local hidden variable theories and the predictions of quantum mechanics are incompatible. The theorem does not rule out the existence of non-local hidden variables as theorized by David Bohm.

DeBroglie-Bohm hidden variables interpretation A non-relativistic, non-linear quantum theory with a universal wave function and no requirement for the presence of an observer. It treats particles as guided on classical trajectories by probability or "pilot" waves.

Classical physics The mechanistic physics of Newton and Maxwell in which a theoretical knowledge of all initial conditions of a system allows for the determination of subsequent conditions along a continuum of cause and effect.

Collapse postulate John von Neumann's axiom that a measurement of a superposed quantum system causes the system to "jump" into but one element of the superposition.

Commutation relation In the quantum formalism, when two matrices of complex numbers (A & B) are multiplied together, they do not necessarily commute. There are cases in which $A \times B \neq B \times A$. The difference is a tiny number based upon Planck's constant. This phenomenon reflects fundamental features of quantum mechanics (see: Heisenberg uncertainty principle).

Complementarity Niels Bohr's philosophy that reality is composed of mutually exclusive phenomena—such as waves and particles. Arbitrarily partitioning the universe between classical and quantum realms, it declares that the microscopic world must be interpreted solely in terms of classical physics.

Correspondence principle Bohr's method of analogizing events in the indeterministic quantum world to events described by the laws of classical physics.

Consistent histories interpretation This approach treats the universe quantum mechanically and assigns probabilities to both microscopic and macroscopic states before measurement-type interactions take place. It considers classical physics to be an approximate representation of quantum reality.

Containment The geopolitical strategy articulated after the Second World War by George F. Kennan that called for controlled isolation of the socialist states, short of threatening war.

Copenhagen interpretation An amalgam of various philosophical approaches popularly attributed to Bohr, von Neumann, Werner von Heisenberg, and others. It posits that we can only speak about quantum mechanics experimentally and by reducing it to classical concepts through wave function collapse. Because this approach allowed physicists to "shut up and calculate," it was the prevailing interpretation in the West during the much of the 20th century. Socialist physicists, however, tended to consider it a positivistic and non-materialist explanation.

Cosmology The study of the origin and evolution of the universe.

Cybernetics As coined by the polymath, Norbert Wiener, cybernetics treats information as a physical quantity subject to the laws of entropy and probability. It recognized the role of "feedback loops," i.e. random output that becomes corrective input.

Decision theory Similar to game theory, it parameterizes the branching process of decision-making in terms of utilities, probabilities, and considerations of "rationality."

Decoherence In practice, the irreversible entanglement of a quantum system with its environment.

Degrees of freedom The number of variables in an equation. The more opportunities an object has to change, the more variables are required to reflect this freedom.

Deterrence In the context of the Cold War: threatening perceived enemies with destruction if they attack while trying to moderate the threat enough so that they do not attack preemptively, as both sides are mirroring each other.

Dualism Involves a metaphysical commitment to two distinct, but often causally related phenomena. In quantum physics, the idea that a quantum object behaves like a wave and, also, a particle. In philosophy, the separation of mind and body, spirit and matter.

Entanglement, correlation When two or more quantum systems interact their properties may become correlated, linked, *entangled*. This relation may persist for objects that become superluminally separated. In terms of observation: if you know something about one of the systems, you also know something about its partner. Entangled systems can be viewed as a single system with its own wave function.

Entropy A measure of disorder.

Fallout Pulverized material stirred up and irradiated by thermonuclear explosions and distributed by the prevailing winds.

Fourier analysis A mathematical method that works in classical mechanics to break down complex oscillations into simpler parts. In quantum mechanics, the method decomposes a wave function into its constituent wave functions.

Game theory A mathematical method of modeling complex systems (economics, evolution, warfare) in which choices are assigned probabilities and utility values.

Generalized Lagrange Multiplier Method An improvement upon Lagrange's optimization method that brought cost-benefit analysis into the computer age. Everett's GLM algorithm assigns "prices" to the use of specific resources while weighing the constraints inhibiting the utilization of those resources and delineating optimal performance paths in complex situations affected by hundreds of variables.

General relativity Einstein's theory that gravity is embedded in a geometrized space-time.

Heisenberg uncertainty principle Some properties of quantum objects (e.g. time and energy, position and momentum) cannot be simultaneously measured to a degree of accuracy greater than Planck's constant.

Hidden variables interpretation The idea that quantum mechanics is an incomplete description of the world.

Hilbert space The complex number-valued space used to represent quantum mechanical states. States are treated as vectors in this mathematical system, which describes superposition, interference, and entanglement with a precision that classical language cannot touch.

Idealism The idea that reality is an idea.

Information Everett viewed information as a measure of the spread of a probability distribution.

Interference A property of waves in which the positive and negative values represented by crests and troughs add when combined. In quantum mechanics, the values are complex numbers.

Irreversibility The Schrödinger equation is reversible, i.e. the evolution of quantum states works forwards and backwards in time. There are, however, some physical processes that do not appear to be reversible—the thermodynamical "arrow of time," for example. Or, in practice, quantum entanglement.

Landscape The equations of the string theory Landscape require 10^{500} universes to encompass a full range of solutions.

Lebesque measure A measure of length, area, or volume. It can be applied to Hilbert space as well as classical (phase) space.

Linearity The Schrödinger equation evolves each element of a superposed wave function deterministically.

Locality, non-locality One notion of locality is the relativistic rule that information cannot be transferred between any two points in space faster than the speed of light. Non-locality is the apparent failure of that rule, as evidenced by the experimental verification of quantum entanglement in which information appears to be instantly communicated between separated but entangled objects. Copenhagen-type quantum mechanics cannot explain non-locality, but the MWI deals with it easily by asserting that as all outcomes occur, there is no need for non-locality, or "action at a distance."

Matrix mechanics Heisenberg conceived the use of matrices (arrays) of complex numbers to depict static relationships between quantum states. The method is basically equivalent to Schrödinger's wave mechanics, but particularly suited to treating such continuously valued properties as the position of a particle.

Measurement problem The paradox that the Schrödinger equation evolves quantum superpositions through time with each element of the superposition intact, and yet we experience only one element of a superposition in our classical space.

Momentum Velocity times mass.

Mutual assured destruction The game-Theoretic doctrine of war preparation that underlay the expansion of nuclear weapons systems during the Cold War by both superpowers. Its premise was that superiority in first strike capabilities would deter an opponent from attacking, except out of fear of attack by an opponent fearing attack from an opponent fearing...

Multiverse The collection of all universes. In the Everettian model, it is described as a universal wave function.

Negentropy The opposite of entropy or disorder. A state with high negentropy has low probability, i.e. it contains more information relative to a state of higher probability or lower negentropy.

Ontological argument A supposedly logical construct proving the existence of a god.

Path integrals, sum over histories A method invented by Richard Feynman for calculating the motion of particles in which entire trajectories (histories) have amplitudes.

Planck's constant: h Also called the quantum of action, it is a fundamental constant of nature. It is the ratio of the energy of a quantum of energy to its frequency. Very tiny (at 6.626×10^{-34} Joules), it governs the exchange of quanta, i.e. the minimum amount of energy by which an electromagnetic (or gravitational) system can change.

Positivism The philosophy of science that reality can only be described as a record of experimental results. We cannot talk meaningfully about anything that cannot be empirically validated.

Probability calculus A set of rules that define the properties of probability. It allows us to make predictions based upon the relative frequency of past events.

Probability distribution In quantum mechanics: as constructed from a series of experimental observations on a microscopic system, the distribution contains the results of such measurements, translated as a range of classical probabilities that add to unity (100 percent).

Psycho-physical parallelism Articulated by von Neumann as the requirement that a scientific theory should correspond to our experience of physical reality.

Quantum state A collection of data sufficient to allow us to make predictions about transformations of the data; represented as a vector in Hilbert space.

Rationality The definition appears to be fluid. According to von Neumann-Morgenstern's *Theory of Games and Economic Behavior*, "It may safely be stated that there exists, at present, no satisfactory treatment of the question of rational behavior.... The chief reason for this lies, no doubt, in the failure to develop and apply suitable mathematical methods to the problem; this would have revealed that the maximum problem which is supposed to correspond to the notion of rationality is not at all formulated in an unambiguous way. ..... [It] is much more

level is far more rational (as shown by evolutionary biology) than the
utilitarian pursuit of narrowly perceived self-interests as iconized by game theory and capitalist economics.

Realism In the philosophy of science: that an objective reality exists independent
of human or divine agency or perception.

Recursion A feedback loop. A simple computer program instruction, "add 1 to the
result just obtained," is recursive, for instance. It allows a series of operations to
advance and conclude without having to write a new instruction at every step.
But the recursive method can bog down in paradox. It is not possible to resolve
the truth of the recursive statement, "All reporters are liars. I am a reporter."

Schrödinger equation The time-dependent partial differential equation invented
by Erwin Schrödinger in 1926. It is the standard tool for calculating changes in
quantum states viewed as wave-like entities.

Single Integrated Operating Plan (SIOP) The United States' targeting plan for
nuclear war. As late as 1998, the SIOP targeted the former Soviet Union. It is
extraordinarily complex and ultra-secret and is seemingly unaffected by disarmament talks.

Solipsism My idea that reality is my idea.

Special relativity A theory based upon the assumption that the speed of light is
constant and that the laws of physics are the same for all observers. Features of
Einstein's theory are that there is no privileged frame of reference in four-
dimensional space-time; that energy and mass are equivalent; and that objects
become infinitely massive and time-dilated when moving at the speed of light
(photons do not have mass).

Spin The intrinsic (non-orbital) angular momentum of an electron, nucleus, or
elementary particle at rest. It comes in discreet quantities and is affected by
magnetic fields.

Superposition principle For any two or more possible quantum mechanical states,
the superposition of these states is also possible. The wave function of an object
contains data that, in sum, list all of the configurations of all of the possible
properties of the quantum object. Each datum or configuration or element of a
wave function is also a wave function and, therefore, the seed of a probability
value. The Schrödinger equation deterministically evolves each element in a
wave function through time *linearly* as if each element is equally and potentially
real without singling out any element as more probable or more real than any
other. A *coherent* superposition allows for these elements to interfere with each
other, i.e. to continuously evolve as combined wave functions. But interference
effects are absent in an *incoherent* superposition of quantum states (or wave
functions or state vectors) and each element in an incoherent superposition has
statistical properties, meaning that predictions regarding its future coordinates
in classical reality (phase space) can be made (using the Born rule). The phenomenon of superposition applies, in theory, to macroscopic objects as well as

microscopic, but due to decoherence effects we do not normally see such items as cannonballs or the Eiffel Tower in superpositions. As a metaphor: consider a double- or triple- or infinitely exposed photograph: a superimposition of images representing real things.

Tautology In logic, a proposition that is true by virtue of its abstract form—its conclusion is identical to its premise. In philosophy, a tautological statement conflates cause and effect. In language, a tautological statement is self-referential—it lacks meaningful content or useful definition, i.e. "If A is true, then not-A is false." Or: "A dog is a canine." Or: "There is no god but God."

Unitary As it evolves elements of a superposition through time, the Schrödinger equation preserves these amplitudes so that, when squared, their sum will always equal 1 (100 percent).

Variable A quantity or force that changes in value.

Vector A mathematical object that represents both magnitude and direction. For example: an arrow in flight.

Wargasm Cold War operations researchers often used sexualized language to describe the doctrine of launching the entire arsenal of nuclear weapons in a first or second strike.

Wave function A complex number representing data gathered by observing and recording such properties as position and momentum of quantum objects. It can be described as a vector in an infinite-dimensional mathematical space (Hilbert space) that obeys certain laws. Squaring a wave function (or amplitude) gives the classical probability that the same wave function will reoccur under identical circumstances.

Wigner's friend A famous thought experiment that asks what happens when an observer of a quantum measurement is part of the system observed.

Acknowledgments

Special thanks:

Mark Everett for trusting me with his father's papers and his family archive; and for commenting on draft chapters.

Stacey L. Evans for reading raw drafts and making many helpful suggestions and for her's and our son Miles' good will in the face of late nights and lost weekends and minor car accidents caused by day dreaming about probabilities.

Simon Saunders, H. Dieter Zeh, and Jeffery A. Barrett for many stimulating conversations about Everett, probability, decoherence, and the philosophy of quantum mechanics. David Deutsch for several talks.

Dr. George B. Wesley for research assistance and comments on early drafts.

Thank you to physicists:

I first learned about Everett's existence from Stephen H. Shenker, who gave me moral support and solid advice during the several years it took to produce this book. Kenneth Ford gave unstintingly of his experience and knowledge and read draft chapters, kindly pointing out errors and making oodles of helpful suggestions. Wojciech H. Zurek and H. Dieter Zeh not only commented on draft chapters, but they painstakingly explained decoherence to me in lay terms. Additional physics pedagogy was received from James B. Hartle, Leonard Susskind, Max Tegmark, Matt Bellis, Don Eigler, David Briggs, Charles Misner. Any errors are purely mine.

Philosophers of science:

Simon Saunders of Oxford University's faculty of philosophy invited me to Oxford, twice, to learn about Everettian quantum mechanics, and we had scores of conversations by email; he also reviewed several draft chapters. Jeffery A. Barrett of University of California, Irvine patiently explained elementary concepts in quantum physics to me and made many helpful comments on the draft book. David Wallace reviewed several chapters and was generous with his conversational time. Harvey Brown made several pithy remarks that illuminated the way. Sean Boocock reviewed the chapters on Bohr and the measurement problem and made many helpful remarks. Arthur Fine honored me with two very informative discussions about the measurement problem. Olival Freire, Stefan Osnaghi, and Fabio Freitas provided primary source material on the "Everettian heresy." Jim Baggott commented on several draft chapters.

Others:

Many thanks to Gary Lucas for long talks about Everett's personal and professional lives and for explaining Generalized Lagrange Multipliers. Donald Reisler for in-depth conversations about his years as Everett's friend and colleague. Charles and Susanne Misner for personal recollections. Misner for comments on draft chapters and writing a foreword. George E. Pugh for his memoir of Everett on file at the American Institute of Physics. Eugene Shikhovtsev of Kostromo, Russia for being the first biographer to tackle Everett: he inspired me and generously offered access to his source materials.

Gregg Herken looked over rough drafts of the Cold War chapters and made useful remarks. Finn Aaserud and Anja Skaar Jacobsen provided historical materials, as did Cecile DeWitt-Morette. Conversations with Harold W. Kuhn greatly informed the game theory chapters. Eugen Merzbacher told me about the existence of the Xavier transcript. Joanna Frawley provided information and read draft chapters on weapons systems research. I learned a lot from attending the Everett@50 confer-ence at Oxford University and the Everett conference at the Perimeter Institute for Theoretical Physics which were packed with wonderful speakers. Useful talks were also had with Shelly Goldstein, Adrian Kent, Wayne Myrvold, Hilary Carteret, Leon N Cooper, Leon Lasden, Paul Davies, Laura Mersini-Houghton and every one of the dozens of people interviewed for this book (named in the bibliography). Thanks to non-academic friends who made useful comments on early drafts: Deborah Hayden, Debbie Hupp, Patrice Gelband, John Morganthaler.

Special thanks to Spencer Weart, Martha and Chris Holler, Matt Isaacs. Also Everett's cousins: Robert Everett, Jean Everett, Edward Everett, and Jim Everett; Everett's high school friends, Ralph Mohr and Fred Wilson; Everett's colleagues Jan Lodal, Ivan Selin, Charles Rossotti, Ken Willis, Tom Green, Elaine Tsiang, Paul Flanagan.

Very special thanks to Sonke Adlung and April Warman of Oxford University Press.

I am indebted to the editors of *Scientific American* for commissioning a magazine profile on Everett in 2007, and to Louise Lockwood and the British Broadcasting Corporation for including me in the making of *Parallel Worlds, Parallel Lives*, and for WGBH NOVA for allowing me to help create a great Website to accompany the award-winning film.

Grant funding to support this project was provided by the American Institute of Physics and the Foundational Questions Institute. Grant funding to support an anthology of Everett's papers (to be published by Princeton University Press, edited by Jeffrey A. Barrett and Byrne) was provided by the National Science Foundation. That grant also supports the creation of a public Web site to hold Everett's scientific papers and related biographical materials.

Thanks to: Scheffel Music Corp for permission to quote from the song *Stranger in Paradise* by George Forrest and Robert Craig Wright; Tom Lehrer for permission to quote from his song *Wernher von Braun*. Grove Press for permission to quote from "The Garden of Forking Paths" by Jorge Luis Borges. Henry Holt & Company for permission to quote from Erich Fromm's *The Sane Society*. Bantam Books, a division of Random House, Inc. for permission to quote from *The Coming of the Quantum Cats* by Frederick Pohl.

Photographs are courtesy of Mark Everett, Emilio Segre Visual Archives, Macmillan Publishing Ltd., Press Association Images, Donald Reisler.

Finally, my parents, Jeb and Beverly Byrne: both writers.

Bibliography

Books referenced

Albert, D. Z. (1992). *Quantum Mechanics and Experience*. Harvard University Press.

Ball, D. (1980). *Politics and Force Levels—The Strategic Missile Program of the Kennedy Administration*. University of California Press.

Barrett, J. A. (1999). *The Quantum Mechanics of Minds and Worlds*. Oxford University Press.

Barrow, J. D., Davies, P. C. W., Harper, C. J. (2004). *Science and Ultimate Reality—Quantum Theory, Cosmology, and Complexity*. Cambridge University Press.

Bell, J. S. (1987). *Speakable and Unspeakable in Quantum Mechanics*. Cambridge University Press, 2004.

Beller, M. (1999). *Quantum dialogue: The making of a revolution*. University of Chicago Press.

Bird, K. and Sherwin, M. J. (2005). *American Prometheus—The Triumph and Tragedy of J. Robert Oppenheimer*. Alfred A. Knopf.

Bohm, D. (1951) *Quantum Theory*. Prentice-Hall.

Bohm, D. and Hiley, B. J. (1993). *The Undivided Universe*. Routledge.

Bohr, N. (1934). *Atomic theory and the description of nature*. Cambridge University Press.

Bohr, N., Rosenfeld, L., *et al.* (1999). *Collected works. 10. Complementarity beyond physics (1928–1962)*. North-Holland Publ. Co.

Borges, J. (1962). *Ficciones*. Grove Press.

Brown, J. (2000). *Minds, Machine, and the Multiverse*. Simon & Schuster.

Carr, B. (2007). *Universe or Multiverse?* Cambridge University Press.

Chalmers, D. J. (1996). *The Conscious Mind, In Search of a Fundamental Theory*. Oxford University Press.

Conway, F. and Siegelman, J. (2005). *Dark Hero of the Information Age—In search of Norbert Wiener, the Father of Cybernetics*. Basic Books.

Davies, P. C. W. (1974). *The Physics of Time Asymmetry*. University of California Press.

Davies, P. C. W. (1982). *The Accidental Universe*. Cambridge University Press.

Deutsch, D. (1997). *The Fabric of Reality*. Penguin Books.

DeWitt, B. and Graham, N. eds. (1973). *The Many-Worlds Interpretation of Quantum Mechanics*. Princeton University Press.

DeWitt, C. M. and Wheeler, J. A. (1968). *Battelle Rencontres, 1967 Lectures in Mathematics and Physics*. W. A. Benjamin.

Dewitt-Morette, C. (2009). *The Pursuit of Quantum Gravity, Memoirs of Bryce DeWitt from 1946–2004*, Springer-Verlay, Heidelberg, (in press).

Dickson, P. (1971). *Think Tanks*. Atheneum.

Dickson, P. (2001). *Sputnik—The Shock of the Century*. Walker Publishing.

Donne, J. (1621). *Poems of John Donne. vol I*. E. K. Chambers, ed. London: Lawrence & Bullen, 1896.

Dresher, M. Tucker, A. W. and Wolfe, P. (1957). *Contributions to the Theory of Games, Volume III*. Princeton University Press.

Eddington, A. S. (1929). *The Nature of the Physical World: Gifford Lectures, 1927*. Macmillan.

Edwards, P. N. (1996). *The Closed World—Computers and the Politics of Discourse in Cold War America*. The MIT Press.

Einstein, A. (1954). *Ideas and Opinions*. Dell Publishing.

Eliot, T. S. (1936). *The Complete Poems and Plays*. Harcourt, Brace & World.

Everett, M. (2008). *Things the Grandchildren Should Know*. Little Brown.

Feynman, R. (1965). *The Character of Physical Law*. Random House, 1994.

Feynman, R., *et al.* (1965). *The Feynman Lectures on Physics—Volume III—Quantum Mechanics*. Addison-Wesley Publishing Co.

Finkbeiner, A. (2006). *The Jasons, The Secret History of Science's Postwar Elite*. Viking.

Frank, P. (1949). *Modern Science and its Philosophy*. Harvard University Press.

Frank, P. (1954). *The Validation of Scientific Theories*. The Beacon Press.

French, A. P. and Kennedy, P. J. eds. (1985). *Niels Bohr: a centenary volume*. Harvard University Press.

Fromm, E. (1955). *The Sane Society*. Henry Holt, 1990.

Fryklund, R. (1962). *100 Million Lives, Maximum Survival in Nuclear War*. Macmillan.

Gell-Mann, M. (1994). *The Quark and the Jaguar*. W. H. Freeman.

Gleick, J. (1992). *Genius—The Life and Science of Richard Feynman*. Random House, 1993.

Ghamari-Tabrizi, S. (2005). *The Worlds of Herman Kahn*. Harvard University Press.

Gold, T. (1967). *The Nature of Time*. Cornell University Press.

Good, I. J. (1962). *The Scientist Speculates: An Anthology of Partly-Baked Ideas*. Basic Books.

Graham, N. (1973). "The Measurement of Relative Frequency," DeWitt, B. and Graham, N. (ed.) (1973).

Heims, S. J. (1980). *John von Neumann and Norbert Wiener—From Mathematics to the Technologies of Life and Death*. The MIT Press.

Herken, G. (1987). *Counsels of War*. Oxford University Press.

Jammer, M. (1974). *The Philosophy of Quantum Mechanics—the interpretations of quantum mechanics in historical perspective*. Wiley.

Jeffress, L. ed. (1951). *Cerebral Mechanisms in Behavior*. Hafner Publishing. (1967)

Kahn, H. (1960). *On Thermonuclear War*. Princeton University Press.

Kaplan, F. (1983). *The Wizards of Armageddon*. Stanford University Press, 1991.

Kruger, L., Gigerenzer, G., and Morgan, M., eds. (1987) *The Probabilistic Revolution Volume 2: Ideas in the Sciences*. The MIT Press.

Kuhn, H. and Tucker, A. (eds) (1950). "Contributions to the Theory of Games I, Annals of Mathematics Studies 24." Princeton University Press.

Kuhn, H. and Tucker, A. (eds) (1953). "Contributions to the Theory of Games II, Annals of Mathematics Studies 28." Princeton University Press.

Kuhn H. (1953). *Lectures on the Theory of Games*. Princeton University Press.

Kuhn, H. ed. (1997). *Classics in Game Theory*. Princeton University Press.

Leinster, M. (1978). *The Best of Murray Leinster*. Ballentine Books.

Lewis, D. (1986). *On the Plurality of Worlds*. Blackwell Publishing.

Lloyd, S. (2006). *Programming the Universe*. Alfred A. Knopf.

Luce, D. R. and Raffia, H. (1957). *Games and Decisions*. John Wiley & Sons.

Margenau, H. (1950). *The Nature of Physical Reality*. McGraw Hill.

Mehra, J. (1973). *The Physicist's Conception of Nature*. D. Reidel Publishing Company.

Miller, W. M. and Greenberg, M. H. (1983). *Beyond Armageddon*. Donald J. Fine.

Mills, C. W. (1956). *The Power Elite*. Oxford University Press, 2000.

Myrvold, W. C., Christian, J. (2009). *Quantum Reality, Relativistic Causality, and Closing the Epistemic Circle*. Springer.

Newton, I. (1730). *Opticks: or a Treatise of the Reflections, Refractions, Inflections and Colours of Light*, 4th edition. William Innes at the Weft-End of St. Pauls.

Nasar, S. (1998). *A Beautiful Mind*. Touchstone.

Pais, A. (1982). *"Subtle is the Lord..." The Science and Life of Albert Einstein*. Oxford University Press.

Pais, A. (1991). *Niels Bohr's Times in Physics, Philosophy and Polity*. Oxford University Press.

Petersen, A. (1968). *Quantum Physics and the Philosophical Tradition*. The MIT Press.

Pohl, F. (1986). *The Coming of the Quantum Cats*. Bantam Spectra.

Popper, K. (1982). *Quantum Theory and the Schism in Physics*. Routledge, 1992.

Poundstone, W. (1992). *Prisoner's Dilemma*. Anchor Books.

Pullman, P. (2008). *His Dark Materials Trilogy*. Scholastic.

Pugh, G. E. (1977). *The Biological Origin of Human Values*. Basic Books.

Rapoport, A. (1964). *Strategy and Conscience*. Schocken Books, 1969.

Rhodes, R. (1995). *Dark Sun: The Making of the Hydrogen Bomb*. Simon & Schuster.

Rhodes, R. (2007). *Arsenals of Folly*. Alfred A. Knopf.

Sakharov, A. (1990). *Memoirs*. Alfred A. Knopf.

Saunders, S., Barrett, J., Kent, A., and Wallace, D. eds. (2010) *Many Worlds? Everett, Quantum Theory and Reality*. Oxford University Press.

Savage, L. J. (1954). *The Foundations of Statistics*. Dover Publications, 1972.

Schelling, T. (1984). *Choice and Consequence—Perspectives of an Errant Economist*. Harvard University Press.

Schrödinger, E. (1995). *The Interpretation of Quantum Mechanics, Dublin seminars (1949–1955) and other unpublished essays*. Oxbow Press.

Schilpp. P. A. ed. (1951). *Albert Einstein: Philosopher-Scientist*. Tudor Publishing.

Shurkin, J. N. (2006). *Broken Genius—The Rise and Fall of William Shockley—Creator of the Electronic Age*. Macmillan.

Susskind, L. (2005). *The Cosmic Landscape—String Theory and the Illusion of Intelligent Design*. Little, Brown.

Tauber, G. E. (1979). *Albert Einstein's Theory of Relativity*. Crown Publishers.

von Neumann, J. and Morgenstern, O. (2004). *Theory of Games and Economic Behavior*. Princeton University Press (60th anniversary edition).

Wang, J. (1999). *American Science in an Age of Anxiety—Scientists, Anticommunism, & the Cold War*. The University of North Carolina Press.

Wiener, N. (1948). *Cybernetics or Control and Communications in the Animal and the Machine*. The MIT Press, 1961.

Wiener, N. (1950). *The Human Use of Human Beings—Cybernetics and Society*. Da Capo Press, 1954.

Wheeler, J. A. and Zurek, W. H. (1983). *Quantum theory and measurement*. Princeton University Press.

Wheeler, J. A. and Ford, K. (1998). *Geons, Black Holes & Quantum Foam—A Life in Physics*. W.W. Norton & Company.

York, H. (1970). *Race to Oblivion*. Simon & Schuster.

Zurek, W. H. (ed.) (1990). *Complexity, Entropy and the Physics of Information*. Addison-Wesley Publishing Company.

Books used for background research

Baggott, J. (2004). *Beyond Measure—Modern Physics, Philosophy and the Meaning of Quantum Theory*. Oxford University Press.

Berlinski, D. (2000). *The Advent of the Algorithm*. Harcourt.

Brown, J. R. (2001). *Who Rules in Science? An Opinionated Guide to the Wars*. Harvard University Press.

Bruce, C. (2004). *Schrödinger's Rabbits—the Many Worlds of Quantum*. Joseph Henry Press.

Cushing, J. T. (1994). *Quantum Mechanics—Historical Contingency and the Copenhagen Hegemony*. University of Chicago Press.

Dick, P. (1962). *The Man in the High Castle*. G. P. Putnam Sons.

Feyerabend, P. (1993). *Against Method*. Verso Books.

Feynman, R. (1985). *QED—The Strange Theory of Light and Matter*. Princeton University Press.

Fine, A. (1986). *The Shaky Game—Einstein, Realism and the Quantum Theory*. University of Chicago Press.

Galbraith, J. K. *The Affluent Society*. Houghton Mifflin. 1958.

Greenberger, D., Hentschel, K. and Weinert, F. eds. (2009). *Compendium of Quantum Physics*. Springer.

Hacking, I. (2001). *An Introduction to Probability and Inductive Logic*. Cambridge University Press.

Isaacson, W. (2007). *Einstein—His Life and Universe*. Simon & Schuster.

Killian, J. R. (1977). *Sputnik, Scientists and Eisenhower—A Memoir of the First Special Assistant to the President for Science and Technology*. The MIT Press.

Kuhn, T. S. (1962). *The Structure of Scientific Revolutions,* The University of Chicago Press.

Lopes, J. and Paty, M. eds. (1977). *Quantum Mechanics a Half Century Later.* D. Reidel Publishing Company.

Lens, S. (1970). *The Military Industrial Complex.* Pilgrim Press.

Leopold, E. (1999). *Breast Cancer. Women, and Their Doctors in the Twentieth Century.* Beacon Press.

Leopold, E. (2009). *Under the Radar—Cancer and the Cold War.* Rutgers University Press.

Morse, P. (1977). *In the Beginnings: A Physicist's Life.* The MIT Press.

Omnes, R. (1999). *Quantum Philosophy—Understanding and Interpreting Contemporary Science.* Princeton University Press.

Penrose, R. (2004). *The Road to Reality—A Complete Guide to the Laws of the Universe.* Random House.

Readers' Guide to Periodical Literature. H. W. Wilson, vols 1925–1962.

Riesman, D. *The Lonely Crowd: a study of the changing American character.* Yale University Press. 1950.

Vilenkin, A. (2006). *Many Worlds in One—The Search for Other Universes.* Farrar, Straus and Giroux.

Weinberg, S. (1992). *Dreams of a Final Theory—The Scientist's Search for the Ultimate Laws of the Universe.* Random House.

Wilson, A. (1968). *The Bomb and the Computer—A Crucial History of War Games.* Delacorte Press.

Referenced and background papers and articles

Aaserud, F. (1995). "Sputnik and the 'Princeton Three:' the national security laboratory that was not to be," *Historical Studies in the Physical and Biological Sciences.* **25**: 2.

Analog. (1976). "Quantum Physics and Reality: The Garden of the Forking Paths," Vol. 96, No. 12. December, 1976.

Aguirre, A., *et al.* (2007). "Towards observable signatures of other bubble universes," arXiv:0704.3473v3.

Ashby, W. R. (1952). "Can a Mechanical Chess Player Outplay its Designer?" *The British Journal for the Philosophy of Science.* **3**(9): 44–57.

Ballentine, L. E. (1973). "Can the Statistical Postulate of Quantum Theory be Derived?—A Critique of the Many-Universes Interpretation," *Foundations of Physics.* **3**: 2.

Bell, J. S. (1964). "On the Einstein-Podolsky-Rosen paradox," *Physics 1.* 195–200. Reprinted in Bell, J. S. (1987).

Bell, J. S. (1966). "The moral aspect of quantum mechanics," (with M. Nauenberg) in *Preludes in Physics,* ed. by A. De Shalit, H. Feshbach, and L. Van Hove. North Holland, Amsterdam. 279–281. Reprinted in Bell, J. S. (1987).

Bell, J. S. (1976). "The measurement theory of Everett and de Broglie's pilot wave." Reprinted in Bell, J. S. (1987).

Bell, J. S. (1981)."Quantum mechanics for cosmologists." Reprinted in Bell, J. S. (1987).

Belinfante, F. J. (1970). "Experiments to disprove that nature would be deterministic," mimeograph, Purdue University, quoted in Jammer, M. (1974). 330.

Belinfante, F. J. (1975). *Measurements and Time Reversal in Objective Quantum Theory*. Pergamon Press.

Bignami, L. (2008). "Contatto." *Focus*. 3/2008.

Bohm, D. (1952). "A suggested interpretation of the quantum theory in terms of 'hidden' variables, I and II," in Wheeler, J. A. and Zurek, W. H. (1983). 369.

Bohr, N. and Rosenfeld, L. (1933). "On the Question of the Measurability of Electromagnetic Field Quantities," in Wheeler, J. A. and Zurek, W. H. (1983).

Bohr, N. (1949). "Discussion with Einstein on Epistemological Problems in Atomic Physics," in Wheeler, J. A. and Zurek, W. H. (1983).

Bohr, N., (1956). *Atoms and Human Knowledge*. Mimeograph copy in basement.

Bohr, N. (1957). *Atoms and Human Knowledge*. Radio address at University of Oklahoma in Bohr, N., L. Rosenfeld, *et al.* (1999). The text is different from Bohr, N. (1956).

Bohr, N. (1958). *Atomic Physics and Human Knowledge*. Wiley.

Born, M. (1926). *Zeitschr. Phys.*, **37**: 863 in Wheeler, J. A. and Zurek, W. H. (1983). 54.

Born, M. (1926B). *Zeitschr. Phys.*, **38**: 803.

Born, M. (1953). "The Interpretation of Quantum Mechanics," *The British Journal for the Philosophy of Science*. **4**: 14.

Brown, H. (1979). "Review" of Belinfante (1975) in *The British Journal for the Philosophy of Science*. 30: 187–209.

Camilleri, K. (2008). "A History of Entanglement: Decoherence and the Interpretation Problem, Studies in History and Philosophy of Modern Physics," presented at the second international conference of the History of Quantum Physics in Utrecht, July 17, 2008.

Camilleri, K. (2009). "Constructing the Myth of the Copenhagen Interpretation," *Perspectives on Science* **17**(1): 26–57.

Cartwright, N. (1987). "Philosophical Problems of Quantum Theory: The Response of American Physicists," in Kruger *et al* (1987). 417–435.

Chang, S., M. Kleban, *et al.* (2008). "When worlds collide," *JCAP* **804**(034): 0712.2261.

Chown, M. (2007). "Into the Void: Could the vast empty hole in space be the imprint of another universe?" *New Scientist*, 10/24/07.

Cooper, L. and van Vechten, D. (1969). "On the Interpretation of Measurement within the Quantum Theory," *American Journal of Physics* **37**: 1212–1220. Reprinted in DeWitt, B. S. and Graham, N. (1973).

Cooper, L. (1976). "How Possible Becomes Actual in the Quantum Theory," *Proceedings of the American Philosophical Society*. **120**(1): 37–45.

Dennett, D. C. (1991). "Real Patterns," *Journal of Philosophy*. **87**: 27–51.

Deutsch, D. (1985). "Quantum theory, the Church-Turing principle and the universal quantum computer," *Proceedings of the Royal Society of London*. A **400**: 97–117.

Deutsch, D. (1985A). "Quantum Theory as a Universal Physical Theory," *International Journal of Theoretical Physics*, 24(1): 1–41.

Deutsch, D. (1996). "Comment on…Lockwood," *The British Journal for the Philosophy of Science*. 47: 222–228.

Deutsch, D. (1999). "Quantum theory of probability and decisions," *Proceedings: Mathematical, Physical and Engineering Science*. 3129–3137.

Deutsch, D. (2002). "The structure of the multiverse," *Proceedings: Mathematics, Physical and Engineering Sciences*. 2911–2923.

DeWitt, B. S. (1967). "Quantum Theory of Gravity. I. The Canonical Theory," *Physical Review*. **160**(5): 1113–1147.

DeWitt, B. S. (1968). "The Everett-Wheeler Interpretation of Quantum Mechanics," in DeWitt, C. M. and Wheeler, J. A. (1968).

DeWitt, B. S. (1970). "Quantum Mechanics and Reality," *Physics Today*. **23**(9): Sept. 1970. Reprinted in DeWitt, B. and Graham, N. (1973).

DeWitt, B. S. (1971). "The Many Universes Interpretation of Quantum Mechanics," *Foundations of Quantum Mechanics*. Academic Press. Reprinted in DeWitt, B. and Graham, N. (1973).

DeWitt, B. S. and Graham, N. (1971A). "Resource Letter on the Interpretation of Quantum Mechanics," *American Journal of Physics*. July 1971.

DeWitt, B. S. *et al*. (1971B). Letters to Editor. *Physics Today*. **24**(10): Oct. 1971.

DeWitt, B. S. (2004). "The Everett interpretation of quantum mechanics," in Barrow, J. D. *et al*. (2004).

DeWitt, B. S. (2005). "God's Rays," *Physics Today*. Jan. 2005.

DeWitt, B. S. (2008). "Quantum Gravity, Yesterday and Today," arXiv:085.2935v1.

Dirac, P. A. M. (1963). "The Evolution of the Physicist's Picture of Nature," *Scientific American*. May 1963.

Dowker, F. and Kent, A. (1996). "On the consistent histories approach to quantum mechanics." *Journal of Statistical Physics*. **82**(5): 1575–1646.

Einstein, A. (1918). "Principles of Research," in Einstein, A. (1954).

Einstein, A., Podolsky, B. and Rosen, N. (1935). "Can Quantum-Mechanical Description of Physical Reality Be Considered Complete?" *Physical Review*. **47**: May, 15, 1935.

Einstein, A. and Rosen, N. (1935A). "The Particle Problem in the General Theory of Relativity," *Physical Review*. **48**: July 1, 1935.

Einstein, A. (1949). "Reply to Criticisms," in Schilpp, P.A. (1951).

Ellsberg, D. (1957). "Risk, Ambiguity, and the Savage Axioms," RAND corporation.

Everett, H III. (1957). " 'Relative State' Formulation of Quantum Mechanics," *Reviews of Modern Physics*. **29**(3): 454–462.

Everett, H III. (1957A). "Recursive Games" in *Contributions to the Theory of Games III*. Princeton University Press. 47–78.

Everett, H III. and Pugh, G. E. (1958). "Simple Formulas for Calculating the Distribution and Effects of Fallout in Large Nuclear Weapon Campaigns (With Applications)." 1/9/58. Basement. Also known as WSEG Research Memorandum No. 3 (top secret) and No. 5 (sanitized). The version in the basement was sanitized.

Everett, H. III. and Pugh, G. (1959). "The Distribution and Effects of Fallout in Large Nuclear Weapon Campaigns," *Operations Research*. 7(2): 226–248.

Everett, H. III. (1962). "Generalized Lagrange Multiplier Method for Solving Problems of Optimum Allocation of Resources," WSEG Research Memorandum No. 25.

Everett, H. III. (1963). "Generalized Lagrange multiplier method for solving problems of optimum allocation of resources." *Operations Research*. 399–417.

Fock, V. A. (1957). "The Journey to Copenhagen," *Vestnik Akademii Nauk SSSR*. 27(7): 54–7. An English translation of Fock's article is in Everett's files.

Freire, O. (2009). "Quantum dissidents: research on the foundations of quantum theory circa 1970," unpublished draft.

Geftner, A. (2009). "Dark flow: Proof of Another Universe?" *New Scientist*, 1/23/09.

Geroch, R. (1984). "The Everett Interpretation," *Nous*. 18: 617–633.

Ghamari-Tabrizi, S. (2000). "U.S. Wargaming Grows Up," http://www.strategypage.com.

Gleason, A. M. (1957). "Measures on the Closed Subspaces of a Hilbert Space," *Journal of Mathematics and Mechanics*. 6: 885–893.

Graham, N. (1970). *The Everett Interpretation of Quantum Mechanics*. PhD thesis, University of North Carolina, Chapel Hill.

Greaves, H. (2004). "Understanding Deutsch's probability in a deterministic multiverse," *Studies in History and Philosophy of Modern Physics*. 35: 423–456.

Greaves, H. and Myrvold, W. (2007). "Everett and Evidence," in Saunders *et al.* (2010).

Hartle, J. B. (1968). "Quantum Mechanics of Individual Systems," *American Journal of Physics*. 36(8): 704–712.

Hartle, J. B. (2008). "The quasiclassical realms of this quantum universe," in Saunders, S., Barrett, J., Kent, A., and Wallace, D. eds. (2010).

Halliwell, J. (1991). "Quantum Cosmology and the Creation of the Universe," *Scientific American*. 76: Dec. 1991.

Healey, R. A. (1984). "How Many Worlds?" *Nous*. 18: 591–616.

Heisenberg, W. (1927). "The Physical Content of Quantum Kinematics and Mechanics," in Wheeler, J. A. and Zurek, W. H. (1983).

Heisenberg, W. (1927B). "Principle of Indeterminism," in Wheeler, J. A. and Zurek, W. H. (1983).

Hellman, G. (2009). "Interpretations of Probability in Quantum Mechanics…" in Myrvold, W. C., Christian, J. (2009).

Helmer, O. (1957). "The Game-Theoretical Approach to Organization Theory," *Report of the Third Conference on Games, March 11–12, 1957*. Princeton University. Also in Helmer, O. (1963). "The game-theoretical approach to organization theory," *Synthese* 15(1): 245–253.

Holman, R., Mersini-Houghton, L., Takahashi, T. (2006). "Cosmological Avatars of the Landscape II: CMB and LSS Signatures," http://arxiv.org/abs/hep-th/0612142.

Howard, D. (2004). "Who Invented the Copenhagen Interpretation? A Study in Mythology," *PSA 2002*. Part II, *Symposium Papers*. Proceedings of the 2002 Biennial Meeting of the Philosophy of Science Association, Milwaukee, Wisconsin, November 7–9, 2002. A special issue of *Philosophy of Science* 71 (2004): 669–682.

Jacobsen, A. S. (2007). "Leon Rosenfeld's Marxist defense of complementarity," *Historical Studies in the Physical and Biological Sciences.* 37(Suppl.): 3–34.

Jacobsen, A. S. (2008). "The Complementarity Between the Collective and the Individual—Rosenfeld and Cold War History of Science," *Minerva*, 46: 195–214.

Jaynes, E. T. (1957). "Information Theory and Statistical Mechanics," *Physical Review.* 106(4): May 15, 1957.

Kahn, H. and Mann, I. (1957). "Military Planning in an Uncertain World, Part One, Techniques of Systems Analysis." RAND, RM-1829-1-PR, June 1957.

Kahn, H. and Mann, I. (1957A). "Ten Common Pitfalls," RAND, RM-1937, July 1957.

Kashlinksy, A., *et al.* (2009). "A new measurement of the bulk flow of X-ray luminous clusters of galaxies," Oct. 27, 2009. arXiv:0910.4958

Kent, A. (1990). "Against many-worlds interpretations," *International Journal of Modern Physics A.* 5(9): 1764–1762.

Kent, A. (2009). "One world versus many: the inadequacy of Everettian accounts of evolution, probability, and scientific confirmation," in Saunders, S., Barrett, J., Kent, A., and Wallace, D. eds. (2010).

Kobakhidze, A. and L. Mersini-Houghton (2006). "Birth of the universe from the landscape of string theory," *The European Physical Journal C.* 49(3): 869–873.

LANL History: "50th Anniversary Article: Evolving from Calculators to Computers," Los Alamos National Laboratory. http://www.lanl.gov/history/atomicbomb/computers.shtml

Lacey, H. M. (1969). *The British Journal for the Philosophy of Science.* 20(1): May, 1969.

Lévy-Leblond, J. (1976). "Towards a Proper Quantum Theory (Hints for a Recasting)," *Dialectica.* 30(2/3). Basement; also in Lopes, J. and Paty, M. eds. (1977)

Lockwood, M. (1996). "'Many Minds' Interpretations of Quantum Mechanics," *The British Journal of the Philosophy of Science.* 47.

Lodal, J. M. (1969). Office of the Assistant Secretary of Defense for Systems Analysis presentation at conference on Application of Mathematical Programming Techniques, Cambridge, U.K., The English Universities Press.

London, F. and Bauer, E. (1939). "The Theory of Observation on Quantum Mechanics," in Wheeler, J. A. and Zurek, W. H. (1983). 219–220.

Margenau, H. (1956). "Philosophical Problems Concerning the Meaning of Measurement in Physics," parts of which were presented at the Symposium of Measurement sponsored by the Philosophy of Science Association in New York on Dec. 29, 1956, later published in *Philosophy of Science* 25: 23–33 (1958).

Margenau, H. (1963). "Measurements and Quantum States: Part II," *Philosophy of Science.* 30(2): 138–157.

McNamara, R. S. (1967). "The Dynamics of Nuclear Strategy," Department of State Bulletin, 59: 10/9/67.

Mermin, N. D. (2009). "What's bad about this habit," *Physics Today.* May 2009. 8–9.

Merali, Z. (2009). "Cosmic Collision," *Discover.* Oct. 2009.

Mersini-Houghton, L. (2005). "Wavefunction of the Universe on the Landscape," ArXiv preprint hep-th/0512304.

Mersini-Houghton, L. and R. Holman. (2009). " 'Tilting' the universe with the landscape multiverse: the dark flow," 2: 006.

Misner, C. (1957). "Feynman Quantization of General Relativity," *RMP*. 29(3): 497.

Misner, C. (1969). "Quantum cosmology. I.," *Physical Review* 186(5): 1319–1327.

Morgenstern, O. (1957). "In memoriam: John von Neumann," *Report of the Third Conference on Games held at Princeton University, March 11 and 12, 1957*. Basement.

Mott, N. (1929). "The Wave Mechanics of the α-Ray Tracks," *Proceedings of the Royal Society*, in Wheeler, J.A. and Zurek, W. H. (1983).

Osnaghi, S., Freitas, F. and Freire, O. (2009). "The Origin of the Everettian Heresy," *Studies In History and Philosophy of Science Part B: Studies In History and Philosophy of Modern Physics.* 40(2): May 2009.

Pauling, L. (1963). Nobel Peace Prize for 1962 Lecture. 12/11/63. http://nobelprize. org/nobel_prizes/peace/laureates/1962/pauling-lecture.html.

Petersen, A. (1963). *Niels Bohr's Philosophy, two talks given on Danish Radio*. April, 1963. Niels Bohr Collection, University of Chicago Library, Series II, Box One, Folder 10.

Rosenfeld, L. (1963). "Niels Bohr's contributions to epistemology," *Physics Today*. 16(10): 47–54.

Saunders, S. (1993). "Decoherence, relative states, and evolutionary adaptation." *Foundations of Physics.* 23(12): 1553–1585.

Saunders, S. (1995). "Time, quantum mechanics, and decoherence," *Synthese*. 102(2): 235–266.

Saunders, S. (1998). "Time, quantum mechanics, and probability," *Synthese*. 114(3): 373–404.

Saunders, S. (2005). "What is Probability?" in A. Elitzur, S. Dolev, N. Kolenda, eds. (2005). *Quo Vadis Quantum Mechanics?* Springer Verlag.

Saunders, S. (2010). "Many Worlds: An Introduction," in Saunders, S. *et al.* (2010).

Schlosshauer, M. and A. Fine (2005). "On Zurek's Derivation of the Born rule," *Foundations of Physics.* 35(2): 197–213.

Schrödinger, E. (1935). "Discussion of Probability Relations Between Separated Systems," *Proceedings of the Cambridge Philosophical Society.* 31: 555–563.

Schrödinger, E. (1935A). "The Present Situation in Quantum Mechanics," in Wheeler, J. A. and Zurek, W. H. (1983).

Schrödinger, E. (1952). "Are There Quantum Jumps? Part I," *The British Journal for the Philosophy of Science.* 10: 109–123.

Schrödinger, E. (1952A). "Are There Quantum Jumps? Part II," *The British Journal for the Philosophy of Science.* 11: 240–242.

Schrödinger, E. (1952B). "July 1952 colloquium," in Schrödinger, E. (1995).

Shannon, C. E. (1948). "A Mathematical Theory of Communication," *The Bell System Technical Journal.* 27:July–Oct 1948. 379–423, 623–656.

Shikhovtsev, E. (2003). "Biographical sketch of Hugh Everett III." space.mit.edu/ home/tegmark/everett/everettbio.pdf.

Shimony, A. (1963). "Role of the Observer in Quantum Theory," *American Journal of Physics*, 31: 755–773.

Simmons, G. J. (1963). "Mathematical Model for an Associative Memory," A Sandia Corporation Monograph.

Stapp, H. (2002). "The basis problem in many-worlds theories," *Canadian Journal of Physics*. **80**(9): 1043–1052.

Stein, H. (1984). "The Everett Interpretation of Quantum Mechanics: Many Worlds or None?" *Nous*. **18**: 635–652.

Stern, A. (1944). "Letter to editor," *Science*. **100**(2599).

Stern, A. (1945). "Letter to editor," *Science*. **101**(2611).

Stern, A. (1949). "A Trend in Contemporary Physics," *Physics Today*. **2**(5): May 1949.

Stern, A. (1953). "Some Concepts in Modern Physics," *American Journal of Physics*. **21**(8): 629–40.

Tegmark, M. and Wheeler, J. A. (2001). "100 Years of the Quantum," *Scientific American*. Feb. 2001.

Tegmark, M. (2003). "Parallel Universes." *Scientific American*. **28**(5): 40–51.

Tegmark, M. (2008). "Many Worlds in Context," in Saunders, S. *et al.* (2010). Also at arxiv.org/abs/0905.2182.

von Neumann, J. (1932). "Measurement and Reversibility," in Wheeler, J. A. and Zurek W. H. (1983).

von Neumann, J. (1951). "The General and Logical Theory of Automata," in Jeffress, L. ed. (1951). 1–41.

Wallace, D. (2001). "Worlds in the Everett Interpretation." *Studies in the History and Philosophy of Modern Physics Part B*. **33**(4): 637–661.

Wallace, D. (2002). "Quantum Probability and Decision Theory Revisited," arXiv: quant-ph/0211104v1.

Wallace, D. (2003). "Everettian Rationality: Defending Deutsch's Approach to Probability in the Everett Interpretation," *Studies in the History and Philosophy of Modern Physics*. **34**: 415–439.

Wallace, D. (2003A). "Everett and Structure," *Studies in History and Philosophy of Modern Physics*. **34**(1): 87–105. See also: Wallace, D. "Everett and Structure," arXiv:quant-ph/0107144v2. July 22, 2005.

Wallace, D. (2007). "Quantum Probability from Subjective Likelihood: Improving on Deutsch's Proof of Probability," *Studies In History and Philosophy of Science Part B: Studies In History and Philosophy of Modern Physics*. **38**(2): 311–332.

Wells, H. G. (1922). "Men Like Gods," *Hearst's International*. Nov. 1922.

Wilson, D. S. and Wilson, E. O. (2007). "Rethinking the Theoretical Foundation of Sociobiology," *The Quarterly Review of Biology*. **82**(4): Dec. 2007.

Wheeler, J. A. (1956). "A Septet of Sibyls," *American Scientist*. **44**(4): 360–377.

Wheeler, J. A. (1957). "Assessment of Everett's 'Relative State' Formulation of Quantum Theory," *Reviews of Modern Physics*. **29**(3): 463–5. Reprinted in DeWitt, B. and Graham, N. (1973).

Wheeler, J. A. (1957A). "No Fugitive and Cloistered Virtue," *Physics Today*. **16**: 30. Jan. 1963.

Wheeler, J. A. (1962). "Three Dimensional Geometry," in Gold, T. (1967).

Wheeler, J. A. (1973). "From Relativity to Mutability," in Mehra, J. (1973).

Wheeler, J. A. (1977). "Include the Observer in the Wave Function?" in Lopes, J. and Paty. M. (1977).

Wheeler, J. A. (1979). "Mercer Street and Other Memories," in Tauber, G. E. (1979).

Wheeler, J. A. (1979A). "Law Without Law," in "Frontiers of Time" in *Proceedings of the International School 'Enrico Fermi,' Course 72, Problems in the Foundations of Physics*, Toraldo di Franco, G. ed. North-Holland Publishing Company, Amsterdam. 395–425.

Wheeler, J. A. (1985). "Bohr's 'Phenomenon' and 'Law Without Law,'" in *Chaotic Behavior in Quantum Systems: Theory and Applications*, Casati, G. ed. Plenum Press. (1985).

Whitrow, G. T. and Bondi, H. (1954). "Is Physical Cosmology a Science?" *The British Journal for the Philosophy of Science.* 4(16): 271–283.

Wiener, N. and Siegel, A. (1955). "The Differential-Space Theory of Quantum Systems," *Supplemento Al Volume II, Series X, Del Nuovo Cimento.* No 4.

Wigner, E. (1961). "Remarks on the Mind-Body Question," in Wheeler, J. A. and Zurek, W. H. (1983). Also in Good, I. J. (1962).

Wigner, E. (1963). "The Problem of Measurement," in Wheeler, J. A. and Zurek, W. H. (1983).

Wigner, E. (1973). "Epistemological perspective," *Contemporary Research on the Foundations and Philosophy of Quantum Theory*, Proceedings of a conference held at the University of Western Ontario, London, Canada, by C.A. Hooker, (ed.). D. Reidel Publishing Company. Volume 2. 369–385.

Wigner, E. (1981). "Interpretation of Quantum Mechanics," in Wheeler, J. A. and Zurek, W. H. (1983).

Zeh, H. D. (1970). "On the interpretation of measurement in quantum theory," *Foundations of Physics.* 1: 69–76.

Zeh, H. D. (2000). "The problem of conscious observation in quantum mechanical description," *Foundations of Physics Letters.* 13(3): 221–233.

Zeh, H. D. (2008). "Feynman's Quantum Theory," http://arxiv.org/abs/0804.3348.

Zeh, H. D. (2009). "Quantum discreteness is an illusion," ArXiv preprint quant-ph/0809.2904.

Zhang, R. and Huterer, D. (2009). "Disks in the Sky: A reassessment of the WMAP 'cold spot,'" arXiv:0908.3988v2.

Zurek, W. H. (1981). "Pointer basis of quantum apparatus: Into what mixture does the wave packet collapse?" *Physics Review D.* 24(6): 1516–1525.

Zurek, W. H. (1991). "Decoherence and the transition from quantum to classical," *Physics Today.* 44(10): 36–44.

Zurek, W. H. (1998). "Decoherence, Einselection, and the Existential Interpretation," *Philosophical Transactions: Mathematical, Physical and Engineering Sciences.* 1793–1821.

Zurek, W. H. (2002). "Decoherence and the Transition from Quantum to Classical—Revisited." *Los Alamos Science.* No. 27.

Zurek, W. H. (2003). "Environment-assisted invariance, entanglement, and probabilities on quantum physics," *Phys. Rev. Lett.* 90: 120404.

Zurek, W. H. (2005). "Probabilities from entanglement, Born's rule from envariance," *Phys. Rev. A.* 71:052105.

Zurek, W. H. (2007). "Relative States and the Environment: Einselection, Envariance, Quantum Darwinism, and the Existential Interpretation," arXiv:0707.2832v1.

Unpublished papers and reports

American Institute of Physics, National Registry of Scientific and Technical Personnel, survey, 5/14/57, basement.

Decision Science Applications Report 39/1038. (1995). Lucas archive.

DeWitt, C. M. (1957). "Conference on the Role of Gravitation in Physics (transcript)," National Technical Information Service, AD118180.

DeWitt, B. S. (2008A). Referee report (unpublished) for *American Journal of Physics* on Yoav Ben-Dov's "Everett's Theory and the 'Many Worlds' Interpretation," circa 1988, courtesy of Cecile DeWitt-Morette.

Dunham, C. L. to Henshaw, P. S. (1957). "Possible Biologic Consequences of Nuclear Warfare," 1/4/57. U.S. DOE Archives, 326 U.S. Atomic Energy Commission, Collection DBM, Box 3368, Folder 35. Available online at DOE.

Enthoven, A. (1966). "Analysis in Defense Decisions," 9/10/66, basement.

Everett, H III. (1956A). "Probability and Wave Mechanics." Basement. There are several different drafts of this paper. This is the typed version handed to Wheeler in late 1956.

Everett, H III. (1956B). "Quantitative Measure of Correlation," basement.

Everett, H III. (1956C). "Objective and Subjective Probability," basement.

Everett, H III. (1957B). "On the Foundations of Quantum Mechanics," basement.

Everett, H III. (1957C). "Wave Mechanics Without Probability," basement.

Everett, H III, (1957D). Princeton student file, 4/15/57.

FBI. (1971). Everett file, Jan. 1971. FOIA document.

Horn, H. O. (1951). English Lit. term paper on "The Hollow Men" by T. S. Eliot. (5/14/51) by Everett's WSEG colleague, Herbert O. Horn, basement.

Institute for Defense Analysis memo, 3/5/63, basement.

Institute for Advanced Study. (1954). *Report to Director.*

Lambda Report 3. (1966). "An Optimization Study of Blast Shelter Deployment," Sept. 1966, basement.

Lambda Report 5. (1967). "Defense Models III," March 1967, Lucas archive.

Lambda Paper 6. (1967). "Defense Models IV, Some Mathematical Relations for Probability of Kill," March 1967, Lucas archive.

Lambda Paper 17. (1968). "Defense Models XI," April 1968, Lucas archive.

Lambda Paper 34. (1969). "Analysis of Cross Targeting," 9/10/69, basement.

National Security Council. (1958). "Discussion at the 387th Meeting of the National Security Council," 11/20/58. Eisenhower Library.

Princeton University. (1957). *Report of the Third Conference on Games, March 11–12, 1957*, basement.

Pugh, G. E., Lucas, G. and Gorman, G. (1978). "Value Driven Decision Theory: Application to Combat Simulations—Decision Science Applications Report to Air Force Office of Scientific Research," May 31, 1978, Lucas archive.

Pugh, G. E. (2005). "Recollections of Hugh Everett III," American Institute of Physics.
Report of autopsy, Chief Medical Examiner, No. Va. District, 7/30/82, basement.
U.S. Congress, Office of Technology Assessment. (1995). *A History of the Department
of Defense Federally Funded Research and Development Centers*, OTA-BP-ISS-157
(Washington, DC: U.S. Government Printing Office, June 1995).
U.S. National Archives, Record Group 77, Records of the Chief of Engineers,
Manhattan Engineer District, Harrison-Bundy File, folder #76.
WSEG/IDA reports available at Department of Defense Reading Room: http://
www.dod.mil/pubs/foi/wseg/: WSEG Report 45; WSEG Report 50, with enclo-
sures; WSEG Report 159; Ponturo, J. (1979). *Analytical Support for the Joint Chiefs
of Staff: The WSEG Experience 1948–1976*, IDA Study S-507.
Werner, F. G. (1962). *Conference on the Foundations of Quantum Mechanics*. Xavier
University, 2002.

Newspapers

Alexandria Gazette, 9/21/54
Princeton Alumni Weekly, 3/12/03.
The Cincinnati Enquirer, 10/2/62, 34.
The New York Times, 7/7/2005, A23.
The New York Times, 4/14/2008.
Washington Post, 1/8/55.

Letters, correspondence

Anderson, S. E. Lt. Gen. USAF to Local Board Number 42, 7/18/56.
Barry. J. emails to Shikhovtsev, E., 2002.
Bohr, N., letter to H. P. E. Hansen, 20 July 1935, Niels Bohr Archive, quoted in Pais
(1991). 446.
Bohr to Wheeler, 4/12/57, basement.
Boone to Taylor, 4/17/53, Mudd.
Everett, H Jr. to Thrift, 10/11/35, basement.
Everett, H Jr. to commanding officer, March–May, 1952, basement.
Everett, H III to Petersen, June 1956, basement.
Everett, H III to DeWitt, 5/31/57, basement.
Everett, H III to Frank, 5/31/57, basement.
Everett, H III to Petersen, 5/31/57, basement.
Everett, H III to American Institute of Physics, 6/1/57, basement.
Everett, H III to NSF, Fellowship Report for 1955–56, 6/24/57, basement.
Everett, H III to RMP, July 1957, basement.
Everett, H III to Podolsky, 3/12/59, basement.
Everett, H III to Jammer, 9/19/73, basement.
Everett, H III to Bill Harvey, 5/20/77, basement.
Everett, H III to Leblond, 11/15/77, basement
Everett, H III to Raub, 4/7/80, AIP, basement
Everett, H III to Wheeler, 3/21/79, AIP, basement.

Everett, H III to Wiener, 5/31/57, basement.
Everett, Liz to Mark Everett, 10/15/91, basement.
Everett, Katharine Kennedy to Hugh Everett III, 5/11/56, basement.
Everett, Katharine Kennedy to Nurse Gantley, 11/22/62, basement.
Everett, Nancy to Hugh Jr and Sara Everett, 9/23/59, basement.
Everett, Nancy to Mark Everett, 9/27/82, basement.
Everett, Nancy to Mark Everett, 4/26/96, basement.
Frank to Everett, H III, 8/3/57, basement.
Deutsch email to Shikhovtsev, 12/6/2000.
DeWitt to Wheeler, 5/7/57, basement.
DeWitt to Wheeler, 4/20/67, American Philosophical Society, Wheeler file.
DeWitt to Wheeler, 9/25/79, American Philosophical Society.
Groenwald to Everett H III, 4/11/57, basement.
Jammer to Everett H III, 8/28/73, basement.
Jaynes to Everett, H III, 6/17/57, basement.
Larregsoffore to Everett, H Jr., 2/14/48, basement.
Leblond to Everett H III, 9/17/78, basement.
Margenau to Everett H III, 4/8/57, basement.
Petersen to Wheeler, 5/26/56, basement.
Petersen to Everett H III, cable and letter, 5/28/56, basement.
Petersen to Everett H III, 4/24/57, basement.
Rockman to Everett H III, 3/2/57, basement.
Rosenfeld to Bergmann, 12/21/59, Niels Bohr Archive, Copenhagen.
Rosenfeld to Belinfante, 6/22/72, Niels Bohr Archive, Copenhagen.
Rosenfeld to Belinfante, 8/24/72, Niels Bohr Archive, Copenhagen.
Rosenfeld to Belinfante, 10/31/72, Niels Bohr Archive, Copenhagen.
Rosenfeld to Bell, 11/30/71, Niels Bohr Archive, Copenhagen.
Siegel to Everett, H III, 4/16/59, basement.
Stern to Wheeler, 5/20/56, basement.
Tsiang to Nancy Everett, 7/22/82, basement.
Wheeler to Everett, H III, 9/21/55, basement.
Wheeler to Dennison, 1/21/56, basement.
Wheeler to Bohr, 4/24/56, basement.
Wheeler to Bohr, 5/24/56, basement.
Wheeler to Dees, 5/24/56, American Philosophical Society.
Wheeler-1 to Everett, H III, 5/22/56, basement.
Wheeler-2 to Everett, H III, 5/22/56, basement.
Wheeler to Everett, H III, 5/25/56, basement.
Wheeler to Stern, 5/25/56, basement.
Wheeler to Shenstone, 5/28/56, basement.
Wheeler to Everett, H III, 9/17/56, basement.
Wheeler to Everett, H III, 7/23/57.
Wheeler to Everett, H III, 10/30/57, basement.
Wheeler to Everett, H III, undated, circa 1958, basement.
Wheeler to Everett, H III, with mimeographed enclosure, circa 1958, basement.

Wheeler to Benioff, 7/7/77, basement, AIP.
Wheeler to Benioff, 9/7/77, basement.
Wheeler to NSF, 1/13/55. American Philosophical Society.
Wheeler to Scalapino, 7/12/79, basement.
Wheeler to Mr. & Mrs. Hugh Everett, 7/12/79, basement.
Wheeler to Jammer, 3/19/92, American Philosophical Society.
Wiener to Wheeler and Everett, H III, 4/9/57, basement.
Zeh to Wheeler, 10/30/80, John Wheeler papers, series II, Box Wo-Ze, American
    Physical Society, quoted in Camilleri (2008).

Tape recorded interviews with author

Jeffrey A. Barrett, June 2008.
Matt Bellis, July 2008.
Paul Benioff, July 2005.
Joseph Clifford, July 2009.
Leon N Cooper, Feb. 2009.
Ann Dean, Feb. 2008.
Dr. Patrick Devlin, Apr. 2008.
Cecile DeWitt-Morette, Aug. 2006.
David Deutsch, Nov. 2004 and Mar. 2006.
Don Eigler, June 2008.
Charles Everett, Dec. 2007.
Jean Everett, Dec. 2007.
Mark Everett, Mar. 2006.
Robert Everett, Aug. 2007.
Paul Federbush, June 2005.
Arthur Fine, Apr. 2008.
Paul Flanagan, Dec. 2007.
Kenneth Ford, April 2006.
Fred Forman, Nov. 2007.
Joanna Frawley, Dec. 2007.
Tom Greening, Nov. 2007.
James B. Hartle, Apr. 2006 and June 2008.
Matthew Kleban, Aug. 2009,
Harold Kuhn Feb. 2008 and Apr. 2008.
Leon Lasden, Feb. 2008.
Jan Lodal, Nov. 2007.
Keith Lynch, Jan. 2008.
Gary Lucas, Oct. 2008.
Juan Maldacena, Apr. 2006.
David Mermin, Apr. 2006.
Charles and Susanne Misner, Nov. 2004.
Ralph Mohr, Apr. 2008.
Louis Painter, Nov. 2007.
Donald Reisler, Apr. 2006 and Feb. 2009.

Charles Rossotti, Dec. 2007.
Simon Saunders, May 2009.
Ivan Selin, Nov. 2007.
Leonard Susskind, Apr. 2006.
Max Tegmark, Apr. 2006.
Hale Trotter, May 2007.
Elaine Tsiang, Apr. 2006.
Deborah van Vechten, Feb. 2009.
Ken Willis, Dec. 2007.
Elaine Tsiang, Apr. 2006.
David Wallace, Apr. 2006.
Fred Wilson, May 2008.
W. Zurek, Apr. 2006.

Interviews with others

Mark Everett interviews with Harvey Arnold, Charles Misner, Susanne Misner, Hal
    Trotter, June 2007.
DeWitt-Morette interview by Kenneth Ford, 2/28/95, AIP.
Wheeler interview with Ken Ford, May 2006.
Everett and Misner cocktail party tape, 1977, basement.
Interview with Zeh recorded by Fabio Freitas 7/25/08. AIP.
Wheeler/Ford, Transcript X, AIP, 3/15/94.
Wheeler/Ford, transcripts I–IX, AIP, Oct. 1994–April 1995.

Unpublished basement documents

Everett, Hugh III. A dozen boxes of papers, photographs, and memorabilia, base-
    ment, including handwritten drafts of thesis, thesis notes, thesis abstract, etc.
    Including "Random Notes on QM thesis" file and "Footnotes" file.
Everett, Nancy Gore. Diaries, correspondence, financial records, basement.
Everett, Katharine Kennedy. *Music in Morning*, short stories, correspondence,
    basement.
Everett, Elizabeth. Letters, basement.
Everett, Hugh Jr. and Everett, Sara Thrift, correspondence, military and financial
    records, basement.

Available at National Security Archive (NSA) web site: http://www.gwu.
edu/~nsarchiv/

National Security Archive Electronic Briefing Book No. 130. (2004). "The Creation
    of SIOP-62."
Blackburn, P. P., Jr. "Office Memorandum U.S. Navy Eyes Only," 10/26/60. NSA
    No. 130.
Blouin, F. J. "JCS memo SM-679-60," 7/15/60. NSA No. 130.
Burke, A. "Memorandum to Chairman of JCS from Office of Chief of Naval
    Operations," 9/30/59. NSA No. 130.

Burke, A. "Admiral Burke's conversation with Secretary Franke," 8/12/60. NSA No. 130.

Burke, A. "NAVAL MESSAGE TO CINCPAC," 11/22/60. NSA No. 130.

Blackburn, P. P Jr. & Burke, A. "EXCLUSIVE U.S. NAVY EYES ONLY," 11/24/60. NSA No. 130.

Burke, A. "Special Edition Flag Officers Dope," 12/4/60. NSA No. 130.

Kistiakowsky, G. "Annex" to "J.C.S.2056/208," 1/27/61. NSA No. 130.

"Memorandum for the Record, Secretary McNamara's visit to JSTPS," 2/4/61. NSA No. 130.

Parker, Adm. "NAVAL MESSAGE EXCLUSIVE FOR ADMIRAL BURKE," 2/6/61. NSA No. 130.

Shoup, D. "J.C.S. 2056/220," 2/11/61. NSA No. 130.

Twining, N. F. "J. C. S. 2056/131," 8/20/59. NSA No. 130.

National Security Archive Electronic Briefing Book 173. (2005). "The Nixon Administration, The SIOP, and the Search for Limited Nuclear Options, 1969–1970."

Wainstein, *et al.* (1975). "The Evolution of U.S. Strategic Command and Control, and Warning, 1945–1972," Institute for Defense Analyses Study S-467, June 1975. Partial text, original classification: top secret.

Other resources

American Philosophical Society: John Archibald Wheeler Papers.

American Institute of Physics, Niels Bohr Archive: Everett files.

Niels Bohr Archive, Copenhagen: Wheeler, Bohr, Rosenfeld files.

Seeley G. Mudd Manuscript Library, Princeton University: Everett's transcripts and student file.

Index

Scholastic Aptitude Test
(SAT), 44
Schönborn, Christoph, 385
Schrodinger, Erwin, 55, 83,
350
Bohr and, 87
complementarity and, 110
EPR paradox and, 98
many worlds theory and,
144, 169
mentalist approach and,
101
on model limitations, 195
Mott experiment and,
99–100
Popper and, 362, 362–63
wave function collapse
and, 99–101
Schrödinger equation, xii, 5,
373
Albert-Loewer
interpretation and, 374
Belinfante and, 317–18
Bohr and, 87
causal relation and, 111
Chalmers and, 376
collapse theory and, 103
Copenhagen
interpretation and,
110–12
decoherence theory and,
367–68
Feynman and, 181–82
many worlds theory and,
137, 145–46, 164
measurement problem
and, 83
superposition principle
and, 84
unitary property and,
96n5, 256n22, 333n29,
394
Schrödinger's cat, 172
Schrödinger's jellyfish,
99–101, 315
Schwinger, Julian, 127, 179,
328
science
anthropic principle and,
350, 386
atomic bombs and, 10
cosmic religious feeling
and, 10–11
House Un-American
Activities Committee
and, 21, 124–25, 266

militarization of, 124
National Science
Foundation and, 39
Nobel Prize and, 65, 66n13,
94, 114, 117, 122, 124n20,
192, 302, 366, 371n42
priestly caste of, 10–11
scientists as businessmen
and, 125–27
Vatican attacks and, 385
science fiction, 31, 303,
319–20, 359–62
Scientific American, 30, 196,
350, 355, 356, 372
*Scientist Speculates, The: An
Anthology of Partly-
Baked Ideas* (Good), 301
Second Law of
Thermodynamics, 39
security clearances
Everett and, 4, 35–36, 188,
212, 264, 266
Oppenheimer and, 57
Q level, 4, 268
WSEG requirements and,
192
Selin, Ivan, 294, 297
Semi-Automatic Ground
Environment (SAGE),
75, 190
"Septet of Sibyls, A"
(Wheeler), 161–62, 167
servomechanisms, 37, 66,
148–49, 164, 190, 348
Shannon, Claude, 37, 39, 144,
149, 173, 199
Shapley, Lloyd S., 56, 66, 75
Sheckley, Robert, 360–61
Shenstone, Allen, 167
Shimony, Abner, 254, 257,
259, 301–2
Shockley, William, 190, 192
Shoup, David, 245
Sideways in Time (Leinster),
360
Silent Language, The (Hall),
20
"Simplification of the
Procedure of
Determining the Basic
Solutions of Matrix
Games, A" (Everett), 38
Single Integrated Operating
Plan (SIOP), 195, 393
Army/Navy issues and,
242–43

Burke and, 243–45
China and, 238
city blasting and, 240–41
counterforce proponents
and, 241
damage function and,
241–42
Eisenhower and, 238, 240
first strike capability and,
239–40
Joint Chiefs and, 238–39,
242
kill ratios and, 241–42
Lambda Corporation and,
268, 278, 283
McNamara and, 245
Marine Corps and, 245
Pentagon and, 238
revision of, 246
secrecy of, 239
Shoup and, 245
Soviet Union and, 238–46
Wargasm and, 239, 245
Site R, 344
socialism, 49
sociobiology, 298
solipsism, 41, 114, 147–48,
259, 341, 393
"So Long, Suckers" (game),
75–77
"Some Mathematical
Relations for Probability
of Kill" (Everett and
Galiano), 283
Southern Railway Company,
13, 15
Soviet Union. *See also* Cold
War
arms race and, 74, 129,
193–94, 197–99, 213–15,
227–37
economic growth rate of,
231–32
first strike and, 195–97,
227–46, 281–83, 292–93
Fock and, 216–18
missile gap and, 193–94,
213–15, 228, 235
nuclear warfare and,
199–200 (*see also*
nuclear warfare)
Report 50 and, 215, 227–37
Single Integrated
Operating Plan (SIOP)
and, 238–46
Sputnik and, 194

Peter Byrne is an investigative reporter and science writer based in Northern California. His work has appeared in *Scientific American, Physics World, Mother Jones, Salon.com, SF Weekly, San Francisco Bay Guardian, North Bay Bohemian* and many other magazines and newsweeklies. He has received national recognition for his investigative reporting, including from Investigative Editors & Reporters and Project Censored. He is a member of the Foundational Questions Institute and the National Association of Science Writers. He consulted for (and appeared in) the BBC4/NOVA production about Everett, *Parallel Words, Parallel Lives*. Byrne and Jeffrey A. Barrett co-edited *The Everett Interpretation of Quantum Mechanics, Collected Works 1955–1980 with Commentary* (Princeton University Press, 2012).

Printed and bound by CPI Group (UK) Ltd, Croydon, CR0 4YY